GitHub for Next-Generation Coders

Build your ideas, share your code, and join a community of creators

Igor Irić

GitHub for Next-Generation Coders

Copyright © 2024 Packt Publishing

All rights reserved. No part of this book may be reproduced, stored in a retrieval system, or transmitted in any form or by any means, without the prior written permission of the publisher, except in the case of brief quotations embedded in critical articles or reviews.

The author acknowledges the use of cutting-edge AI, such as ChatGPT, with the sole aim of enhancing the language and clarity within the book, thereby ensuring a smooth reading experience for readers. It's important to note that the content itself has been crafted by the author and edited by a professional publishing team.

Every effort has been made in the preparation of this book to ensure the accuracy of the information presented. However, the information contained in this book is sold without warranty, either express or implied. Neither the author, nor Packt Publishing or its dealers and distributors, will be held liable for any damages caused or alleged to have been caused directly or indirectly by this book.

Packt Publishing has endeavored to provide trademark information about all of the companies and products mentioned in this book by the appropriate use of capitals. However, Packt Publishing cannot guarantee the accuracy of this information.

Group Product Manager: Preet Ahuja
Publishing Product Manager: Prachi Rana
Book Project Manager: Ashwin Dinesh Kharwa
Senior Editor: Mudita S
Technical Editor: Rajat Sharma
Copy Editor: Safis Editing
Proofreader: Mudita S
Indexer: Tejal Soni
Production Designer: Prafulla Nikalje
DevRel Marketing Coordinator: Rohan Dobhal

First published: July 2024
Production reference: 1050724

Published by Packt Publishing Ltd.
Grosvenor House
11 St Paul's Square
Birmingham
B3 1RB, UK

ISBN 978-1-83546-304-8

www.packtpub.com

For my beloved son, Vuk, who inspires every page of this book and every day of my life.

– Igor Irić

Contributors

About the author

Igor Irić is a senior Azure solutions architect expert and GitHub trainer. With over 15 years of experience, he's aided major companies in building cutting-edge solutions. As a mentor, Igor leads developer teams and enhances their skills through transformative training. His engagements with global enterprises have refined his expertise in technology strategies. As a reliable guide for clients facing technical hurdles, Igor provides invaluable advice and training. Passionate about sharing knowledge, he actively discusses cloud best practices, microservices, the cloud, artificial intelligence, DevOps, and software innovation.

Many thanks to my wife, Sanja, whose love and patience have supported me at every step of my career. I also want to thank my parents for inspiring my passion for IT by providing me with my first computer.

About the reviewers

Pavle Davitkovic is a software engineer with over five years of experience. His focus is on the .NET stack, where he has developed applications ranging from web to mobile. He holds a bachelor's degree in modern computer technologies from the College of Applied Technical Sciences in Niš, Serbia. In his spare time, he is a tech content creator.

Florian Holzapfel has been working as a software developer for more than 20 years. Currently, he is working as a senior software developer at glueckkanja AG where he has been contributing to various projects on the product development team using technologies such as C++, Go, Typescript, and React. Over the years, he has had the opportunity to create, maintain, and improve software solutions that impact users worldwide. He is passionate about solving complex problems and staying up to date with the latest technologies.

Sorin Pasa is a seasoned cloud architect and Microsoft Certified Trainer with nearly 20 years of experience in the IT industry. Well known for his expertise in designing and implementing scalable cloud solutions, Sorin has consistently driven innovation and efficiency across various organizations and industries, in multiple countries. His deep knowledge of Microsoft technologies, combined with a passion for teaching, has empowered countless professionals to enhance their skills. Sorin's commitment to excellence and his ability to navigate complex technical landscapes make him a pivotal figure in the cloud computing community.

Soundarya Srikanta is a passionate software engineer with a wealth of experience in the tech industry. An active GitHub contributor, she believes it's the ultimate tool for developers. With a master's degree in computer science from Northeastern University, she's conquered industry giants such as AWS and TCS. Relentlessly innovative, she scoffs at repetitive tasks, automating them effortlessly. Beyond coding, Soundarya's fitness regime and love for animals are the envy of all. She's always eager to learn new things and enjoys being part of a team, bringing fresh ideas and energy to every project.

I thank my family for their unwavering support and belief in my abilities, my dear friends for their cheers and support to lift my spirits higher than the stars, and my well-wishers for their wisdom, guidance, and warm wishes that have sculpted me into the unstoppable force I am today.

With heartfelt gratitude, thank you all for standing by me through thick and thin. Your presence makes every step of this journey meaningful and unforgettable.

-Soundarya Srikanta

Table of Contents

Preface ... xv

Part 1: Getting Started with GitHub

1

Introduction to Version Control and GitHub ... 3

Exploring the benefits of GitHub to young coders	4
Understanding version control in GitHub – keeping track of your changes	5
Here's where "Git" version control and GitHub shine together!	5
Understanding the types of VCS	6
How does this work for coding?	7
What are deltas in version control?	8
How do deltas work in coding?	8
Understanding distributed version control systems	10
What are snapshots in version control?	11
Getting started with Git and GitHub	**13**
Setting up your GitHub account and account types	**16**
What is a user account?	18

Organization accounts – team headquarters	21
Navigating the GitHub interface – a guide for beginners	22
So, what is the Code section about?	22
And Issues, what's that?	23
Pull requests – sounds important?	24
Discussions, are they like meetings?	24
Actions, do they do my work?	25
Projects – that sounds serious?	25
Wiki, is that like Wikipedia?	26
What's in the Security section?	26
Insights, do they tell the future?	26
Settings – what can I set there?	27
Summary	**28**
Quiz	**28**

2

Mastering GitHub Essentials　31

Technical requirements	31	Getting your tools ready – downloading and installing Git	43
How Git interacts with GitHub – Git commands and GitHub workflows	32	Creating and managing personal access tokens	47
Git clone – cloning your first GitHub repository (the welcome kit)	32	How to create a PAT	48
Git add – adding files to your repository, the local way and the GitHub way	34	Keeping your token safe	51
Creating files directly on GitHub	35	Using your token with Git	51
Git commit – saving your snapshot	37	Why a good README file is like a welcoming front door on GitHub	52
Git push – the delivery of the changes to GitHub	38	Editing a README file right on GitHub (hands-on)	55
Git fetch – the update checker	38	Summary	57
Git pull – bringing it all together	39	Quiz	57
Network interaction	39	Challenge – Launching your space adventure!	59
What are repositories on GitHub?	39	Steps to success	59
The GitHub neighborhood	40	Bonus challenge	60
Creating and managing a repository on GitHub	40		

Part 2: Collaborative Development Using GitHub

3

GitHub Features for Collaborating on Projects　63

Technical requirements	64	Creating an issue	66
Setting up your team – inviting collaborators on GitHub	64	Discussing the issue	68
		Assigning the issue	68
Finding the invitation spot	64	Linking to a pull request	69
Awaiting their RSVP (répondez s'il vous plaît – please respond)	65	Introduction to pull requests	70
		What are pull requests?	70
Checking the guest list	65	Using pull requests in your website project	70
Understanding GitHub issues – what's troubling our project?	66	Review and feedback	71
		Merging the pull request	72

Why are pull requests important?	72	Setting up a project for your one-page portfolio website	81
What's buzzing? Managing GitHub notifications	**72**	Why use a Kanban board?	86
		How can I change my project settings?	86
Spotting notifications	73	**Understanding wikis on GitHub**	**87**
Tuning channels	73	How can wikis help?	87
Marking notifications	74	Creating a wiki for your one-page website on GitHub	87
Reducing notification flood with filters	75		
Customizing alerts	76	**What are GitHub repository insights?**	**89**
Emailing updates	76	How to use Insights	89
Exploring GitHub Discussions	**77**	Why are insights helpful?	92
The cool factor in project collaboration	77	**Summary**	**92**
It's like a chat room for your project	77	**Quiz**	**93**
What types of discussions could we have?	77	**Challenge – Teaming up for a stellar mission!**	**94**
How do I start a discussion?	78		
Building a stronger team	81	Your mission	95
Setting up projects on GitHub	**81**	Steps to success	95
Why do projects matter in GitHub?	81		

4

Branching in GitHub and Git — 97

Technical requirements	**98**	**Creating a local copy of the repository (cloning)**	**106**
Branching with GitHub Flow and creating a branch on GitHub	**98**	Cloning the project	106
		Switching to the correct branch	108
Setting up your own space – creating a new feature branch	98	**Adding and committing files to a local repository**	**109**
Making changes and committing	99	Adding your new files to Git	109
Show and tell – creating a pull request	101	Understanding "Check Untracked Files" in Git	111
Teamwork – discussing and reviewing the PR	101	Adding your new file changes to the Git staging area	112
Approving the PR	102		
Making it official – merging the PR	102	**Understanding Git's working area, staging area, and history and pushing your changes to GitHub**	**113**
Cleanup time	102		
Understanding GitHub Flow in practice – two developers, two features, one project	**103**	Working area (or working directory) – what is it?	114

Staging area (or index) – what is it?	114	Quiz	118
History (or repository)	115	**Challenge – Navigating the stars in**	
Pushing your changes to GitHub	116	**Space Explorer!**	**120**
Why are these areas important?	118	Your mission?	120
Summary	**118**	Steps to success	121

5

Collaborating on Code through Pull Requests 123

Technical requirements	**123**	Deleting a GitHub branch	139
Getting familiar with PRs and their importance	**124**	**Enhancing your website with GitHub's easy editing features**	**142**
Creating a new PR	124	Direct editing on GitHub – quick and easy	142
GitHub PR interface – proposing your draft	128	Using github.dev – your full-fledged editor	143
Why is a good PR important?	130	**Summary**	**144**
Reviewing a PR	**133**	**Quiz**	**145**
What's the Files changed tab?	134	**Challenge – Cosmic collaboration in**	
Ensuring quality	136	**Space Explorer!**	**146**
Learning and growing	137	Your mission?	146
Merging the changes	138	Bonus exploration	148

6

Resolving Merge Conflicts – on GitHub and Locally 149

Technical requirements	**150**	Solving conflicts with removed files	164
Understanding merge conflicts and how they occur	**150**	**Summary**	**166**
		Quiz	**166**
What is a merge conflict?	151	**Challenge – Stellar enhancements in**	
How does a merge conflict happen?	151	**Space Explorer!**	**168**
Addressing merge conflicts	**153**	Your mission	168
Merge conflict in action	153	Steps to success	168
Resolving merge conflicts using GitHub's UI	155	Bonus challenge	169
Having multiple merge conflicts	159		
Resolving merge conflicts using the command line like a pro	160		

Part 3: Mastering Git Commands and Tools

7

Git History and Reverting Commits — 173

Understanding Git and GitHub history – tracking changes to your website 174	**Unraveling mysteries with git diff – the tale of the unseen changes** 190
Viewing the history locally in a Git repository 174	What is the diff command? 190
Branches – parallel universes of Git history 179	Other diff commands 191
The Activity page – a chronicle of your repository's journey 179	**Undoing changes with git reset and cherry-picking** 193
Explaining git bisect – finding the needle in the haystack 182	Types of git reset 193
A git bisect mystery – the case of the missing CSS styling 183	Picking specific changes with cherry-picking 193
The short way of using git bisect – let Git do the detective work 187	**Summary** 195
Reverting commits to a previous version 188	**Quiz** 196
	Enhanced challenge – Space Explorer splash screen, score tracking, and mastering Git commands 197

8

Helpful Tools and Git Commands — 199

Crafting shortcuts and keeping it clean – advanced Git commands 200	**Navigating your website project with GitHub Desktop** 208
Streamlining your website work with Sourcetree 202	Setting up GitHub Desktop 209
Why should you use Sourcetree? 203	What can you do with GitHub Desktop? 211
Getting started with Sourcetree 204	**Crafting your website with GitHub Codespaces** 215
Connecting your repository 204	Why would you use Codespaces? 215
Viewing your project timeline 205	How to get started with Codespaces 215
Create, merge, and switch branches 206	Deleting Codespaces 218
Staging and committing changes 207	**Managing your project's buzz with DevHub** 219
Pull, Fetch, and Push 207	

How do you set up DevHub?	220	Quiz	222
Why should you use DevHub in your daily workflow?	221	Challenge – Crafting a Game Over screen for Space Explorer	224
Summary	221		

Part 4: Advanced GitHub Functionalities

9

Leveraging GitHub Actions for Automation   229

Technical requirements	230	Managing secrets and environment variables	243
Understanding a GitHub Actions workflow	230	What are secrets and environment variables?	243
Creating action workflows	233	Using secrets and environment variables for MY ARTICLES	243
Exploring prebuilt actions	234		
Understanding the uses keyword	237	Troubleshooting and optimizing GitHub Actions	246
Seeing it in action	237		
What are reusable workflows in GitHub Actions?	238	Troubleshooting GitHub Actions	247
		Common issues to look for	247
Automated testing and deployment with GitHub Actions	239	Optimizing GitHub Actions	248
		Summary	248
What's automated testing?	239	Quiz	249
Deployment	239		
How do we set this up with GitHub Actions?	239		

10

Enhancing GitHub Security Measures   251

Technical requirements	252	Keeping your website safe – vulnerability scanning with Dependabot	259
Setting up collaboration in your website repo	252		
Setting up 2FA for your portfolio website	255	What is vulnerability scanning?	260
		And what's Dependabot?	260
What is 2FA?	255	Protecting your one-page website with CODEOWNERS	264
How to set up 2FA	255		

What is CODEOWNERS?	264	Planning a secure development strategy	267
How does CODEOWNERS help?	265	Building and deploying securely	277
Applying secure coding practices	**267**	**Summary**	**278**
Why secure development matters	267	**Quiz**	**278**

11

Engaging with the Open Source Community — 281

Technical requirements	282	Choosing a license	292
Exploring open source projects	282	Need more help?	294
Sounds very interesting, but where to start?	282	**Creating a license for your**	
Understanding project pages	283	**GitHub repository**	**294**
Getting started	284	**Open source etiquettes and**	
Making meaningful contributions		**best practices**	**297**
with real examples	**286**	**Summary**	**297**
Navigating licensing and legal		**Quiz**	**298**
considerations	**291**		

Part 5: Personalizing Your GitHub Experience

12

Crafting Your GitHub Profile — 303

Technical requirements	303	Why should you add stats and achievements?	315
Optimizing your GitHub		Showcasing your GitHub trophies	315
profile overview	**303**	**Displaying projects and**	
Showcasing skills and expertise	**306**	**contributions strategically**	**317**
Creating your README profile	306	Adding contributions and highlights	318
What to include in the README file?	308	Using GitHub Pages to show off your work	320
Adding stats and achievements to		**Summary**	**321**
your profile	**312**	**Quiz**	**321**
What are these widgets?	312		

13

GitHub Copilot Aiding Code Creation — 325

Technical requirements	325	Test-driven development practice	339
Understanding GitHub Copilot, your coding assistant	326	Code refactoring	341
How does it work?	326	**Prompt engineering with GitHub Copilot**	**342**
A quick introduction to AI and LLMs	328	How does prompt engineering work?	342
Why is Copilot so cool?	328	Cool examples of prompt engineering with GitHub Copilot	343
What to keep in mind when coding with Copilot	328	Tips for great prompt engineering	344
How to get started	328	**Is GitHub Copilot free for coders in school?**	**345**
Installing the GitHub Copilot extension	332	How to apply for free GitHub benefits	345
Code completion and suggestions with Copilot	**334**	A word of wisdom for students	346
Using Copilot for clean code and best practices	**337**	**Summary**	**347**
Unit testing generation	337	**Quiz**	**347**

Index — 351

Other Books You May Enjoy — 360

Preface

Welcome to *GitHub for Next-Generation Coders*! This book is designed to take young coders on an explorative ride into the world of GitHub, a platform that revolutionizes how people collaborate on software projects. Whether you're starting your first coding project or looking to enhance your skills, this guide will provide you with the knowledge and tools necessary to succeed in the collaborative environment of GitHub.

Throughout this book, we will cover everything from the basics of setting up a GitHub account to advanced features that can help you manage complex projects and contribute to the open source community. By the end of the book, you will not only be comfortable using GitHub; you will also understand how to use its many features to improve your coding projects and collaborate effectively with others around the world.

Who this book is for

GitHub for Next-Generation Coders is ideal for young individuals who are eager to learn about version control and collaboration. Whether you're a middle school student just starting out with coding or a high school student looking to manage group projects more efficiently, this book is written to help you understand the concepts at your own pace and apply them practically.

What this book covers

Chapter 1, *Introduction to Version Control and GitHub*, dives into the essentials of version control and explains why it's crucial for managing changes and collaboration. You will set up your first GitHub account and explore the platform's fundamental features.

Chapter 2, *Mastering GitHub Essentials*, deepens your understanding of GitHub by covering repositories, branches, commits, and merges. You will start managing your projects on GitHub with confidence and ease.

Chapter 3, *GitHub Features for Collaborating on Projects*, covers GitHub's powerful tools for team collaboration, including issues, pull requests, and code reviews. You will learn how these features facilitate effective teamwork.

Chapter 4, *Branching in GitHub and Git*, discusses the concept of branches in GitHub as a means of working on different parts of a project simultaneously without affecting the stable version.

Chapter 5, *Collaborating on Code through Pull Requests*, provides practical guidance on how to work with others on GitHub using branches and pull requests, ensuring smooth collaboration and integration of changes.

Chapter 6, *Resolving Merge Conflicts – on GitHub and Locally*, explains how to handle merge conflicts that may arise when multiple people are editing the same parts of a project, ensuring a seamless merge process.

Chapter 7, *Git History and Reverting Commits*, covers the tools for tracking project history and undoing changes with commands such as `git log` and `git revert`, giving you control over your project's historical changes.

Chapter 8, *Helpful Tools and Git Commands*, explores a range of Git commands and tools that enhance your productivity and project management capabilities on GitHub.

Chapter 9, *Leveraging GitHub Actions for Automation*, explains how to implement automation in your projects using GitHub Actions to streamline workflows for testing, deployment, and more.

Chapter 10, *Enhancing GitHub Security Measures*, covers securing your projects by managing access, using security features, and following best practices to protect your code.

Chapter 11, *Engaging with the Open Source Community*, explains how to contribute to open source projects, discusses the significance of open source, and covers how to make impactful contributions.

Chapter 12, *Crafting Your GitHub Profile*, will help you enhance your GitHub profile to showcase your projects, skills, and professional accomplishments effectively.

Chapter 13, *GitHub Copilot Aiding Code Creation*, explores how GitHub Copilot can help you write better code faster, using AI to assist in your coding tasks.

To get the most out of this book

Here is what to do to get the most out of this book:

Engage with the examples and exercises provided in each chapter

Participate in community discussions and open source projects to practice your skills

Reflect on how you can apply the lessons learned to your own coding projects

Software/hardware covered in the book	Operating system requirements
JavaScript	Windows, macOS, or Linux
HTML 5	Windows, macOS, or Linux
CSS	Windows, macOS, or Linux
YAML	Windows, macOS, or Linux
Markdown	Windows, macOS, or Linux
Json	Windows, macOS, or Linux

If you are using the digital version of this book, we advise you to type the code yourself or access the code from the book's GitHub repository (a link is available in the next section). Doing so will help you avoid any potential errors related to the copying and pasting of code.

Download the example code files

You can download the example code files for this book from GitHub at `https://github.com/PacktPublishing/GitHub-for-Next-Generation-Coders`. If there's an update to the code, it will be updated in the GitHub repository.

We also have other code bundles from our rich catalog of books and videos available at `https://github.com/PacktPublishing/`. Check them out!

Conventions used

There are a number of text conventions used throughout this book.

`Code in text`: Indicates code words in text, database table names, folder names, filenames, file extensions, pathnames, dummy URLs, user input, and Twitter handles. Here is an example: "You can use this command to create a new branch: `git switch -c name-of-your-branch`."

A block of code is set as follows:

```
<!-- Profile Picture -->
<p align="center">
    <img src="https://github.com/[your-username]/biographyii/blob/main/1674712595713-plava2.jpg" alt="Your Name" width="200">
</p>
```

Bold: Indicates a new term, an important word, or words that you see onscreen. For instance, words in menus or dialog boxes appear in **bold**. Here is an example: "Click on the dropdown menu **Branch: main** and type the name of your new branch."

> **Tips or important notes**
> Appear like this.

Get in touch

Feedback from our readers is always welcome.

General feedback: If you have questions about any aspect of this book, email us at customercare@packtpub.com and mention the book title in the subject of your message.

Errata: Although we have taken every care to ensure the accuracy of our content, mistakes do happen. If you have found a mistake in this book, we would be grateful if you would report this to us. Please visit www.packtpub.com/support/errata and fill in the form.

Piracy: If you come across any illegal copies of our works in any form on the internet, we would be grateful if you would provide us with the location address or website name. Please contact us at copyright@packt.com with a link to the material.

If you are interested in becoming an author: If there is a topic that you have expertise in and you are interested in either writing or contributing to a book, please visit authors.packtpub.com.

Share Your Thoughts

Once you've read *GitHub for Next-Generation Coders*, we'd love to hear your thoughts! Scan the QR code below to go straight to the Amazon review page for this book and share your feedback.

https://packt.link/r/1835463045

Your review is important to us and the tech community and will help us make sure we're delivering excellent quality content.

Download a free PDF copy of this book

Thanks for purchasing this book!

Do you like to read on the go but are unable to carry your print books everywhere?

Is your e-book purchase not compatible with the device of your choice?

Don't worry! Now with every Packt book, you get a DRM-free PDF version of that book at no cost.

Read anywhere, any place, on any device. Search, copy, and paste code from your favorite technical books directly into your application.

The perks don't stop there, you can get exclusive access to discounts, newsletters, and great free content in your inbox daily

Follow these simple steps to get the benefits:

1. Scan the QR code or visit the following link:

https://packt.link/free-ebook/9781835463048

2. Submit your proof of purchase.
3. That's it! We'll send your free PDF and other benefits to your email directly.

Part 1: Getting Started with GitHub

In this section, we start from the very beginning, guiding you through the essentials of version control and the GitHub platform. You'll learn how to set up your GitHub account and create your first repository, and you'll discover the fundamentals of version control systems. By the end of this section, you'll be familiar with navigating GitHub, managing files, and beginning to collaborate on projects.

This part contains the following chapters:

- *Chapter 1, Introduction to Version Control and GitHub*
- *Chapter 2, Mastering GitHub Essentials*

1
Introduction to Version Control and GitHub

This first chapter is like a launching pad for your exciting GitHub journey! We're going to introduce you to GitHub and version control, which is a superpower for coders. We'll guide you through step-by-step processes, providing practical tips and easy-to-follow examples.

First off, we'll help you understand the core of GitHub and why version control is so awesome. Think of version control like a magic time-traveling book that lets you see and undo all your changes, keeping your code organized and safe.

Next, we'll dive into the fun stuff! We'll show you how to create your special space on GitHub, called a *repository*. Think of it as your treasure chest where you keep all your coding gems. You'll learn how to add your code, make changes to it, and share it with others.

We'll also explore how GitHub can be a fantastic tool for working together with friends or even with coders from around the world! It's like having a coding party where everyone brings their own unique ideas and creations.

Here's a sneak peek of what we'll cover in this chapter:

- Exploring the benefits of GitHub to young coders
- Understanding version control in GitHub – keeping track of your changes
- Getting started with Git and GitHub
- Setting up your GitHub account and account types
- Navigating the GitHub interface – a guide for beginners

By the end of this chapter, you'll feel at home in the GitHub universe, understand the magic of version control, and be ready to start your own coding adventures and share them with the world. Let's code together and have a blast!

Exploring the benefits of GitHub to young coders

Have you been coding for a little while and are looking to improve your skills further? Whatever stage you're at, GitHub is a platform you'll want to get familiar with.

Let's explore why:

- **Learning from the community**: Have you ever found yourself stuck on a coding problem, not sure how to proceed? On GitHub, you can find countless projects and code snippets. Seeing how others tackle problems can provide a new perspective and help solve your own coding dilemmas.
- **Collaboration**: Ever worked on a group project and found it hard to keep track of who did what? GitHub makes teamwork smoother with its collaboration features. You can work with others easily, track changes, and combine everyone's work seamlessly.
- **Building a portfolio**: Do you dream of landing a cool job in tech? GitHub can act as your portfolio, showcasing your projects to potential employers. It's like your public coding diary, demonstrating your skills and the journey of your progress.
- **Version control**: Ever made a change to your code and realized you liked it better the way it was the day before? Git's version control allows you to go back in time and compare changes so you never lose your work.
- **Receiving feedback**: Want to get better at coding? GitHub allows you to build up the community, which can provide valuable feedback on your projects. Repository and/or community members can point out bugs, suggest improvements, and help you become a better coder.
- **Access to resources**: Looking for a library or a tool to make your coding project easier? GitHub hosts a great number of resources, many of which are free to use.
- **Integration with tools**: Do you use other coding tools? GitHub integrates with many tools and platforms that can improve your workflow, making your coding process more efficient and enjoyable.

GitHub is not just a place to store your code. It's a platform where you can learn, collaborate, showcase your work, and grow as a coder. So, why wait? Dive into GitHub and start exploring the endless possibilities it offers to young coders like you!

In the next section, we're going to learn how GitHub helps you keep track of every change you make to your projects, just like keeping a detailed diary of your work. We'll look at different ways to manage changes with version control systems, understand the concepts of deltas and snapshots, and explore how distributed version control systems keep teams in sync. It's all about making sure you can always see what's been done and by whom, making teamwork smoother and more organized.

Understanding version control in GitHub – keeping track of your changes

Have you ever stopped to ask yourself how developers manage the ever-evolving versions of their software? Think of it like you're writing a story. You write the first draft, then the second, and so on. Each draft is unique, with its own set of changes, yet all of them are interconnected. Now, what if you wanted to go back to a particular draft to revisit some ideas, or share your drafts with fellow writers for a collaborative masterpiece? This is where version control steps in as your time-traveling companion in the digital domain!

In coding, **version control systems** (**VCSs**) are like magical journals that keep a record of every change made to a project. They allow developers to travel back in time to any version of their project, compare different versions, and even merge changes made by different people. It's like having a superpower that lets you not only undo mistakes but also collaborate without a problem.

GitHub is the place where developers come together to share and collaborate on projects. GitHub is built on top of Git, a popular VCS that we will explain later in the *Getting started with Git and GitHub* section, making it a robust platform for team-based projects. It's like a massive online library, but instead of books, it houses code. Every project on GitHub lives in its own repository (or *repo* for short), a nicely ordered digital shelf where all versions of a project are stored and organized.

Curious about how this all comes together?

Let's say you and your friends decide to build a portfolio website. Each one of you has unique ideas to make the website cooler. Alice suggests adding a top menu, Bob wants a gallery, and Charlie is keen on a blog. You all set out to work on your individual parts, but how do you bring it all together without stepping on each other's toes?

Here's where "Git" version control and GitHub shine together!

Each one of you creates a branch from the main branch (a **branch** is a separate workspace where you can make changes and try out new ideas without affecting the main version of your project) and works on your ideas, and once you're satisfied, you propose the changes back to the main branch through a **pull request**. Your friends can review the changes, discuss them, and even suggest tweaks. Once everything looks good, the changes are merged into the main branch. Your collaborative portfolio website now has a gallery, a top menu, and a blog section, all built together without any problems.

And guess what? All the discussions, changes, and versions are nicely recorded, thanks to Git's version control magic and GitHub's collaborative platform. So, if someday you decide that the blog should include news, you know exactly where to go and how to make it happen without disrupting the rest of your website!

This journey of creative collaboration and organized evolution is what makes GitHub very popular among developers. It's not just a platform; it's a community where ideas grow, mix, and come to life, one version at a time.

Understanding the types of VCS

What are centralized version control systems?

Picture a train station at the heart of a large city full of activities. This station is no ordinary one; it's the nerve center where all the city's data trains come and go. Now, what if this city represented a software project, and the train station, a **centralized version control system (CVCS)**?

Interesting, isn't it?

A CVCS is like this central station where all changes to a project are stored and managed. In this digital city, developers are the conductors of data trains, each carrying a unique set of changes or updates. They all meet at the central station, deposit their updates, and pick up changes made by others.

Sounds organized, right? But let's dive a bit deeper.

You're a developer-conductor. Each morning, you pull into the central station, pick up the latest version of the project, and move along through the day making your changes. Once satisfied, you steer your data train back to the central station to deposit your updates. This cycle continues, keeping the project evolving while maintaining a sense of order.

However, the CVCS has its bad sides. Picture a sunny day, our central station full of activity. Suddenly, a storm strikes and the station experiences a blackout, as seen in *Figure 1.1*. Panic sweeps through the city as the heart that held all the project's data is now inaccessible. Developers are stranded with their updates, unable to share or access others' changes.

Figure 1.1 – A CVCS as a central station

This scenario illustrates a significant characteristic (and a potential downside) of a CVCS—its single point of failure. Everything is stored on the central server. If it's down or faces issues, the flow of updates halts, causing a ripple effect across the project.

Now, let's draw our attention to another aspect. Remember how every conductor picks up the latest version of the project from the central station? This illustrates the centralized nature of a CVCS. Every piece of data, every update, and every version resides in one central place.

There's a sense of simplicity and order with a CVCS. It's like having a disciplined, centralized government managing the flow of data. It's easier to administer, control access, and ensure everyone is on the same page. Yet, this centralized nature also brings along a few challenges, such as the potential for a single point of failure and the dependency on network access to the central server.

How does this work for coding?

In coding, a CVCS acts as a central repository where all code and its versions are stored. Developers *check out (pull)* the latest version of the code from this central server, work on their local machines to make changes, and then *check in (push)* their updates back to the central repository. This way, everyone has access to the latest version of the code, and all changes are tracked centrally.

In this setup, the CVCS keeps a record of all changes, who made them, and when. It's an organized way to manage code in a team setting. However, it does have its drawbacks. For instance, if the central server goes down, developers can't share or access the updates from the repository. Also, every operation is dependent on the central server, which can slow down processes if the network is slow or if the team is spread out across different locations. You can get an idea about a CVCS from *Figure 1.2*:

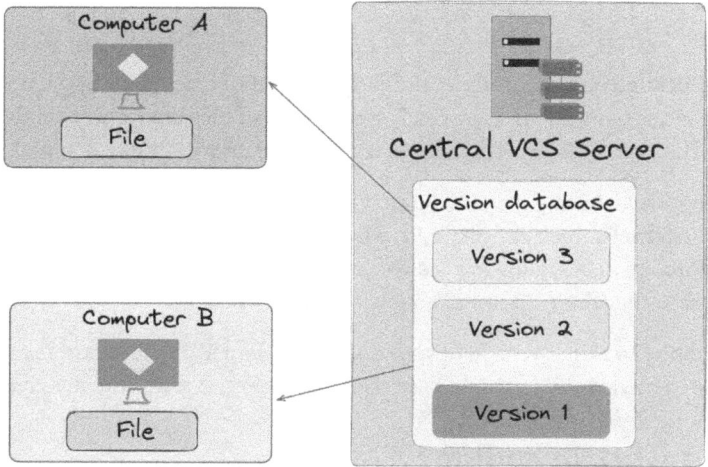

Figure 1.2 – The CVCS flow

The simplicity and centralized control make a CVCS easy to administer and ensure consistency in the code base, but at the cost of potential bottlenecks and a single point of failure.

What are deltas in version control?

Imagine you're an artist, and every day, you add a new layer to your painting, enhancing it with fresh details, colors, and expressions. Each layer modifies just part of the painting, leaving other parts unchanged. As days go by, your artwork becomes a vibrant collection of these layers, each contributing to the overall story of the painting:

Figure 1.3 – Layers of a painting

In the world of version control, a **delta** works similarly. It's a set of changes or updates applied to a file or a group of files. Rather than saving completely new copies of files after every single change, VCSs typically save these changes as deltas. This means they only record what has changed.

For example, let's say you're writing a digital book. Each day, you type up new ideas or edit what you wrote before. Your VCS keeps track of every edit, whether you add a paragraph or fix a spelling error. These changes are stored as deltas, allowing the system to document how your book evolves over time without needing to save the whole document anew with every change.

When you need to look at an earlier version of your document, your VCS uses these deltas to reconstruct it. It's like being able to travel back in time to see your document as it was on any given day.

How do deltas work in coding?

In coding, deltas refer to the differences or changes between two versions of a file or files. These are vital to version control as they help track the evolution of the project without storing complete file copies at each stage.

When you or someone else makes changes to the code, the VCS figures out the delta—what's been added or removed—and saves it. This acts like a detailed record of all edits, allowing for a more manageable and efficient way to handle changes.

These deltas, also called **diffs**, are essential to teamwork. They let everyone see exactly what has changed, who made the changes, and when they were made. By tracking changes this way, the VCS helps manage the project's progress and ensures that nothing gets lost, even as many people collaborate and make continuous changes.

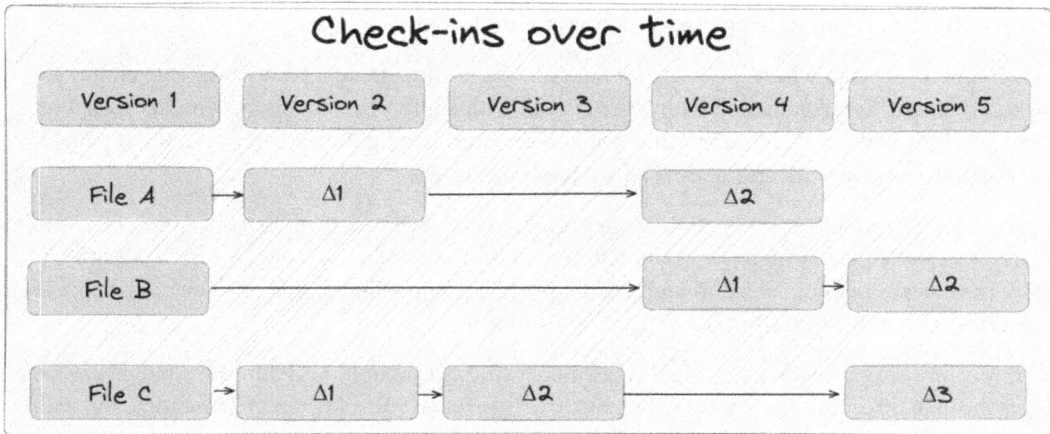

Figure 1.4 – Check-ins over time with deltas

This diagram shows a visual way to track changes to files over time in a project, such as updates to your one-page portfolio website. Each row represents a different file, labeled as **File A**, **File B**, and **File C**. The columns represent different versions of your project, from **Version 1** to **Version 5**:

- In **Version 1**, all three files are the original versions
- By **Version 2**, **File A** and **File C** have been changed ($\Delta 1$), while **File B** is still the original
- In **Version 3**, **File A** hasn't changed from **Version 2**, **File B** has not been changed, and **File C** has been updated from its original ($\Delta 1$)
- Moving on to **Version 4**, **File A** has another change ($\Delta 2$), **File B** has been changed ($\Delta 1$), and **File C** is the same as in **Version 3**
- Finally, in **Version 5**, **File A** is the same as in **Version 4**, **File B** has another new change ($\Delta 2$), and **File C** has yet another update ($\Delta 3$)

The "Δ" symbol stands for a change or "delta." So, $\Delta 1$ on **File A** means the first set of changes to the original version of **File A**, and $\Delta 2$ on **File A** represents a second set of changes.

This kind of tracking allows you to see the history of updates to each file, and by comparing these versions, you can understand how your website evolved over time. It's a bit like looking at a photo album of your website, with snapshots of how each part changed from one version to the next.

Understanding distributed version control systems

A **distributed version control system** (**DVCS**) such as Git gives developers a complete project history on their local machines. This allows them to work independently on code, fixing bugs or adding features without worrying about disrupting the main project.

Now, picture a scenario where a developer named Ada is addressing a tricky bug. She clones the repository, creating her own version within the DVCS. With a full copy of the project on her computer, she works on the code, tests solutions, and resolves the bug. Throughout this process, she is isolated from other developers' work and doesn't affect their progress.

When Ada is confident in her solution, she shares it with the central repository. In the context of GitHub, she creates a pull request, which is a well-documented summary of her changes for her colleagues to review. They assess her contribution and discuss potential adjustments, and once everyone agrees, the changes are merged into the main project repository.

If Ada were working directly with other developers using Git without GitHub, she could share her changes through direct pushes to a shared repository. This would still allow for collaboration and code review, but without the structured workflow of pull requests.

With a DVCS, every developer's input, whether it's a bug fix, a feature, or an optimization, is a valuable contribution that is shared, reviewed, and improved through collaboration, driving the project toward new and exciting possibilities with the combined knowledge and skills of the team.

In the context of using Git with GitHub, you get the best of both worlds: the decentralized power of a DVCS with the collaborative tools that make managing contributions easier and more structured, like some features of a CVCS but with much more flexibility and independence.

Figure 1.5 shows a representation of a DVCS:

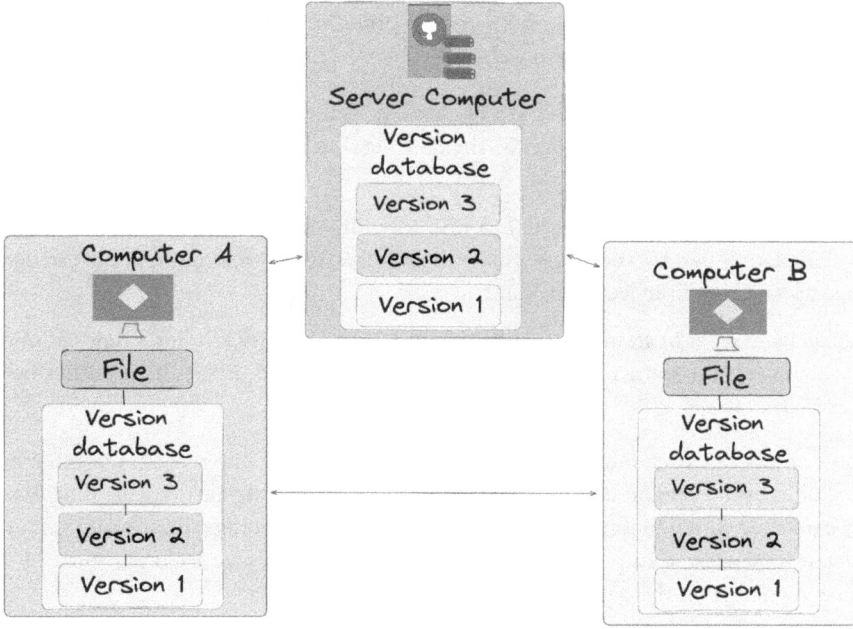

Figure 1.5 – A DVCS

In the DVCS, you are saving your changes like snapshots or like images of everything you have changed over time.

What are snapshots in version control?

Snapshots are like photographs capturing the state of your project at specific moments in time. Each snapshot holds a picture of what your code looked like at that moment.

So, how does this magic happen? Every time you reach a milestone, no matter how minor, you can take a snapshot of your code. In Git, this is called a **commit**. Each commit is a snapshot, a milestone on the development timeline of your project.

As time goes on, your code grows, but sometimes bugs appear. That's where the power of snapshots comes into play. You can go back in time, revisiting each snapshot to figure out when and how the bug entered your code.

Snapshots allow you to compare your current code with past versions, helping you understand how your code has evolved. If necessary, you can even revert to a previous snapshot when everything was working smoothly.

These snapshots are not isolated; they are connected in a chronological chain, each one leading to the next. They tell the story of your project from its beginning to the present day.

Figure 1.6 illustrates how a group of files changes over time through different versions of a project, such as updates to your one-page portfolio website:

- **File A** starts off in **Version 1** without any changes. In **Version 2**, it has a set of changes (noted as **A1**), which stays the same in **Version 3**. Then, in **Version 4**, **File A** sees another set of changes (**A2**), which carries over into **Version 5**.
- **File B** begins unchanged in **Versions 1**, **2**, and **3**, but in **Version 4**, it has a set of changes (**B1**). These changes remain in **Version 4**. But by **Version 5**, **File B** has a new set of changes (**B2**), which is updated again in **Version 5** (**B2**).
- **File C** has its first set of changes (**C1**) in **Version 2**. In **Version 3**, it has a new set of changes (**C2**), which stays the same in **Version 4**. Finally, in **Version 5**, **File C** has another new set of changes (**C3**).

The pattern here shows that each file can be updated independently in different versions. For example, **File C** is updated in every version except in **Version 4**, while **File B** doesn't change until **Version 4**. This way, you can track which files have been updated and when, keeping a clear history of each part of your project as it grows and improves. It's a lot like keeping a diary for each part of your website, noting down each change on the date it was made.

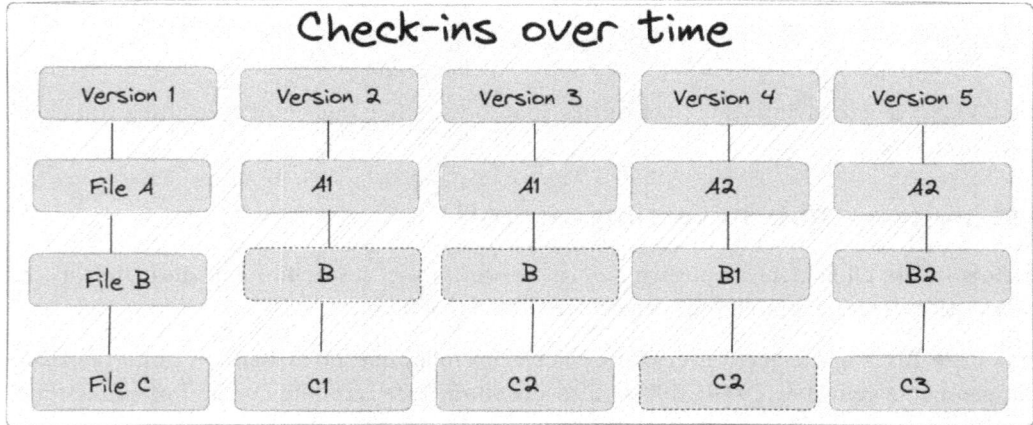

Figure 1.6 – Check-ins over time

When you take a snapshot in version control, it's like capturing the current state of your project. If you make changes later, a new snapshot is taken, preserving the changes while keeping the previous ones intact.

Now that we've covered the basics of version control and how it helps keep track of your changes, let's move forward. In the next section, we'll take these concepts and show you how to apply them practically. You'll learn how to set up Git on your machine, start managing your projects on GitHub, and begin collaborating with others effectively. This step-by-step guide will help you make the most of these powerful tools in your coding projects.

Getting started with Git and GitHub

So, what's Git?

Imagine a lively digital city where code serves as the foundation and versions act as significant points of reference. In this city, Git acts as your reliable companion, a mystical journal documenting each step of your coding adventure, whether you're constructing, revising, or occasionally removing code structures:

Figure 1.7 – Git as a lively digital city

Now, why would a developer need such a record?

Imagine working on a digital sculpture, adding pieces, when suddenly, another piece falls off! Panic sets in as you realize that the work you did a day ago is now damaged. Here's where Git jumps in like a time-traveling machine. With a swift motion, Git turns back the hands of time, and your work is restored!

How does GitHub fit into this journey?

Now, let's investigate the ecosystem of GitHub, a lively metropolis where developers from every corner of the digital domain meet (*Figure 1.8*). But what makes GitHub more than just a meeting point?

Figure 1.8 – GitHub as a bustling metropolis

Picture GitHub as a colossal library, with Git being the great librarian (*Figure 1.9*). Each developer, upon entering, is given an endless shelf to fill with their books of code. Each book, a repository, holds the tale of a project, chronicled with precision by Git.

Getting started with Git and GitHub   15

Figure 1.9 – GitHub as a colossal library

But unlike any library of the ordinary world, this one grows on collaboration. You're not just a silent subscriber, you're a part of a community, writing the epic of innovation, chapter by chapter, line by line.

Now, as you write down your tale, you might stumble upon a chapter written by another writer that sparks a new thought. With a sense of excitement, you decide to build upon it. In GitHub, this act of collaborative branching is encouraged. You duplicate the chapter, add your part, and propose to merge it with the original. The other writer, appreciative of your contribution, accepts, and the tale grows richer.

But GitHub isn't just about the serious business of code crafting. It's a platform where you can collaborate with others, contribute to various projects, and learn from the work of other developers. Your profile serves as a portfolio, showcasing your projects, contributions, and coding skills, making it easier to connect and collaborate with others in the developer community.

In the next section, you'll learn how to set up your GitHub account so you can start managing your projects online. We'll cover two types of accounts:

- **User account**: This is your personal account where you can manage your own projects, contribute to others' projects, and more. Think of it as your personal workspace on GitHub.
- **Organization accounts**: If you're working with a team, this type of account will be useful. It's like a shared space for teams where everyone can collaborate on projects together.

You'll understand how these accounts differ and how you can use each type to keep your projects organized and accessible to the right people. This knowledge will help you make the most of GitHub, whether you're working alone or as part of a team.

Setting up your GitHub account and account types

Now it's time for you to set up your first GitHub account. It's very easy to create a GitHub account. You can set up your free account by following these simple steps:

1. Go to https://github.com/ and click the **Sign up** button in the upper-right corner:

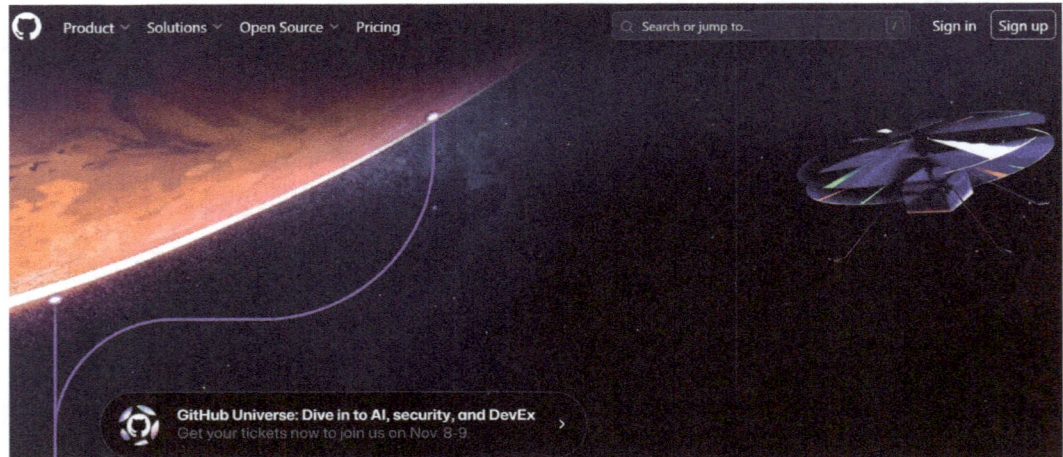

Figure 1.10 – GitHub sign-up

2. You will be redirected to the following screen, where you will need to enter your email address and password:

Figure 1.11 – GitHub account creation

3. You will receive a verification email at the address you provided when creating your GitHub account.
4. Click on the link to complete the verification process and creation of the GitHub account.

If you already have an account, then navigate to `https://github.com/` and click the **Sign in** button in the upper-right corner:

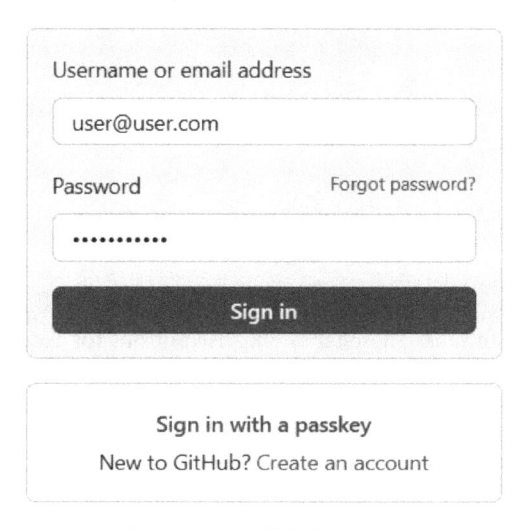

Figure 1.12 – GitHub sign-in

Before we dive into the journey of the GitHub repositories, let's talk about different types of GitHub accounts.

What is a user account?

A user account is your own personal GitHub account. It's where you can work on your projects, either alone or with a few friends. Here's what you can do with a user account:

- **Create repositories**: You can build various projects in your personal space
- **Contribute:** You can visit other people's houses (repositories) and help them with their projects
- **Personal profile**: You have a profile where others can see who you are and what you've worked on, and even follow you to stay updated on your activities

Ready to explore your GitHub user account view? Let's get started.

Your GitHub personal user profile comprises several distinct sections, as shown in *Figure 1.13*:

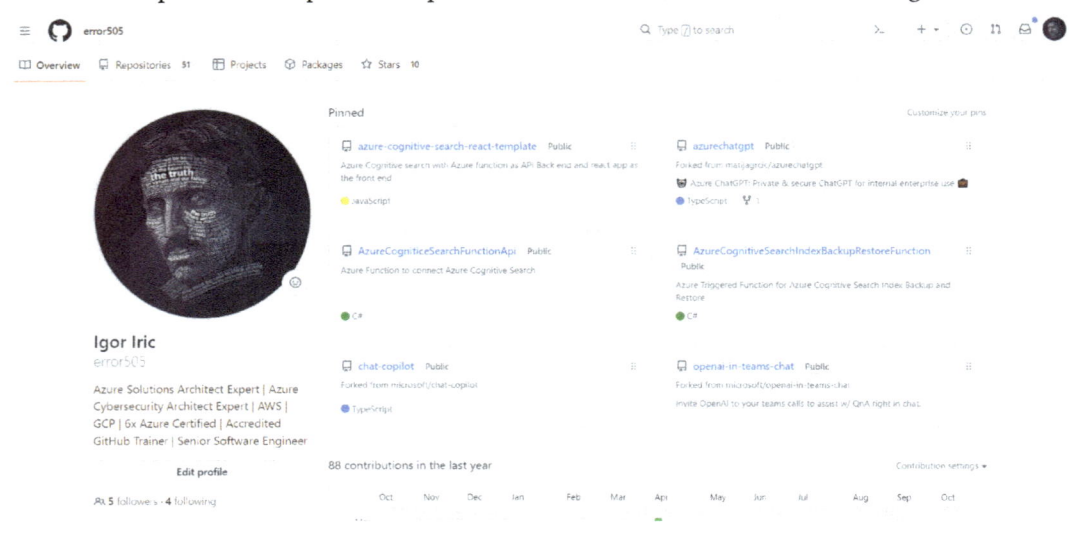

Figure 1.13 – GitHub's user account overview

Let's learn about these sections in detail:

- **User account navigation tabs**: These tabs comprise buttons for accessing various sections of your user account, including **Overview**, **Repositories**, **Projects**, **Wikis**, **Packages**, and more:

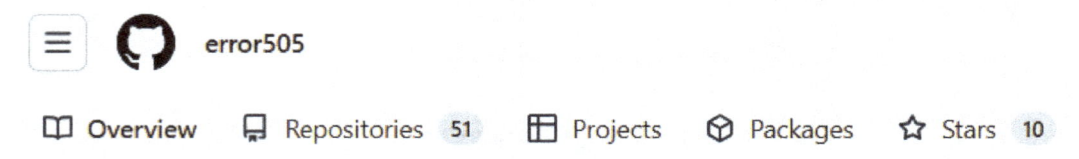

Figure 1.14 – GitHub's user navigation tabs

- **Global navigation**: You can access the global navigation menu by clicking on the *burger menu* icon located in the top-left corner. This menu provides links to essential features such as **Home**, **Issues**, **Pull requests**, **Discussions**, **Marketplace**, **Notifications**, and other useful resources:

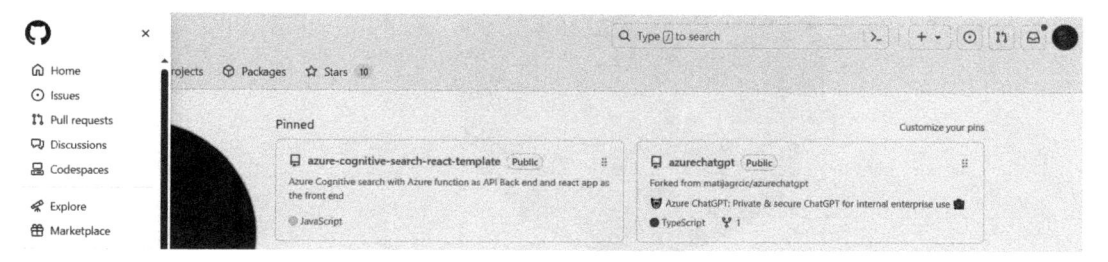

Figure 1.15 – GitHub's global navigation

- **Repository view**: Located at the center of the **Overview** page, the **Repositories view** is where you can easily access and manage your pinned repositories in GitHub. Here, you can conveniently organize and showcase the repositories that matter most to you, allowing for quick and effortless access to your key projects and contributions:

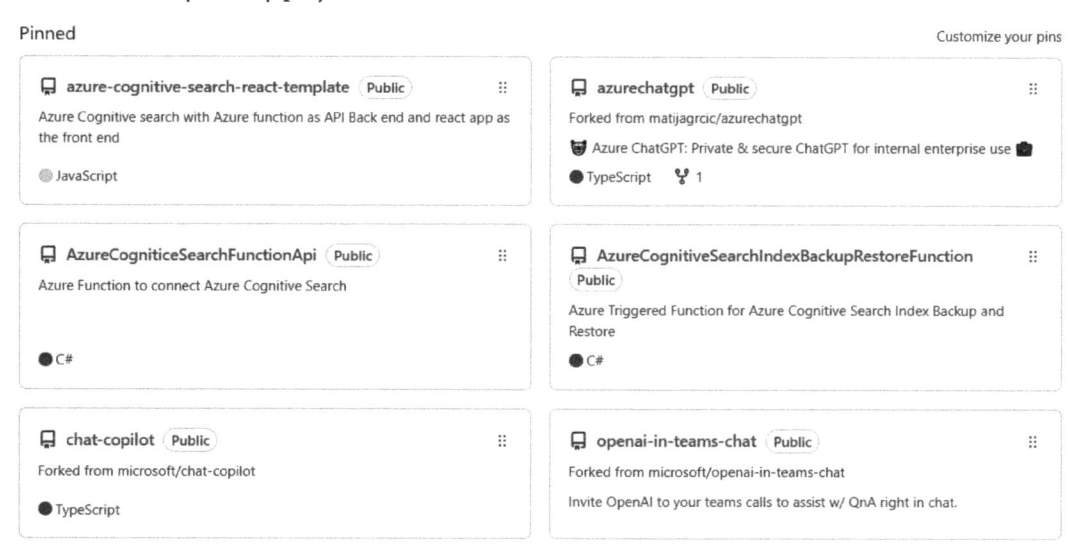

Figure 1.16 – GitHub's Repositories view

- **Biography view**: Positioned on the left side of the page, this is where you can craft a summary about yourself by clicking on the **Edit profile** button. Below your profile, you'll find statistics on the number of users you follow and those who follow you on GitHub. Additionally, you can showcase your achievement badges and list the organizations you're a part of:

20 Introduction to Version Control and GitHub

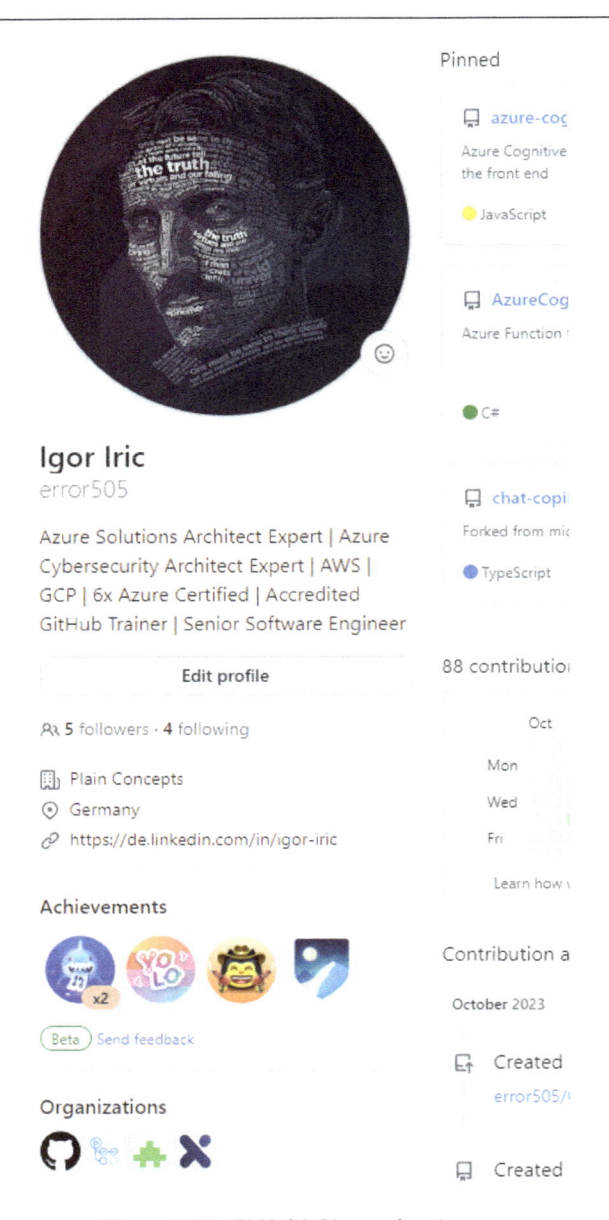

Figure 1.17– GitHub's Biography view

- **Contribution view**: Located at the bottom of the page, this provides valuable insights into all your contributions to repositories over the selected year on GitHub. It showcases your commit activity, pull requests, issues opened, and more, allowing you to track your involvement in open source projects and your development journey. This view helps you visualize your impact and contributions to the GitHub community:

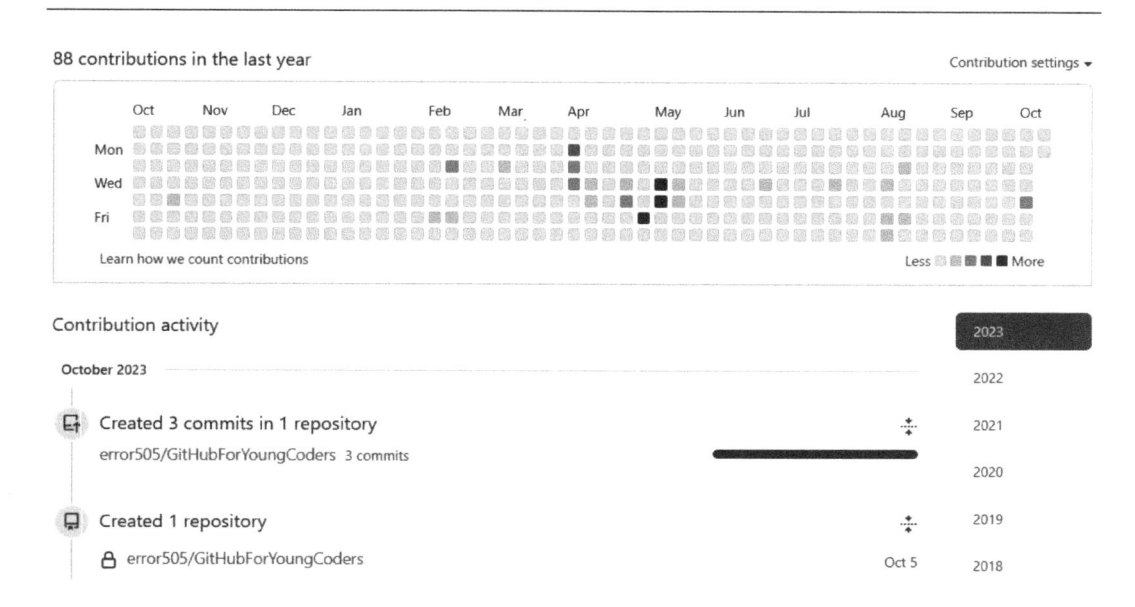

Figure 1.18 – GitHub's Contribution view

Organization accounts – team headquarters

An organization account is like a big office building where a whole team or company can work together. It has more space and extra features to help groups collaborate on multiple projects. Here's the breakdown:

- **Team management**: You can create various teams, assign different projects, and set specific roles for members
- **Multiple repositories**: It's like having several floors in your building, each for a different project
- **More privacy options**: You can have private rooms that only certain members can access
- **Unified billing**: Instead of paying for things individually, all expenses are handled together, making budgeting easier
- **Professional profile**: Showcase your team, your projects, and your achievements in a professional setup

In *Figure 1.19,* you can see what an organization account on GitHub looks like:

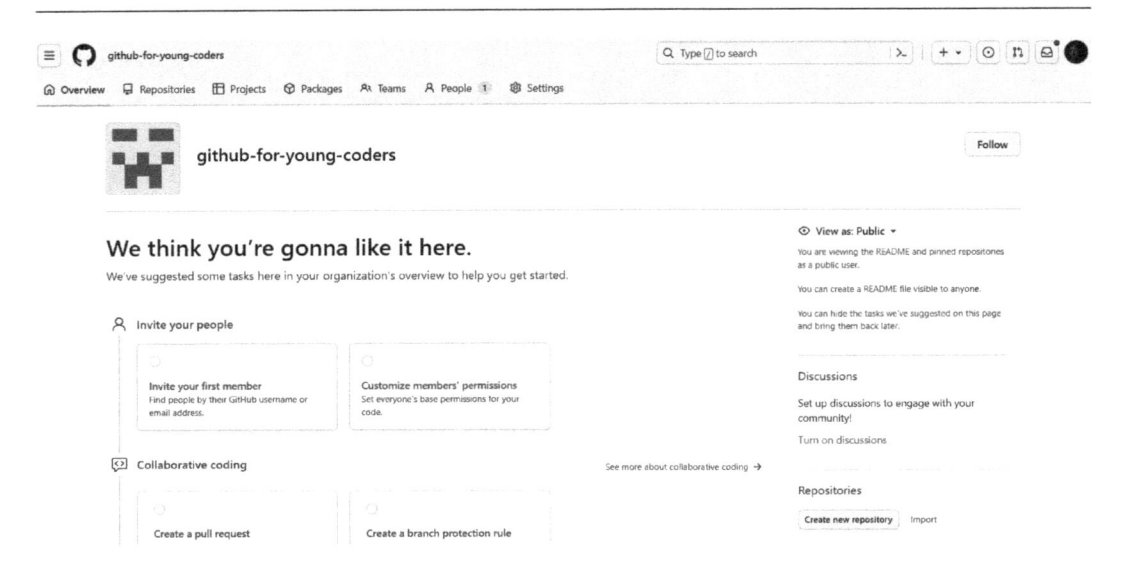

Figure 1.19 – GitHub's Organization view

Now that you're familiar with setting up your GitHub account and the different types of accounts available, let's take a closer look at how you can actually use GitHub in your projects. Next up, we'll guide you through the GitHub interface where you'll learn about key features such as the **Code** section, **issues**, pull requests, and more. This will help you understand where everything is and how to use each feature to manage your projects effectively. Whether you're just starting out or looking to get more from GitHub, this next section will give you the tools you need to navigate and make the most of your GitHub experience.

Navigating the GitHub interface – a guide for beginners

Think of GitHub as a multifunctional toolbox, each tool designed for a specific task in your project repository. Here's a friendly walkthrough of the main tools you'll find in this digital toolbox such as **Code**, **Issues**, **Pull requests**, **Projects**, and so on. You can see how this looks in *Figure 1.20*:

Figure 1.20 – GitHub's sections

So, what is the Code section about?

Think of the **Code** section as the heart of your GitHub project repository. This is where your project lives. It's like the main folder on your computer where you keep all your project files. It contains all the coding files, documents, and resources you need for your project:

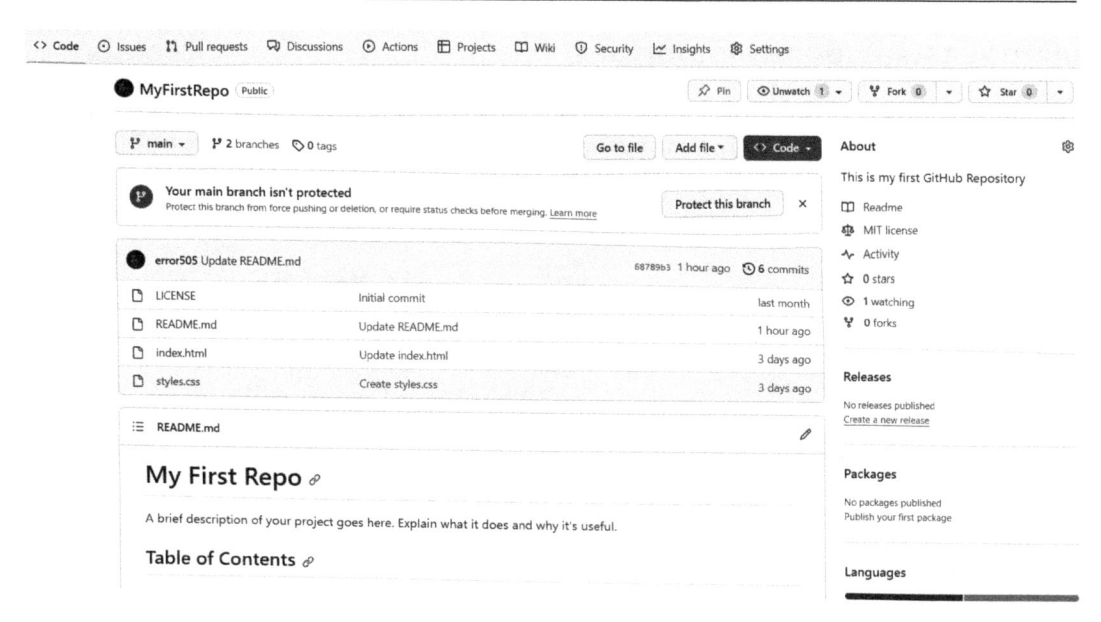

Figure 1.21 – GitHub's Code section

In the next chapter, *Mastering GitHub Essentials*, we'll dive into each aspect of the GitHub **Code** view. We'll also explore how to create and manage a repository on GitHub, giving you a comprehensive understanding of these fundamental tools.

And Issues, what's that?

The **Issues** section is your project's repository's bulletin board. Found a bug? Have an idea? Post it here. It's like leaving sticky notes for your team, saying, "*Hey, let's fix this* or *What about trying this?*" You can see a glimpse of this section in the following figure.

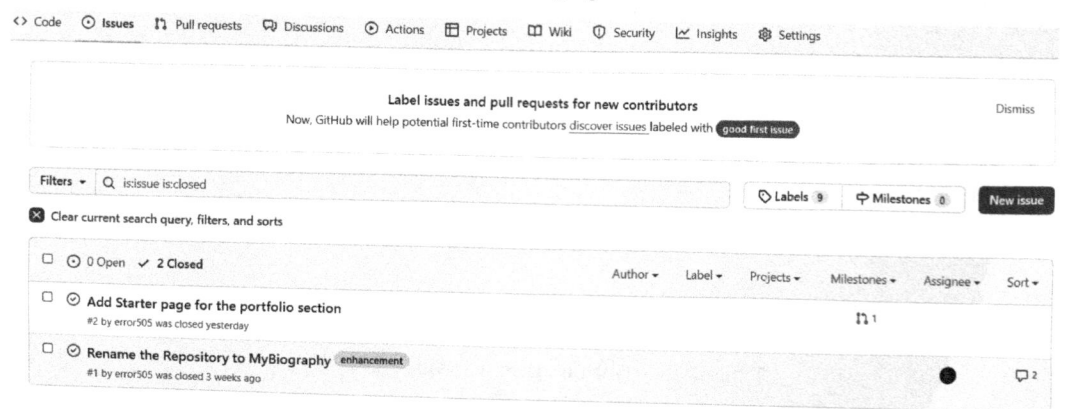

Figure 1.22 – GitHub's Issues section

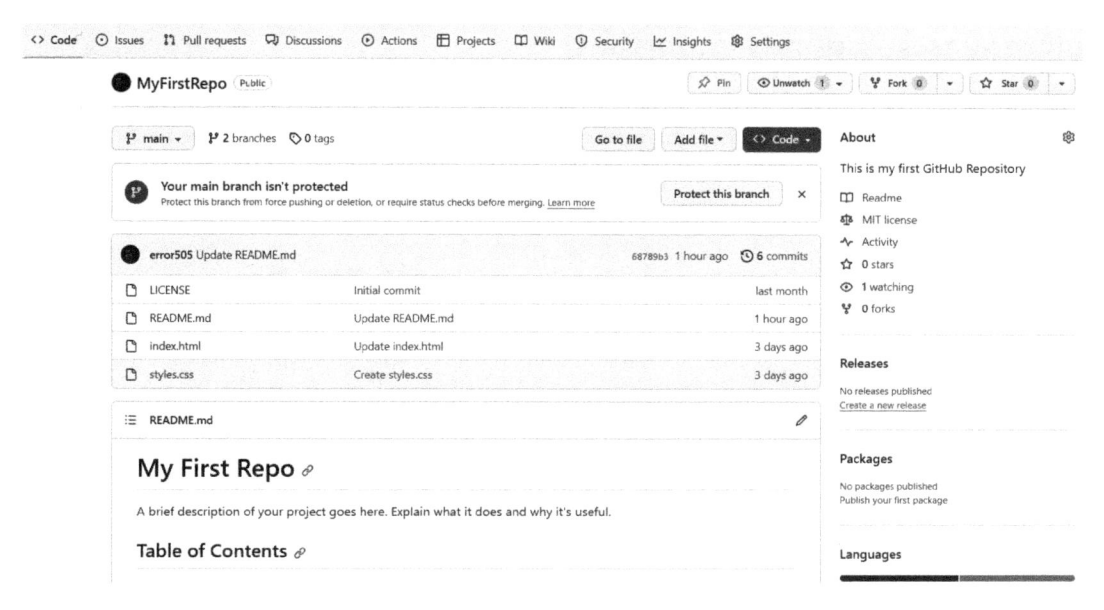

Figure 1.21 – GitHub's Code section

In the next chapter, *Mastering GitHub Essentials*, we'll dive into each aspect of the GitHub **Code** view. We'll also explore how to create and manage a repository on GitHub, giving you a comprehensive understanding of these fundamental tools.

And Issues, what's that?

The **Issues** section is your project's repository's bulletin board. Found a bug? Have an idea? Post it here. It's like leaving sticky notes for your team, saying, "*Hey, let's fix this* or *What about trying this?*" You can see a glimpse of this section in the following figure.

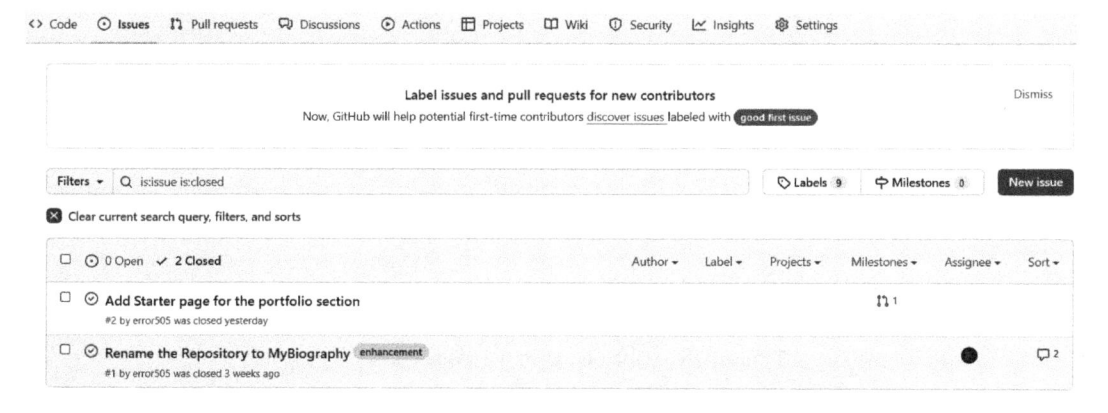

Figure 1.22 – GitHub's Issues section

Pull requests – sounds important?

Pull requests are your way of saying, "*I've made something cool, let's add it to our project!*" It's like showing your work to the team and asking for its thumbs-up before your code joins the party:

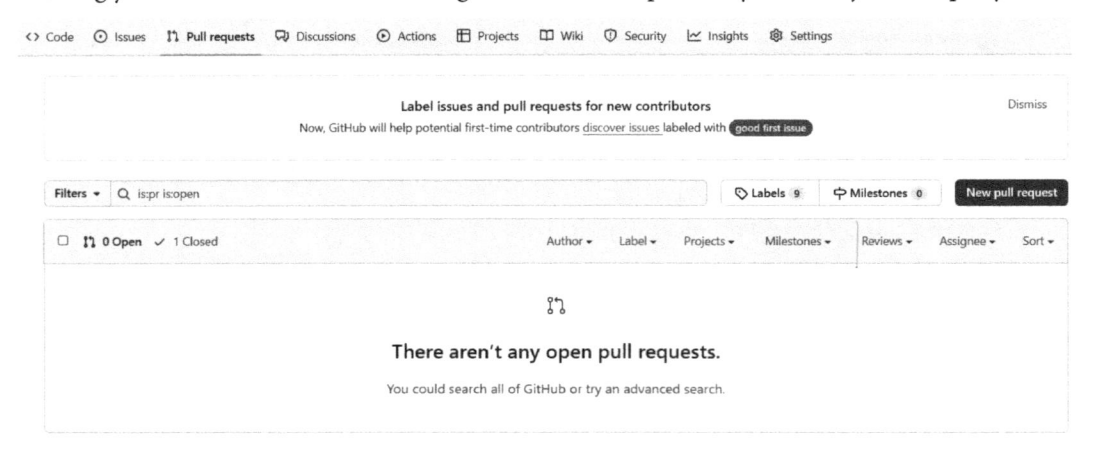

Figure 1.23 – GitHub's PR section

Discussions, are they like meetings?

Exactly! The **Discussions** section is your project's chat room. This is where ideas fly around, questions get asked, and collaborations happen. It's the virtual water cooler of your GitHub project:

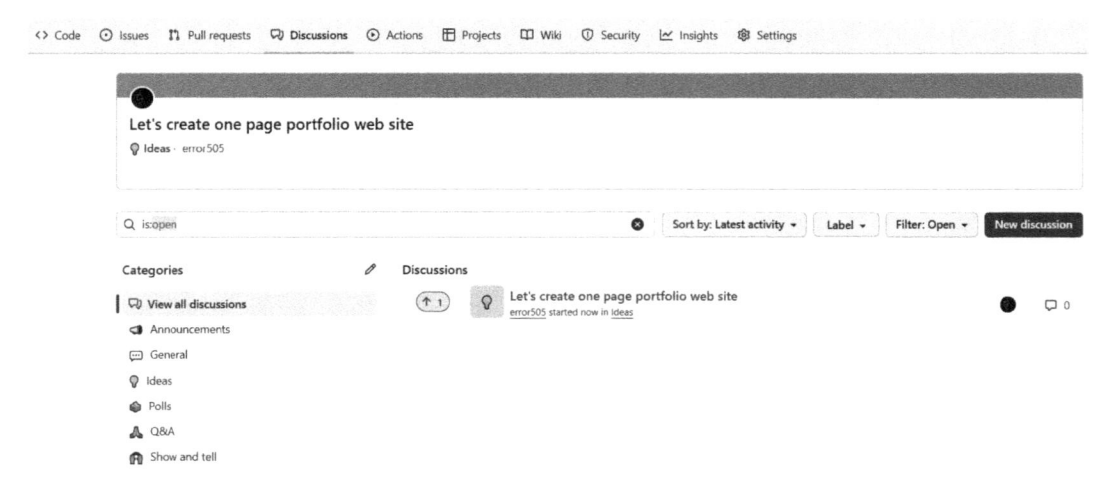

Figure 1.24 – GitHub's Discussions section

Actions, do they do my work?

Well, sort of! **Actions** are your automated assistants. They handle tasks such as testing your code or updating and deploying your website, doing the repetitive work for you so you can focus on the fun stuff:

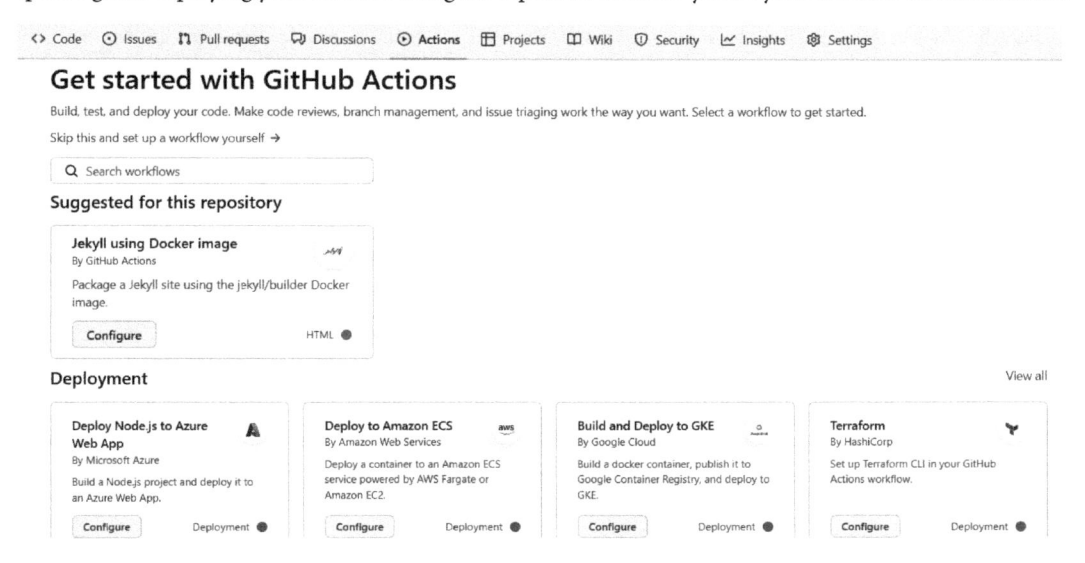

Figure 1.25 – GitHub's Actions section

Projects – that sounds serious?

GitHub's **Projects** section functions as the central hub for managing your work. In this space, you can organize your tasks, monitor progress, and assign responsibilities. It acts as your strategic planning board, ensuring that every aspect of your project stays aligned and on schedule:

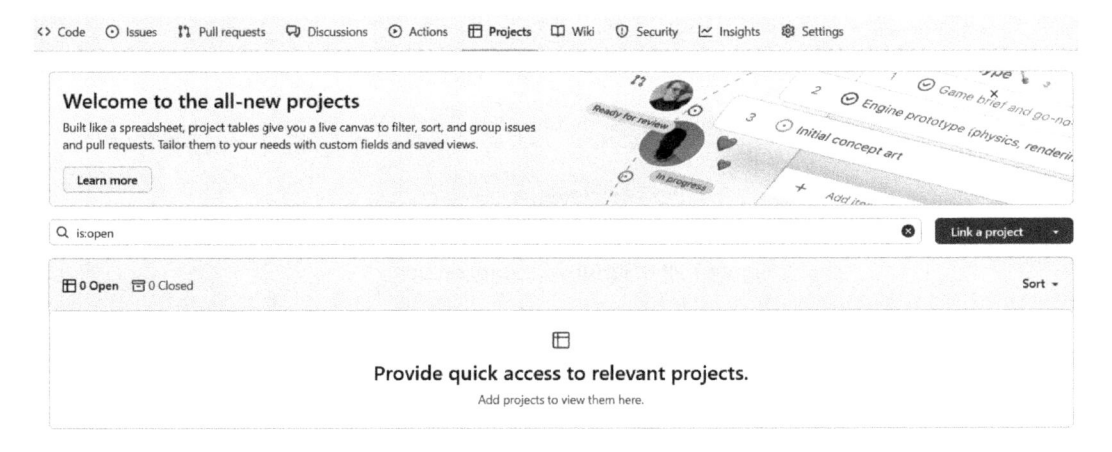

Figure 1.26 – GitHub's Projects section

Wiki, is that like Wikipedia?

Spot on! Your project's **Wiki** section is its knowledge base. It's the go-to place for documentation, how-to, guidelines, and all the wisdom you've accumulated about your project:

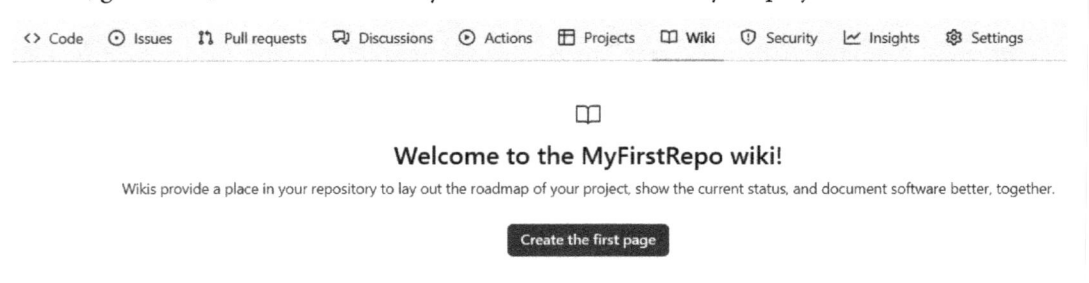

Figure 1.27 – GitHub's Wiki section

What's in the Security section?

Think of **Security** as your project's bodyguard. It helps you find and fix any security weaknesses, keeping your project safe from virtual bumps and bruises:

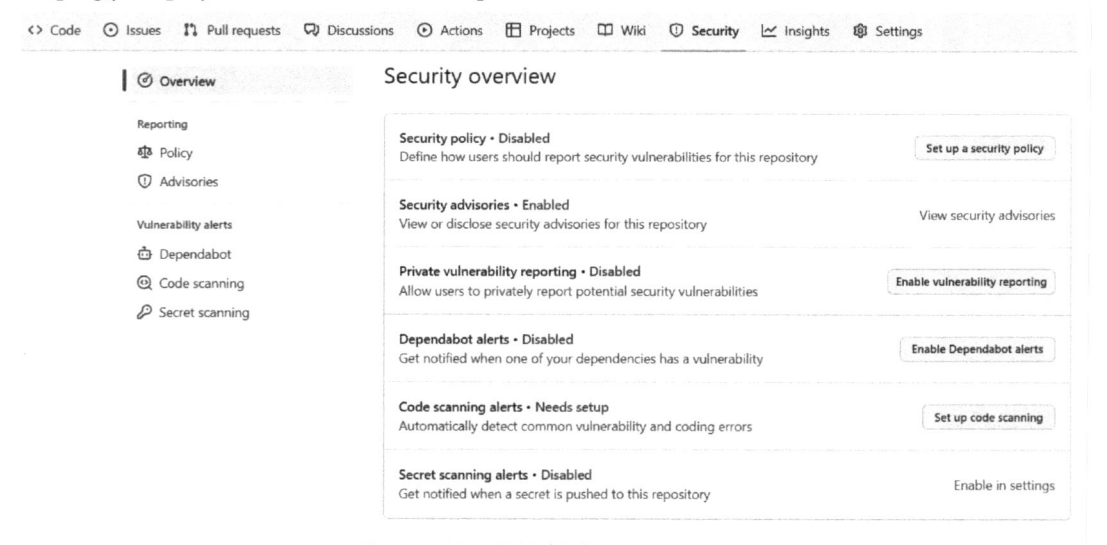

Figure 1.28 – GitHub's Security section

Insights, do they tell the future?

Not quite, but close! The **Insights** section is like a report card showing how your project is doing. It offers stats and graphs on your project's health, progress, and team contributions:

Navigating the GitHub interface – a guide for beginners 27

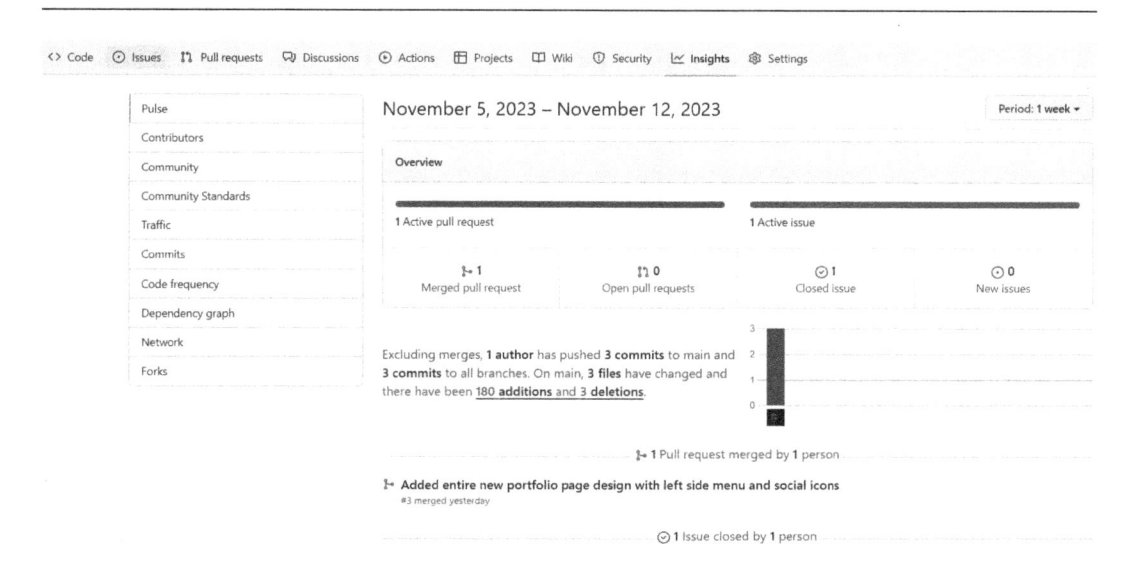

Figure 1.29 – GitHub's Insights section

Settings – what can I set there?

The **Settings** section is your control room. Here, you manage collaborators, set rules, and tweak your project's settings. It's where you customize your GitHub experience to fit your project's needs:

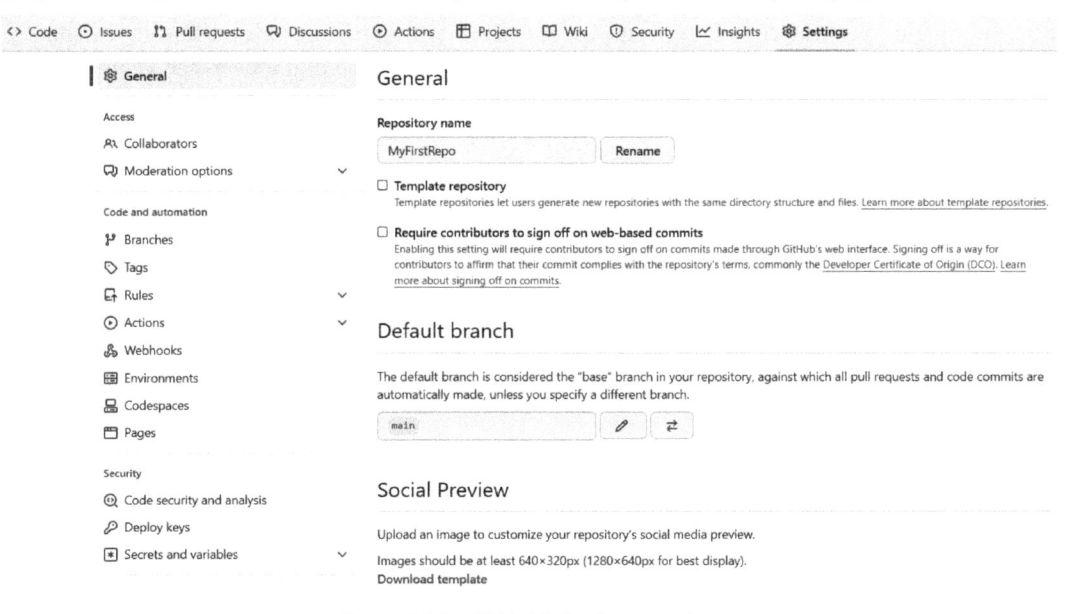

Figure 1.30 – GitHub's Settings section

Now that we've covered the basics of navigating the GitHub interface and explored sections like Code, Issues, Pull Requests, Discussions, Actions, Projects, Wiki, Security, Insights, and Settings, let's see how much you've learned.

You'll find a *Quiz* section designed to test your knowledge and help reinforce what we've discussed. It's a fun way to ensure you understand the key points before moving on to more advanced topics. Ready to test your skills? Let's get started!

Summary

This chapter explained how GitHub uses version control, which is like a magic book that lets you see all the changes you've made to your code and even undo them if you need to. It keeps everything safe and in good order. It then showed you how to create your own space in GitHub, called a *repository*, which you can think of as a treasure chest for your coding projects. You've learned how to add your code, change it, and share it with others, like showing your artwork to the world.

Next, the chapter talked about how GitHub is great for working with others. It's like having a big party where everyone shares their coding ideas. This makes coding more fun and creative.

The benefits of using GitHub are pretty cool. You can see how other people solve coding problems, which can help you learn new ways to solve your own. If you make a change and don't like it, GitHub's version control lets you go back to how it was before. Additionally, you can get advice and tips from other coders, which can help you improve.

In the next chapter, you will be diving into the deep end of the pool, but in a fun way! It's all about getting your hands dirty (in a good way) with Git and GitHub. You're going to learn how to use some powerful tools in GitHub, almost like learning spells to control your coding universe. It's going to be super fun and really useful to your coding adventures!

Quiz

Check your knowledge earned by reading this chapter with these quiz questions:

1. What does GitHub use to help manage and organize code?

    A. Version control

    B. Spell check

    C. Grammar correction

    D. Auto-formatting

    Answer: A. Version control

2. True or false: GitHub is only used by professional coders.

    A. True
    B. False

    Answer: B. False

3. Fill in the blank: Version control in GitHub is like a _____ that lets you see and undo all your changes.

    Answer: magical time-traveling book

4. Why is version control considered a superpower for coders?

    Answer: It allows coders to track and revert changes easily, keeping their code organized and safe.

5. Which of the following is a key benefit of using GitHub for your code?

    A. It increases file size
    B. It makes code less secure
    C. It helps organize and track changes
    D. All of the above

    Answer: C. It helps organize and track changes

6. True or false: Version control can only track changes in text files.

    A. True
    B. False

    Answer: B. False

7. What is the source of the analogy used in this chapter to explain GitHub's functionality?

    A. A toolbox
    B. A time-traveling book
    C. A recipe book
    D. A roadmap

    Answer: B. A time-traveling book

8. Fill in the blank: GitHub is an incredible world of _____ and version control.

    Answer: coding

9. True or false: Every change made in GitHub version control is automatically saved forever.

   A. True
   B. False

   Answer: A. True

10. What is one practical tip mentioned in the chapter for using GitHub version control effectively?

    Answer: Regularly commit changes to track progress effectively

2
Mastering GitHub Essentials

In this chapter, we will dive into the basics of Git and GitHub. You'll learn how to set up your workspace, use Git commands for effective project management, and collaborate with others. This chapter equips you with the knowledge to clone repositories, understand Git's basic commands, manage personal access tokens, and create meaningful README files.

By the end of this chapter, you'll be able to set up projects, save changes, and work collaboratively on GitHub.

In this chapter, we're going to cover the following main topics:

- How Git interacts with GitHub – Git commands and GitHub workflows
- What are repositories on GitHub?
- Getting your tools ready – downloading and installing Git
- Creating and managing personal access tokens

Technical requirements

Before starting, ensure you have Git installed on your machine. You can download it from https://git-scm.com/. Additionally, create a GitHub folder named ch2 to organize any files or projects related to this chapter. You can download the code for this chapter at https://github.com/PacktPublishing/GitHub-for-Next-Generation-Coders/tree/main/Chapter%202.

How Git interacts with GitHub – Git commands and GitHub workflows

Suppose you've just joined a team of developers to work on an exciting new project hosted on GitHub. You're eager to dive in, but first, let's understand how Git commands on your local machine interact with GitHub, making collaboration seamless and efficient.

Git clone – cloning your first GitHub repository (the welcome kit)

When you join a project, the first thing you do is use the `git clone` command to copy the project from GitHub to your local machine. Think of it as receiving a welcome kit on your first day at a new job. This kit contains all the existing work done on the project:

Figure 2.1 – The GitHub welcome kit

So, you would finally like to clone the GitHub repository to your local computer? OK, let's see how you can accomplish that:

1. Navigate to the repository on GitHub you created previously, `MyFirstRepo`, and click on the green **Code** button:

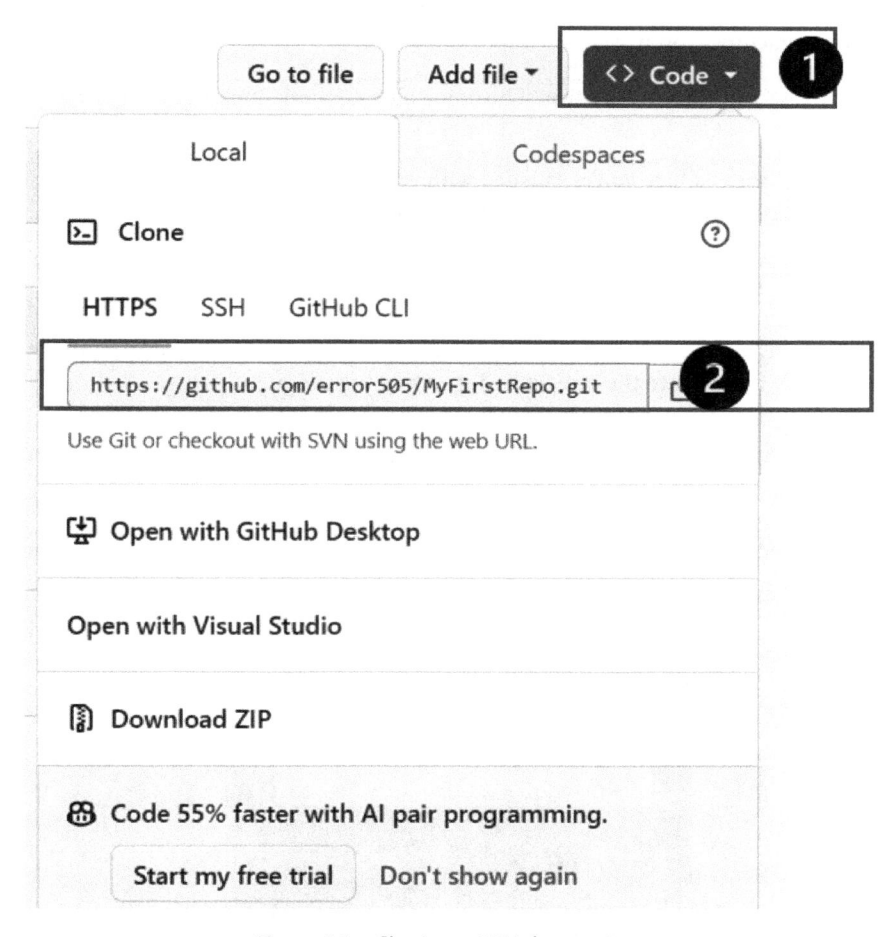

Figure 2.2 – Cloning a GitHub repository

2. You'll see a URL (it should end in `.git`). Copy this URL.

3. Now, go back to your Git Bash or Terminal, navigate to the folder where you want your project to live, and type the following command, replacing URL with the URL you just copied:

    ```
    $ git clone URL
    ```

4. Now you can access your project locally. Once the cloning process is complete, navigate to your project folder that has been created during the cloning process:

    ```
    $ cd MyFirstRepo
    ```

Now, on your local machine, you have three important areas:

- **Working directory**: This is where you'll be doing all your work, making changes, and creating new files.
- **Staging area** (**index**): Once you're satisfied with your changes, you move them to the staging area, kind of like putting your finished draft in a "to be reviewed" folder.
- **Local history**: When you're completely satisfied with your changes, you commit them. This is like saving a final draft in a "completed work" folder.

That's it. You've now set up Git, introduced yourself, and brought your GitHub project to your local machine. You're ready to dive into the project and start making progress.

Git add – adding files to your repository, the local way and the GitHub way

After making some code changes in the working directory, you use `git add` to move them to the staging area, like sorting and packing items into a box, ready to be shipped off to the project's main storage on GitHub:

Figure 2.3 – Adding files to the repository

Imagine your repository is like a package storage. Right now, it's empty, but you're about to add some items (packages or components) to create valuable content (your project). You can either create items directly in your package (locally on your computer) or obtain some pre-made items from the storage (GitHub) and include them in your package. Let's explore both methods!

Follow these steps to create packages (files) locally using `git add` and `touch` commands:

1. **Creating Your Packages (Files)**:

    I. Open your Git Bash or Terminal.

    II. Navigate to your repository's folder using the `cd MyFirstRepo` command.

    III. Now, let's create a file. Type `touch index.html` to create a new file called `index.html`.

2. **Storing Your Packages**: Your package (file) is prepared, but it's not in the storage yet. Type `git add index.html` to inform Git that you're ready to place this package in your storage.

Creating files directly on GitHub

In this section, we're going to learn how to create files directly on GitHub, a useful skill for quickly adding content to your repository without the need for local file management. This process is straightforward and can be done entirely through the GitHub website, making it a convenient option for managing your project files.

Let's dive into each step to ensure you know exactly how to create and manage your files on GitHub:

1. Create your packages (files) on GitHub:

    I. Navigate to your repository on GitHub.

    II. Click on the **Add file** button and choose **Create new file**:

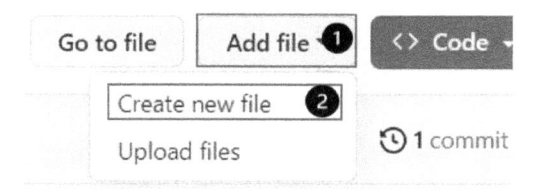

Figure 2.4 – Adding a file using the GitHub interface

2. Give your file a name, `index.html`:

Figure 2.5 – Naming a file

3. You can write something in your file, such as `Hello, World!`, in the big text box below the Edit button. Later, we will start building your own portfolio page that you will host on GitHub for free.

Figure 2.6 – Committing your changes using the GitHub interface

4. In the upper-right corner, click on the green **Commit changes** button to save your new package (file) to the community garden.

5. A new pop-up window will appear, allowing you to define your commit message and provide a description:

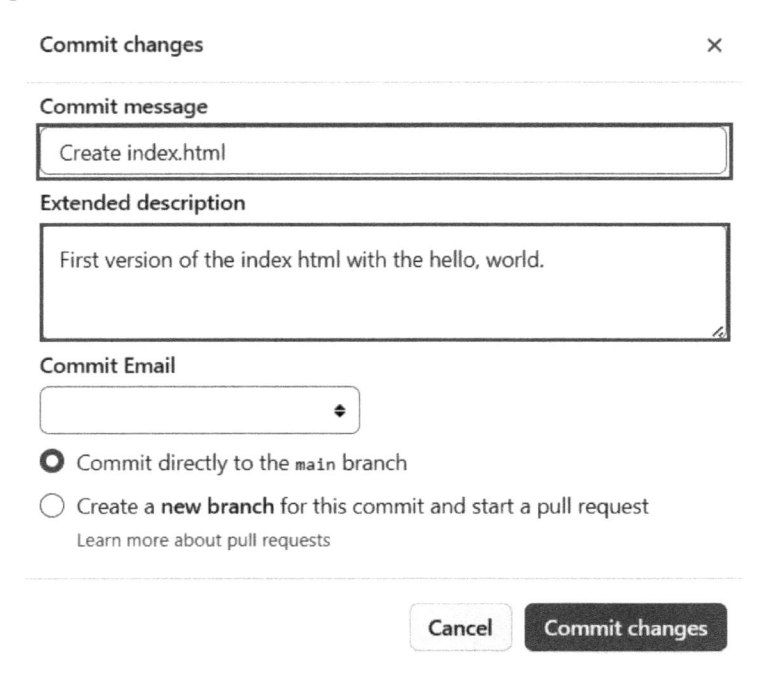

Figure 2.7 – Committing the message and detailed description

Offering meaningful commit messages and descriptions is crucial, as they enable your colleagues to understand the additions made to the repository and their purpose.

Git commit – saving your snapshot

Before shipping off the changes to GitHub while working locally, `git commit` takes a snapshot of the changes in the staging area. It's like having a packing list for the box, detailing what's included (*Figure 2.8*).

Figure 2.8 – Git commit

In this instance, you've made a commit to the main branch, which may not be the ideal approach. However, we will learn more about branches and branching strategies to improve your understanding.

Git push – the delivery of the changes to GitHub

Now it's time to share your work with the team. `git push` is the delivery truck that takes your box of changes from your local machine to GitHub, making it available to everyone on the project:

Figure 2.9 – Git push

Git fetch – the update checker

While you're working, your teammates are also making changes. `git fetch` is your way of checking what's new on GitHub without merging those updates into your local project, like checking for package deliveries without bringing them inside yet:

Figure 2.10 – Git fetch

Git pull – bringing it all together

When you're ready to integrate the latest updates from GitHub to your local machine, `git pull` does the job. It's like bringing in the packages (your teammates' changes) inside and opening them up to add to your project. It combines fetching the latest updates and merging them into your local project in one command.

Network interaction

Among these commands, `git clone`, `git push`, `git fetch`, and `git pull` are the key players that interact with GitHub over the network, bridging the space between your local machine and the online repository on GitHub.

These commands are your toolkit for a collaborative workflow, ensuring you and your team stay in sync while working on the project, no matter where you are located.

You've now learned how Git interacts with GitHub through various commands and workflows. We've covered cloning repositories with git clone, adding files both locally and directly on GitHub, saving changes with git commit, pushing updates with git push, and keeping everything in sync with git fetch and git pull. This foundation helps you manage your code effectively and collaborate with others.

Next, we'll explore what repositories on GitHub are all about. Understanding repositories is essential, as they are the core of your projects on GitHub. We'll look at how to create, manage, and use repositories to keep your projects organized and accessible.

What are repositories on GitHub?

Imagine you've just moved to a new city and you're in search of a place to live—a space that you can call your own, where all your belongings and memories would reside. In development, a repository (or *repo* for short) serves as this home for your project:

Figure 2.11 – The GitHub repository

A **repository** is a centralized location where your project's files and revision history reside. It's like a big filing cabinet where all project-related materials, such as code files, documentation, and more, are stored and organized. Each file has its own drawer in this cabinet, and every change to these files is very well recorded, like a logbook of who changed what and when.

Now, let's dive into the neighborhood of this city *GitHub*, where repositories live.

The GitHub neighborhood

GitHub is like a huge neighborhood with millions of houses (repositories). Each house belongs to different individuals or teams working on various projects. Some houses are full of activity, with many people contributing and making changes, while others might be quiet, with only one or two inhabitants.

In this neighborhood, you can visit other houses, admire their design (code), and even suggest improvements (contribute). If you really like a particular house's design, you can duplicate it (fork) and have your own version to tinker with.

Creating and managing a repository on GitHub

When you decide to build your own house in this neighborhood, you create a new repository on GitHub. This is where you'll store all the plans, designs, and progress of your project. As you work, you'll make many changes—adding new rooms, painting walls, or fixing the plumbing. Each of these changes is recorded in your repository, allowing you (and others) to track the evolution of your project over time.

By now, you should really want to start creating a new repository, so let's start:

1. Click on the + icon in the top-right corner of the GitHub homepage.
2. Select **New repository** from the drop-down menu:

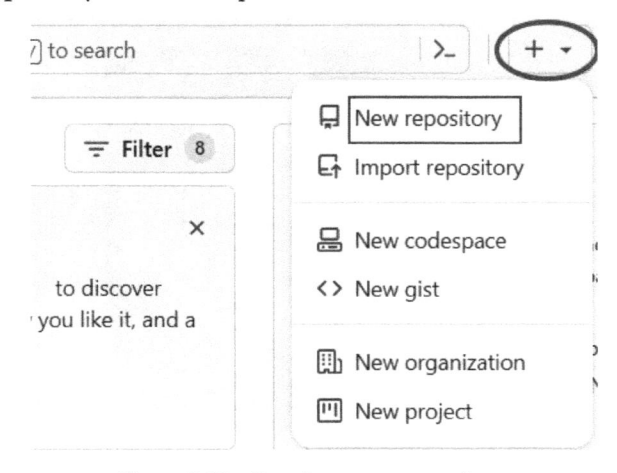

Figure 2.12 – Creating a new repository

3. Now you must fill in all the required fields to create your new repository, but first, let's fill in the owner and the name of the repository:

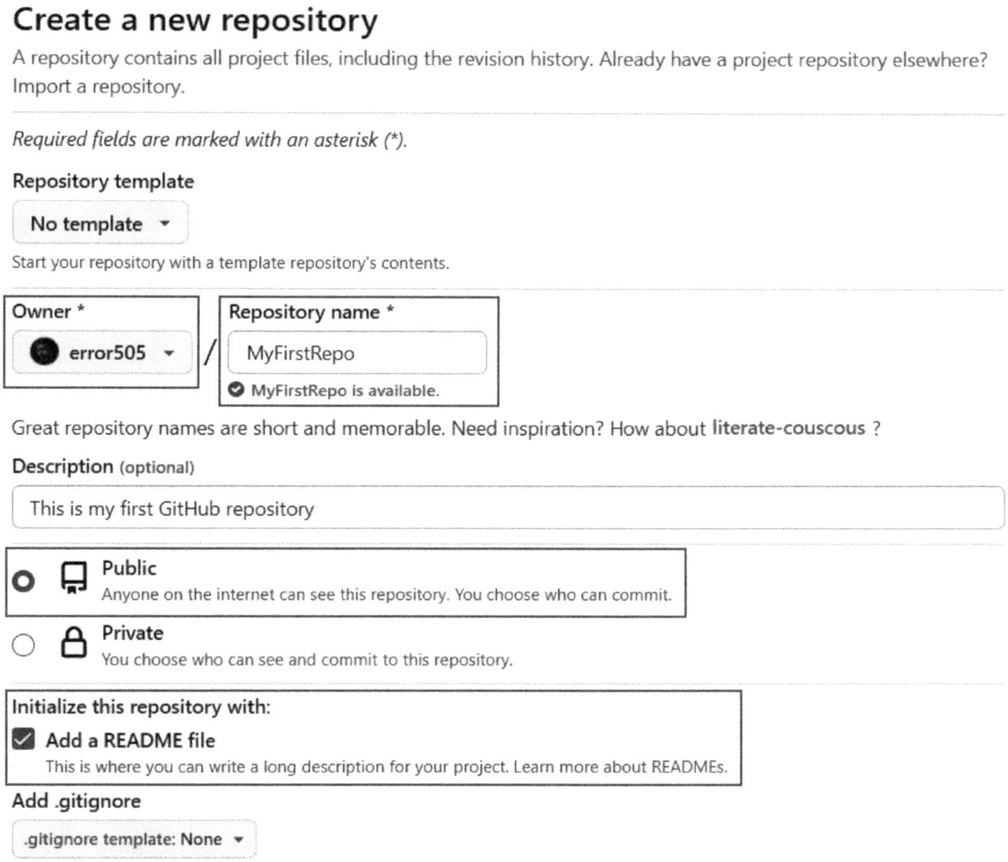

Figure 2.13 – Configuring a repository

4. Lastly, select the license type; in this instance, we will choose **MIT License**. Then, proceed by clicking on the green **Create repository** button:

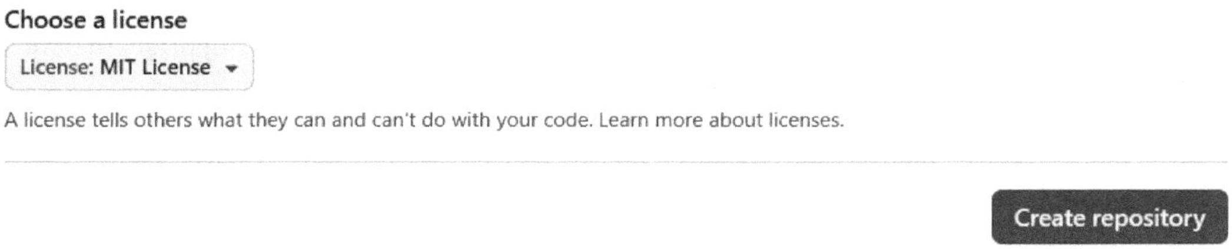

Figure 2.14 – Choosing a license for the repository

The next screen will display the newly created repository along with the GitHub repository view, including the README file:

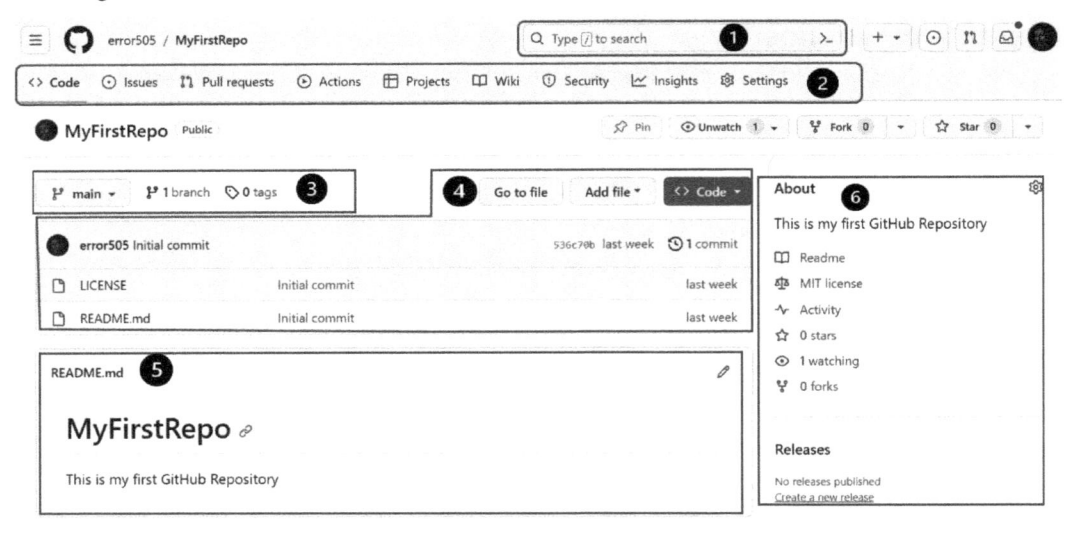

Figure 2.15 – Repository overview

Let's break down the GitHub repository page:

1. **Search bar**: This is where you can type and look for anything on GitHub. Think of it like a Google search, but just for GitHub stuff.

2. **Repo management**: This section is like the control center of your project. You can see different parts of your project, such as issues (problems you want to fix), pull requests (changes you or others want to add), and more.

3. **Branches**: Imagine your project is like a tree. The main part of the tree is the **main** branch. But sometimes, you want to try out new things without changing the main tree, so you make a smaller branch. That's what branches in GitHub are—different versions of your project.

4. **Files**: This area shows you all the files and folders in your project, just like files and folders on your computer!

5. **README**: This is a special file where you can write about your project. It's like a welcome sign that tells people what your project is about and how they can use it. It is very important to have a good README file for each repository you create on GitHub.

6. **Metadata**, **releases**, **packages**, and **contributors**:

    - **Metadata**: This is the **About** section, which is a short description of your project

    - **Releases**: This is where you can put out new versions of your project for others to use

That's a simple breakdown of the GitHub repository page. GitHub is like a big playground for your projects, and these parts help you organize and share your work with others! Next, let's look at how to download and set up Git.

Getting your tools ready – downloading and installing Git

Starting your journey with Git and GitHub is like setting up your workspace before starting on a creative project. The first step is to ensure you have the right tools in place. Here's how you get started:

1. Navigate to the Git website at https://git-scm.com/.
2. Click on the **Download** button and the website will automatically suggest the right version for your operating system—be it Windows, Mac, or Linux:

Figure 2.16 – Downloading Git

3. Once the download is complete, open the installer.
4. Follow the on-screen instructions. It's usually safe to go with the default settings if you're new to Git:

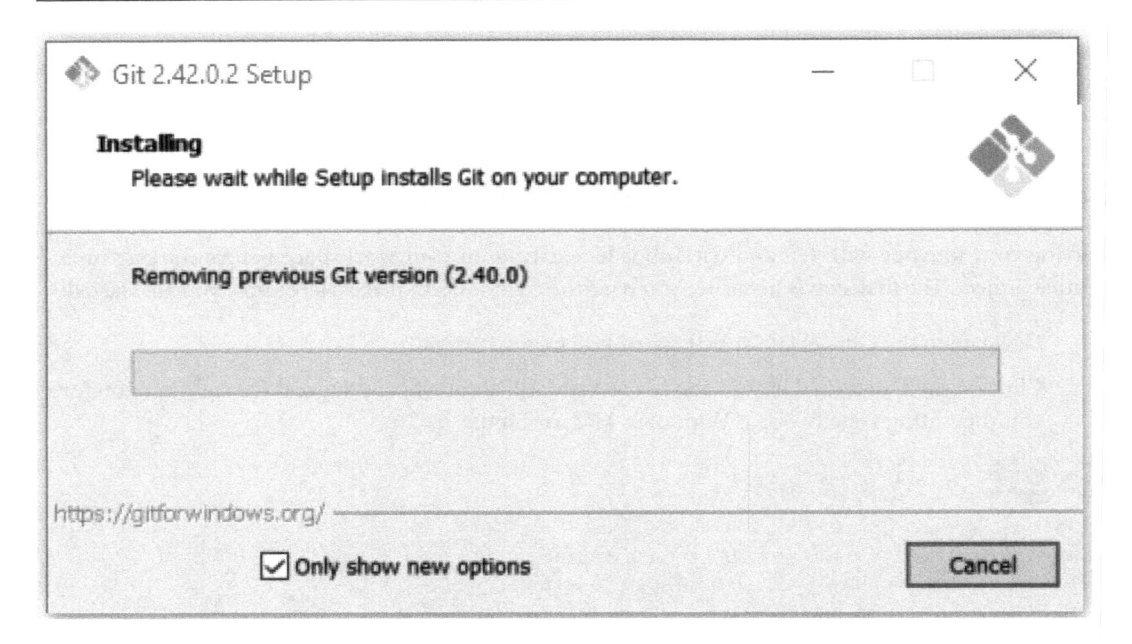

Figure 2.17 – Installing Git locally

Now that you've got Git installed, it's time to introduce yourself to it. Git needs to know who you are so that it can attach your name and email address to the changes you make.

But before we go further, let's understand the following Git configuration levels:

- `--system`
- `--global`
- `--local`

Understanding Git configuration levels

Imagine you live in a big house with many rooms, and you have a favorite tune you like to whistle while you wander around. Now, you decide that each room should have its own unique echo to your tune, and also that everyone in the house should have a common beat they follow, but only when they are in the house. Lastly, when you step outside, there's a neighborhood rhythm that everyone in the area follows.

In Git, setting configurations is a bit like setting the musical rules in this scenario. There are three levels where you can set your preferences: system, global, and local. Let's break them down:

- **System** (`--system`)

  - This is the *neighborhood rhythm* in our story
  - When you set a configuration at the system level, it's like setting a rule for the whole computer, no matter who uses it

  For example, if you set a system-level rule that says `Every filename should be written in capital letters`, then everyone working on any project on that computer will have to follow that rule.

- **Global** (`--global`)

  - This is the *house beat* in our story
  - Global configurations are like setting rules for yourself, but they apply wherever you work on your computer

  For example, suppose you prefer a certain way of naming your files, such as starting with the date. So you set a global rule, and now, no matter which room (project) you're in, you follow this rule.

- **Local** (`--local`)

  - This is the *unique room echo* in our story.
  - Local configurations are the most personal. They only apply to the room (project) you're in.

  For example, in one particular project (room), you decide that you want to name all your files in a special way, such as ending them with `_final`. This rule only applies here, in this project.

So, when you're setting your preferences in Git using `git config`, you decide which level to apply them to, much like deciding where your musical rules apply in our story. Here's how you might set a username at each level:

- **System**: `git config --system user.name "Your Name"`
- **Global**: `git config --global user.name "Your Name"`
- **Local**: `git config --local user.name "Your Name"`

Each of these commands sets your username for different scopes within your computer, from broad to narrow: everyone, just you, or just you in a specific project.

Let's configure Git on your local machine:

1. **Start your Terminal**: Open the Git Bash on Windows or Terminal on Mac/Linux.
2. **Set your username**: Type the following command, replacing Your Name with your actual name:

   ```
   $ git config --global user.name "Your Name"
   ```

3. **Set your email address**: Similarly, set your email address with the following command, replacing you@example.com with your actual email address:

   ```
   $ git config --global user.email you@github.com
   ```

4. **View your configuration settings**: To view your current Git configuration, you can run the following command:

   ```
   $git config --list
   $git config --global --list
   ```

Figure 2.18 shows the result of using the git config --list command. This command lists all the Git configuration settings for the user. It includes information like the **username**, **email**, and other settings that control how Git behaves. For example, it shows the user's name and email address, the text editor used for writing commit messages, and various alias commands to make working with Git easier.

Key Points in the Figure:

- **user.name** and **user.email**: These lines show the configured name and email of the user, which Git uses to label the author of each commit.
- **core.editor**: This specifies the text editor that Git will open when you need to write a commit message.
- **alias. commands**: These are shortcuts for Git commands. For instance, alias.lol is a shortcut for a longer git log command that shows a graph of commits.

```
● PS C:\Users\igor.iric\Desktop\GitHub for Young Coders> git config --list
  diff.astextplain.textconv=astextplain
  filter.lfs.clean=git-lfs clean -- %f
  filter.lfs.smudge=git-lfs smudge -- %f
  filter.lfs.process=git-lfs filter-process
  filter.lfs.required=true
  http.sslbackend=openssl
  http.sslcainfo=C:/Program Files/Git/mingw64/etc/ssl/certs/ca-bundle.crt
  core.autocrlf=true
  core.fscache=true
  core.symlinks=false
  core.editor="C:\\Program Files\\Notepad++\\notepad++.exe" -multiInst -notabbar -nosession -noPlugin
  pull.rebase=false
  credential.helper=manager
  credential.https://dev.azure.com.usehttppath=true
  init.defaultbranch=master
  user.name=Igor
  user.email=igor.iric@        ..com
  difftool.sourcetree.cmd='' "$LOCAL" "$REMOTE"
  mergetool.sourcetree.cmd=''
  mergetool.sourcetree.trustexitcode=true
  credential.helperselector.selected=cache
  alias.lol=log --oneline --graph --decorate --all
  alias.lol2=log --oneline --decorate --all
  alias.lold=git log --oneline --graph --decorate --all
  fetch.prunne=true
  fetch.prune=true
```

Figure 2.18 – Git config list

You've now got your tools ready by downloading and installing git, and configuring your local git setup. With everything set up, you can now start managing your projects efficiently.

Next, let's move on to creating and managing personal access tokens. These tokens are important for securely accessing your GitHub account and automating certain tasks. We'll learn how to create these tokens and use them effectively.

Creating and managing personal access tokens

What's a personal access token? Think of a **personal access token** (**PAT**) as a special password that you can create to access your GitHub account, without using your regular password. It's a way to let certain apps or tools have limited access to your account.

How to create a PAT

Follow these steps to create a PAT:

1. Go to your GitHub account in your web browser and log in. Now, click on your profile picture in the top-right corner and select **Settings** from the drop-down menu:

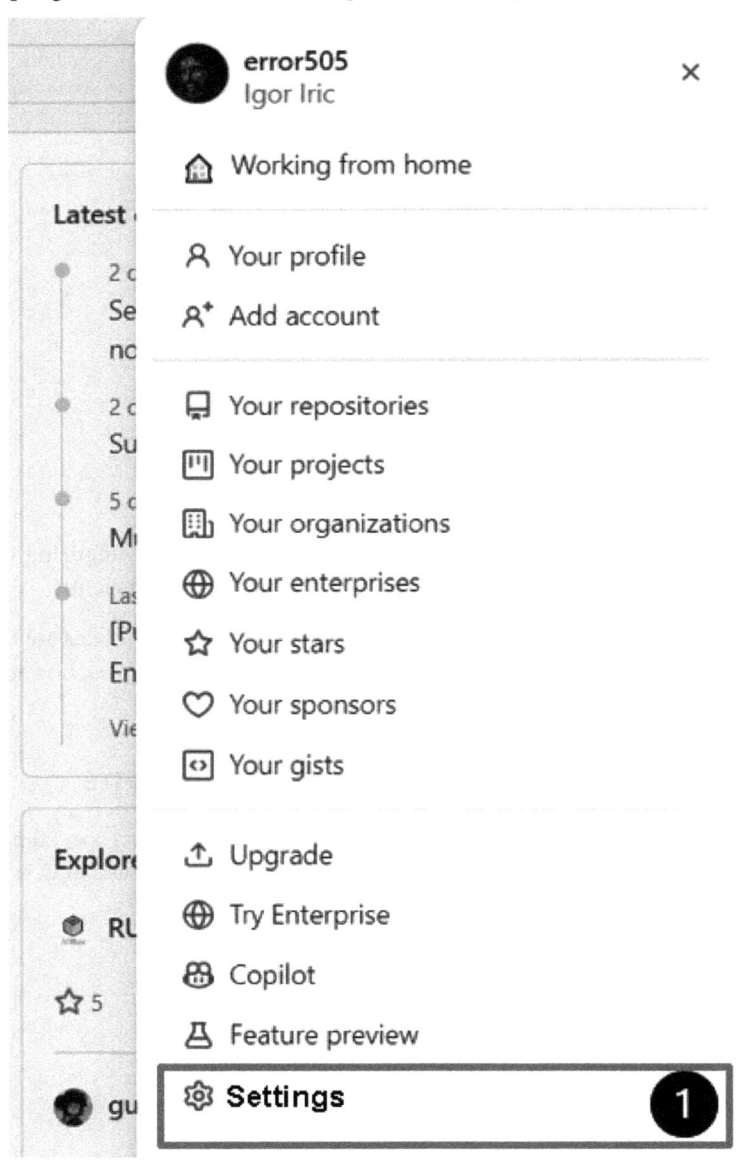

Figure 2.19 – Token creation user settings menu

2. On the left sidebar, click **Developer settings**:

- Code security and analysis

Integrations
- Applications
- Scheduled reminders

Archives
- Security log
- Sponsorship log

- **Developer settings** ❷

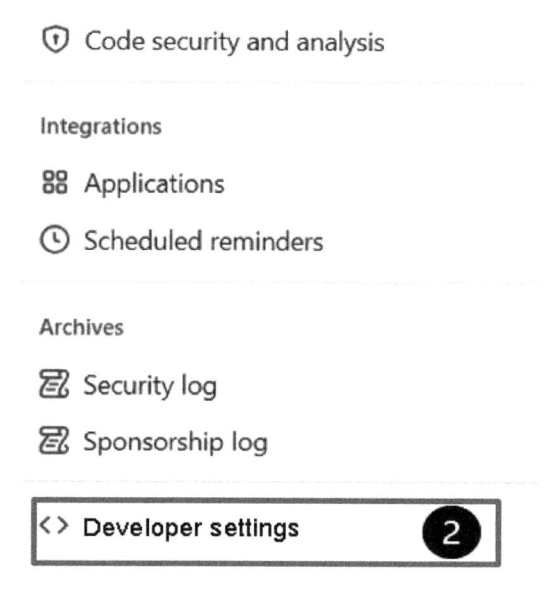

Figure 2.20 – Token creation: developer settings

3. Now click on **Personal access tokens**.
4. Select **Tokens (classic)**:

Settings / Developer Settings

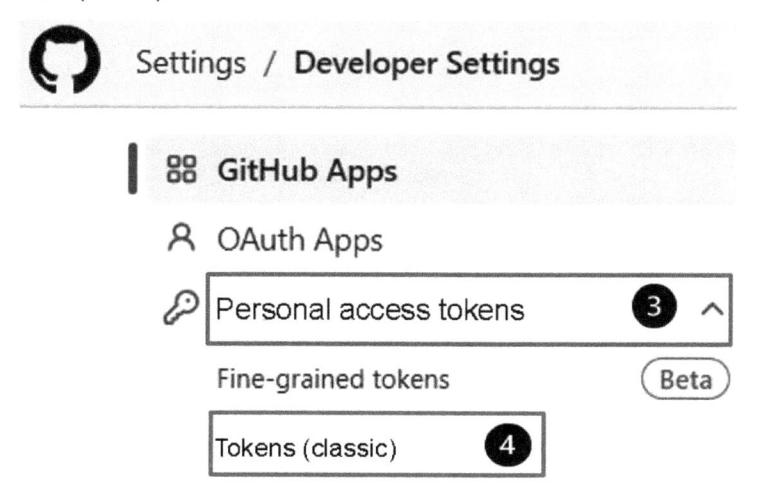

- GitHub Apps
- OAuth Apps
- Personal access tokens ❸
    - Fine-grained tokens (Beta)
    - Tokens (classic) ❹

Figure 2.21 – Token creation: classic token

Now, let's look at how to make your token.

5. Click on **Generate new token**.
6. Select **Generate new token (classic)**:

Personal access tokens (classic)

Tokens you have generated that can be used to access the Git...

gh-developer-training — *repo, user, write:packages*

Generate new token ▼   Revoke all

Generate new token (Beta)
Fine-grained, repo-scoped

Generate new token (classic)
For general use

Delete

Figure 2.22 – Token creation: generating a classic token

7. Name your token something that will help you remember what it's for, such as `token-for-my-app`.
8. Choose the permissions or scopes you want this token to have. For example, to let it access your repositories, check **repo**.
9. Select for how many days you want your token to be valid:

New personal access token (classic)

Personal access tokens (classic) function like ordinary OAuth access tokens. They can be used instead of a password for Git over HTTPS, or can be used to authenticate to the API over Basic Authentication.

Note

token-for-my-app

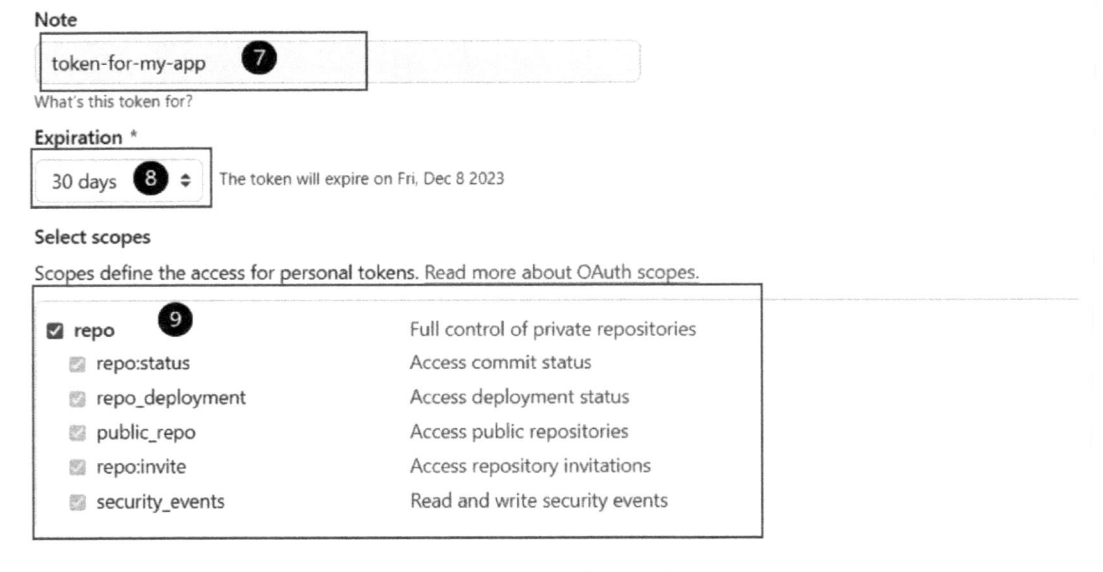

What's this token for?

Expiration *

30 days ⇕   The token will expire on Fri, Dec 8 2023

Select scopes

Scopes define the access for personal tokens. Read more about OAuth scopes.

☑ repo	Full control of private repositories
☑ repo:status	Access commit status
☑ repo_deployment	Access deployment status
☑ public_repo	Access public repositories
☑ repo:invite	Access repository invitations
☑ security_events	Read and write security events

Figure 2.23 – Configuring PAT

10. Click on **Generate token** at the bottom of the page:

☐ admin:gpg_key
  ☐ write:gpg_key
  ☐ read:gpg_key

☐ admin:ssh_signing_key
  ☐ write:ssh_signing_key
  ☐ read:ssh_signing_key

[Generate token] Cancel

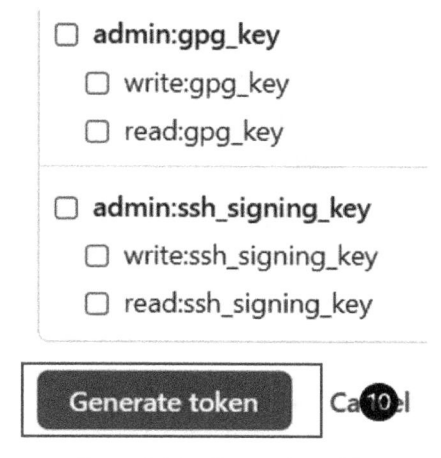

Figure 2.24 – Generating PAT

Keeping your token safe

You'll see your new token now. Make sure to copy it and keep it somewhere safe, because you won't be able to see it again! If you lose it or think someone else might have seen it, you can come back here to make a new one or delete the old one.

Using your token with Git

Now that you have your token, you can use it to log in to GitHub from Git on your computer. Open your terminal on your computer. You'll use a command to tell Git to use this token to log in to GitHub:

1. Type in this command to configure Git to use your GitHub token: `git remote set-url origin https://YOUR-TOKEN@github.com/username/repo.git`.
2. Replace `YOUR-TOKEN` with the token you just made, `username` with your GitHub username, and `repo` with the name of your repository:

```
○ PS C:\Users\igor.iric\Desktop\GitHub for Young Coders> git remote set-url origin https://YOUR-TOKEN@github.com/username/repo.git
```

Figure 2.25 – Configuring your local Git to use your PAT

Now you can use Git to push or pull changes to and from GitHub without needing to type in your GitHub password each time.

Why a good README file is like a welcoming front door on GitHub

When you create a new repository on GitHub for your one-page portfolio website, think of the README file as the front door to your project. It's the first thing visitors see, and it can make a big difference in how people perceive and interact with your project. Here's why it's so important.

The README file – your project's introduction

Just like the front page of your website gives the first impression to your guests, your README file is the first thing people see when they visit your GitHub repository.

A well-written README file is inviting and informative, making visitors more likely to explore your project and understand what it's all about:

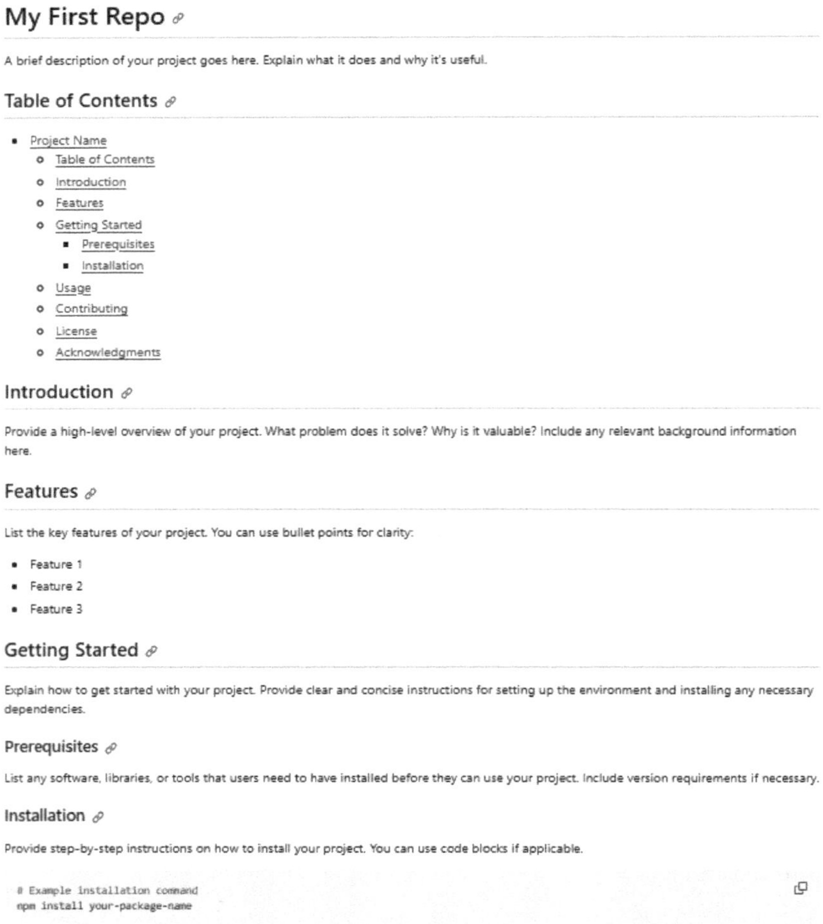

My First Repo

A brief description of your project goes here. Explain what it does and why it's useful.

Table of Contents

- Project Name
  - Table of Contents
  - Introduction
  - Features
  - Getting Started
    - Prerequisites
    - Installation
  - Usage
  - Contributing
  - License
  - Acknowledgments

Introduction

Provide a high-level overview of your project. What problem does it solve? Why is it valuable? Include any relevant background information here.

Features

List the key features of your project. You can use bullet points for clarity:

- Feature 1
- Feature 2
- Feature 3

Getting Started

Explain how to get started with your project. Provide clear and concise instructions for setting up the environment and installing any necessary dependencies.

Prerequisites

List any software, libraries, or tools that users need to have installed before they can use your project. Include version requirements if necessary.

Installation

Provide step-by-step instructions on how to install your project. You can use code blocks if applicable.

```
# Example installation command
npm install your-package-name
```

Figure 2.26 – Example of a good README file

Explaining what your project is

Imagine a friend is visiting your GitHub project. The README file should answer their basic questions, such as *What is this project about?* and *What does this website do?*

It's like having a sign on your front door that says, "*Welcome to my portfolio website project!*"

How to get started

The README file can provide instructions on how to view or use your portfolio website. It's like giving your friend a quick tour of your website and showing them where everything is.

It might include steps such as how to open the website, any special features they should check out, or how to navigate through your project files.

You should encourage collaboration!

If you want others to help you with your project, a good README file is crucial. It's like inviting friends over and giving them a guide on how they can help you decorate or improve your website.

It tells potential collaborators how they can contribute, what kind of help you're looking for, and any important guidelines they should follow.

Can I show my project's status?

A README file can also include information such as how far along your project is (is it just starting, halfway done, or nearly complete?). It's like telling your friends how much of the website is finished and what parts are still under construction.

Tips for a great README file

Some tips for a great README file are as follows:

- **Keep it simple and clear**: Use plain language and be concise, like you're explaining it to a friend who knows nothing about building websites.
- **Update regularly**: Keep the README file updated as your project evolves. It's like keeping your front porch clean and welcoming.
- **Use visuals if necessary**: Sometimes, a picture or a screenshot can help explain things better.

A good README file on GitHub is your project's greeting. It sets the tone for your project and can make the difference between someone just passing by and someone stopping, entering, and maybe even staying to help out.

Here is the Markdown we will use for our first repository to get started with a good README file:

```
# My First Repo
```

A brief description of your project goes here. Explain what it does and why it's useful.

```
## Table of Contents

- [Project Name](#project-name)
  - [Table of Contents](#table-of-contents)
  - [Introduction](#introduction)
  - [Features](#features)
  - [Getting Started](#getting-started)
    - [Prerequisites](#prerequisites)
    - [Installation](#installation)
  - [Usage](#usage)
  - [Contributing](#contributing)
  - [License](#license)
  - [Acknowledgments](#acknowledgments)

## Introduction
```

Provide a high-level overview of your project. What problem does it solve? Why is it valuable? Include any relevant background information here.

```
## Features
```

List the key features of your project. You can use bullet points for clarity:

```
- Feature 1
- Feature 2
- Feature 3
## Getting Started
```

Explain how to get started with your project. Provide clear and concise instructions for setting up the environment and installing any necessary dependencies.

```
### Prerequisites
```

List any software, libraries, or tools that users need to have installed before they can use your project. Include version requirements if necessary.

```
### Installation
```

Provide step-by-step instructions on how to install your project. You can use code blocks if applicable.

```bash
# Example installation command
npm install your-package-name
```

Editing a README file right on GitHub (hands-on)

1. **Go to the GitHub repository we created previously**: This is like opening the folder on your computer where your document is saved.
2. **Look for the pencil (✏) icon at the top right of the file content**: This is the edit button. Click on it:

Figure 2.27 – Editing the README file

3. **Opening File**: The file will now open in an editable text box, similar to a simple text editor. Now, you're ready to make your changes:

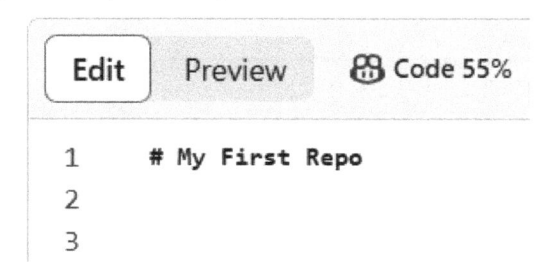

Figure 2.28 – Adding changes to the README file

4. **Make your changes**: Just type in the text box to make your changes to the file. It's like editing a Word document or an email.
5. **Adding your Markdown**: Now use the Markdown code and replace all the contents of the README file.
6. **Review your changes**: You'll see a section called **Preview** (Edit Preview). This shows you what your edits look like. It's like previewing a document before printing it.
7. **Commit your changes**: Once you're happy with your edits, you need to *commit* these changes (in GitHub language, *committing* is like saving your changes):

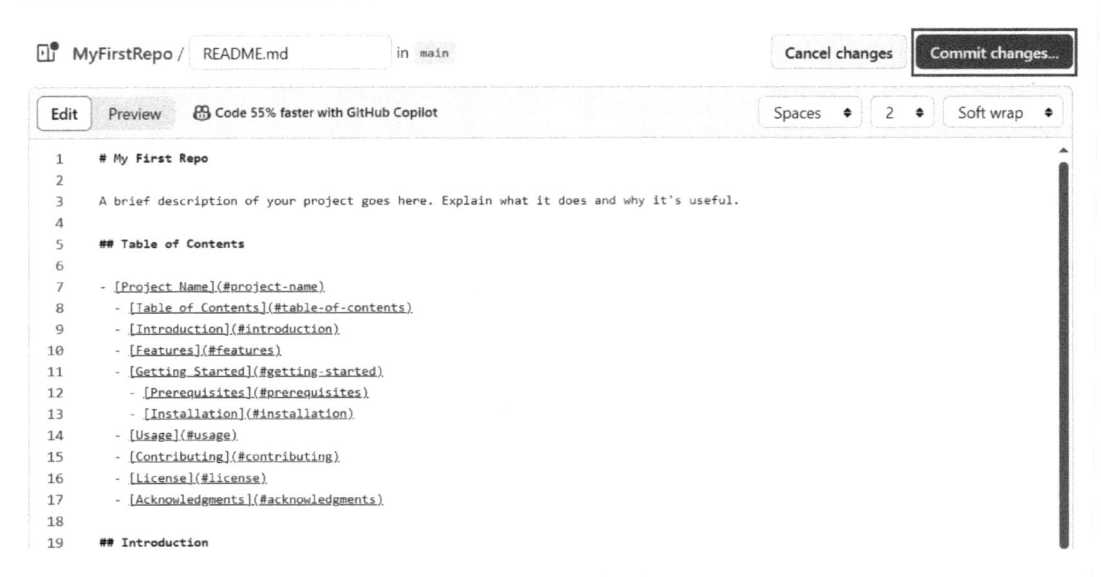

Figure 2.29 – Saving the changes

8. **Adding your Commit Message**: Next, you'll see a box where you can enter a short message explaining what changes you made (this helps you and others understand what was changed and why):

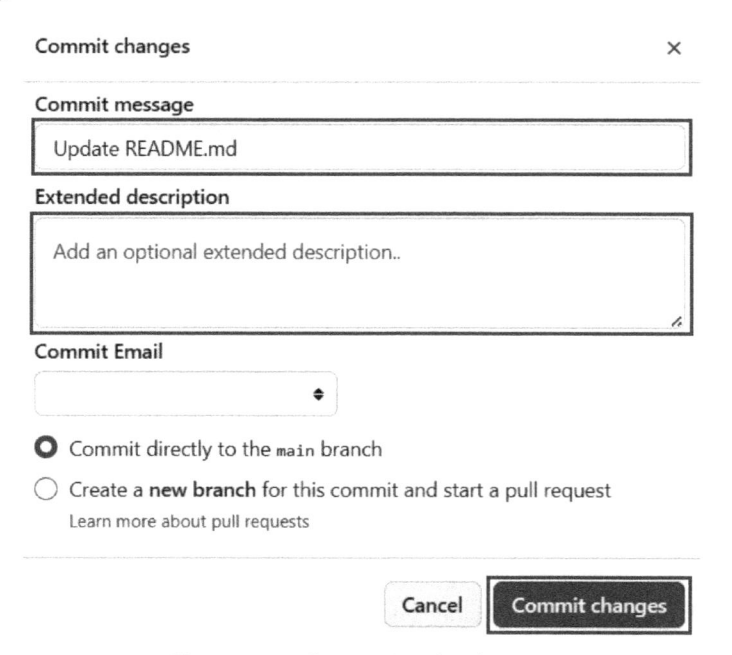

Figure 2.30 – Committing the changes

9. **Finalize the commit**: After adding your message, click on the green button that says **Commit changes**. This is like clicking **Save** on a document.

Your changes are now saved and immediately updated in the file on GitHub.

Summary

In this chapter, we learned the basics of Git and GitHub, including how to clone repositories, use Git commands, save changes, and collaborate with others. These skills are crucial for managing projects and working as a team in coding.

Next, we'll explore GitHub features for project collaboration, such as inviting friends to projects, handling issues, creating pull requests, managing notifications, engaging in discussions, setting up projects, and using insights and charts. This will build on our foundational knowledge, helping us collaborate more effectively.

Quiz

Check your knowledge earned by reading this chapter with these quiz questions:

1. What is a "repository" in GitHub?

    A. A collection of code and project files

    B. A chat room for developers

    C. A type of programming language

    D. A website builder

    Answer: A. A collection of code and project files

2. True or false: Git and GitHub are the same thing.

    A. True

    B. False

    Answer: B. False

3. Fill in the blank: To start using GitHub, you need to create a _____ for your project.

    Answer: repository

4. Why is it important to set up Git on your machine?

    Answer: It allows you to interact with GitHub from your local machine, enabling you to manage and collaborate on projects more effectively.

5. What does "cloning a repository" mean?

   A. Deleting a project
   B. Copying a project to your local machine
   C. Changing a project's name
   D. Sending a project to a friend

   Answer: B. Copying a project to your local machine

6. True or false: You can only use GitHub for coding projects.

   A. True
   B. False

   Answer: B. False

7. What is the purpose of Git commands?

   A. To play games
   B. To communicate with Git and manage your projects
   C. To draw pictures
   D. To write essays

   Answer: B. To communicate with Git and manage your projects

8. Fill in the blank: When you make changes to your code, you need to _____ them to save and update your project on GitHub.

   Answer: push

9. True or false: You can collaborate with others on GitHub without using Git.

   A. True
   B. False

   Answer: B. False

10. Name one basic Git command mentioned in this chapter and its purpose.

    Answer: The `git push` command is used to upload local repository content to a remote repository

Challenge – Launching your space adventure!

Ready to start an out-of-this-world adventure? In this chapter, we've explored the basics of GitHub and version control. Now it's time to put that knowledge into practice with your very own space-themed game: *Space Explorer!*

Your mission?

Launch the `Space Explorer Game` repository on GitHub:

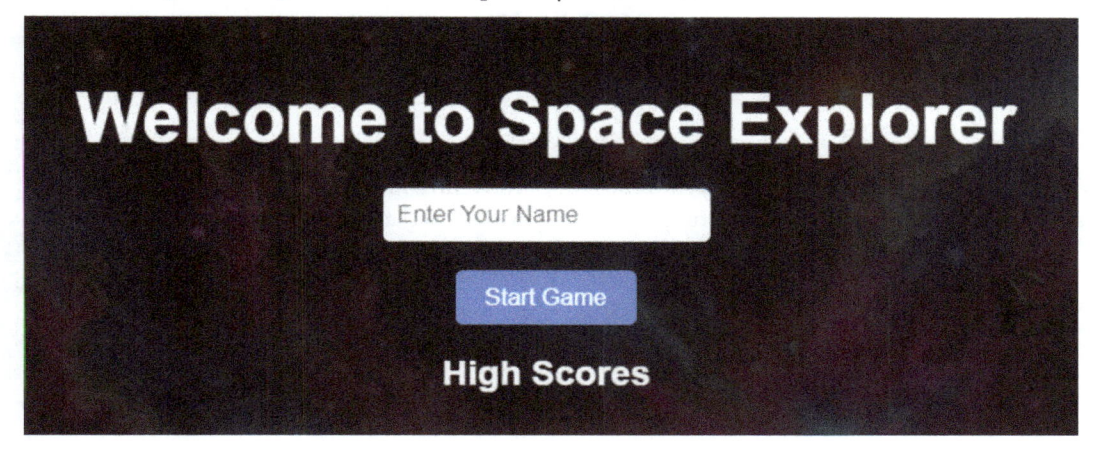

Figure 2.31 – Space Explorer game

Steps to success

Here are your steps to success:

1. **Fork the repository**: Visit the official `Space Explorer Game` repository at `https://github.com/PacktPublishing/GitHub-for-Next-Generation-Coders`.
2. **Click on the Fork button**: This will create a copy of the game in your GitHub account—think of it like cloning an alien spaceship so you have one of your own!
3. **Clone your fork**: Once you've forked the repository, it's time to bring it to your computer. Use the `git clone` command in your terminal to clone the repository. (Need a refresher? Revisit our section on cloning.) You're beaming the spaceship right to your computer!
4. **Explore your local copy**: Open the cloned project in your favorite code editor. Peek inside! You'll find HTML, CSS, and JavaScript files—the building blocks of your space adventure.
5. **Testing the game in your local browser**: Run the HTML file in a browser and see the basic version of the *Space Explorer* game in the `Game Start` folder. It's alive!

6. **Commit to your mission**: Make a small change to start with. How about changing the title in the HTML file to include your name? For example, `<title>[Your Name]'s Space Explorer</title>`.
7. **Saving and commitng the changes**: Save your changes and commit them with a message such as `Personalize game title`. You're marking your territory in this coding cosmos!
8. **Push the changes**: Use the `git push` command to push your changes to your GitHub fork. Your personalized spaceship is now visible in the GitHub universe!
9. **Reflect on your journey**: Look at your GitHub repository online. See your commit there? You've just completed a fundamental GitHub cycle: fork, clone, edit, commit, and push!

Bonus challenge

Write a short "captain's log" entry in your repository's README file, describing your first mission. Use your imagination! What was exciting? What did you discover about GitHub and coding?

Part 2: Collaborative Development Using GitHub

Moving forward, this part delves into more collaborative aspects of using GitHub. You'll explore how to work effectively with others by using branches, managing merge conflicts, and leveraging GitHub's features to enhance collaboration. Practical tips and tricks will help you to manage contributions seamlessly and maintain a clean project history.

This part contains the following chapters:

- *Chapter 3, GitHub Features for Collaborating on Projects*
- *Chapter 4, Branching in GitHub and Git*
- *Chapter 5, Collaborating on Code through Pull Requests*
- *Chapter 6, Resolving Merge Conflicts – on GitHub and Locally*

3
GitHub Features for Collaborating on Projects

In this chapter, we will understand the GitHub features that improve project collaboration, teaching you how to invite collaborators, manage issues, create pull requests, handle notifications, use discussions for team communication, set up projects, and use insights and charts. These tools are designed to simplify teamwork, making it easier to track progress, discuss ideas, and share updates efficiently.

In this chapter, we're going to cover the following main topics:

- Setting up your team – inviting collaborators on GitHub
- Understanding GitHub issues – What's troubling our project?
- Introduction to pull requests
- What's buzzing? Managing GitHub notifications
- Exploring GitHub Discussions
- Setting up projects on GitHub
- Understanding wikis on GitHub
- What are GitHub repository insights?

Technical requirements

A GitHub account and basic familiarity with Git are required. Ensure Git is installed on your computer and set up a GitHub repository for practice. Access the `https://github.com` URL for all activities in this chapter. Create a GitHub folder named `ch3` for organizing your work.

Setting up your team – inviting collaborators on GitHub

Ever been to a party? It's much more fun when you invite friends, right? Now, think of your project on GitHub as a party. It gets better when more people, such as your teammates or friends, join in and contribute. So, how do you invite them to your GitHub project party? Let's walk through it.

Finding the invitation spot

1. Open your web browser and head to your GitHub project page. In our case, it is called `MyFirstRepo`. Now, look for a tab called **Settings** in the top-right corner of the page and give it a click. This is like opening the doors to your party. Consider this as *step zero*.

2. Once you're in the settings, look for a section on the left called **Access**. Click on **Collaborators**. This section is like your guest list.

3. Now, click on the green button that says **Add people**:

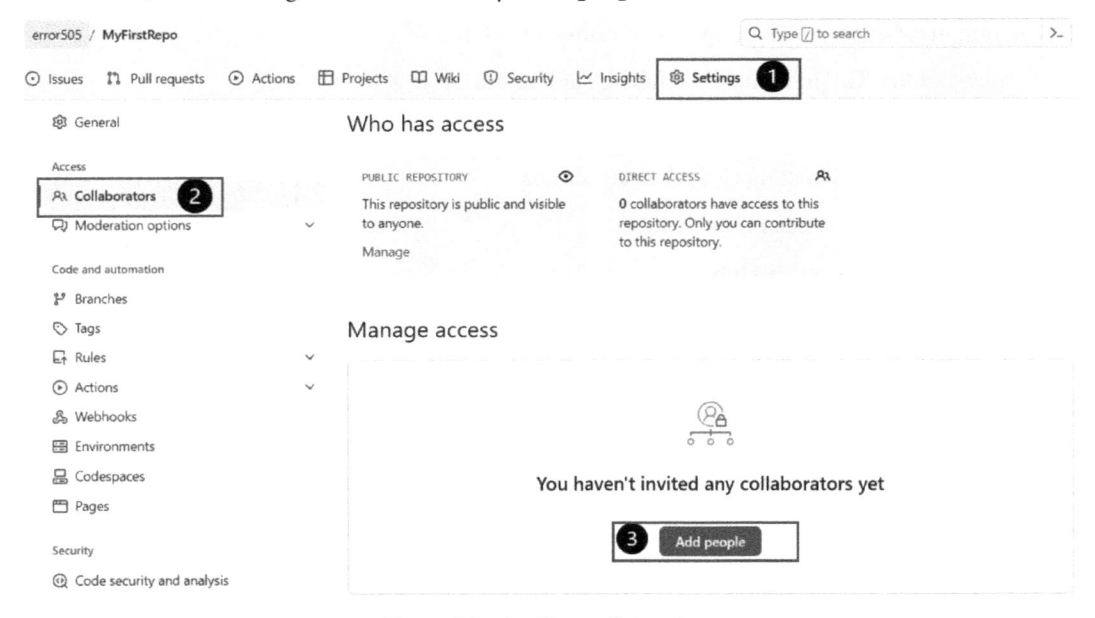

Figure 3.1 – Inviting collaborators

4. Here, you can type in the username or email address of a friend you want to invite. As you start typing, GitHub will help you out by suggesting names:

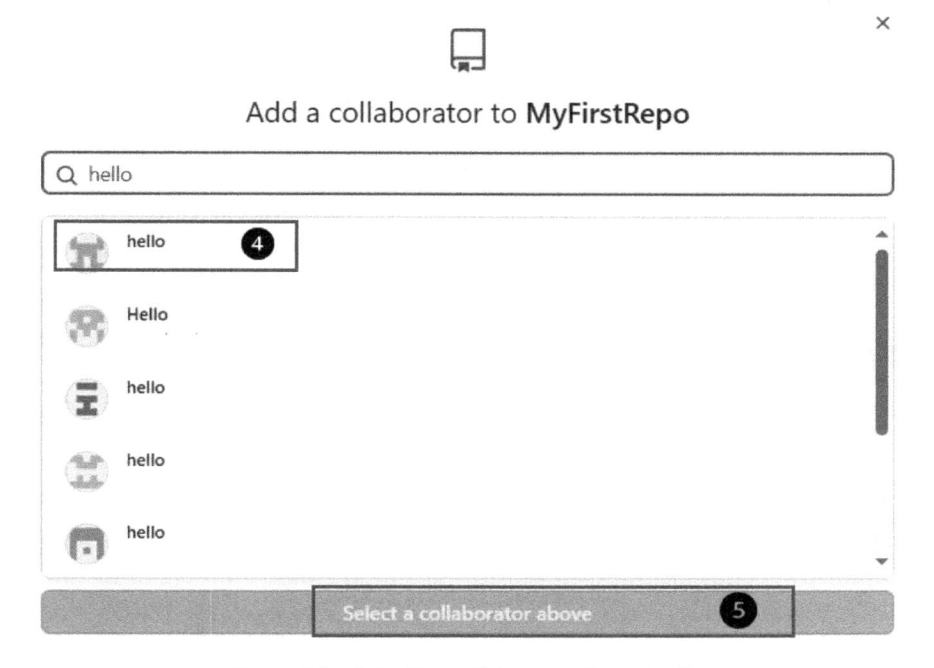

Figure 3.2 – Selecting collaborators from the list

5. Select the friend you want to invite from the list, or finish typing their username or email, and hit the **Select a collaborator above** button to send them an invite. It's like sending out an invitation card to your party.

Awaiting their RSVP (répondez s'il vous plaît – please respond)

Your friends will get an email and a notification on GitHub with your invitation. Once they accept it, they'll be able to join your project, see all the cool stuff you've been working on, and start contributing their own ideas too!

Checking the guest list

At any time, you can go back to the **Collaborators** section in your project settings to see who's accepted your invitation and who's joined the party.

Now, your GitHub project party is set up, your invitations are sent, and as your friends join in, the real fun and creativity begin. Together, you'll create something amazing and maybe learn a thing or two from each other along the way. Next, let's understand an important topic: GitHub issues.

Understanding GitHub issues – what's troubling our project?

Imagine you're working on a big puzzle with a group of friends. You all are placing pieces together, but suddenly, you notice a section of the puzzle looks a bit odd. Maybe a piece is missing, or a couple of pieces don't fit well. You want to let your friends know about this so that everyone is aware and someone can help to fix it. In the collaboration on projects via GitHub, this is where *issues* come into play.

GitHub Issues is like a bulletin board for your project. Whenever you or your teammates find a problem, have a question, or want to suggest an idea, you can post it on the **Issues** board. It's a place where everyone can discuss what's going on and come up with solutions together. For example, if you're working on a website project and the contact form isn't working, you can create an issue such as *Website bug* to get it fixed.

Creating an issue

1. Go to your project repository on GitHub. Click on the **Issues** tab.
2. Next, click the green **New issue** button:

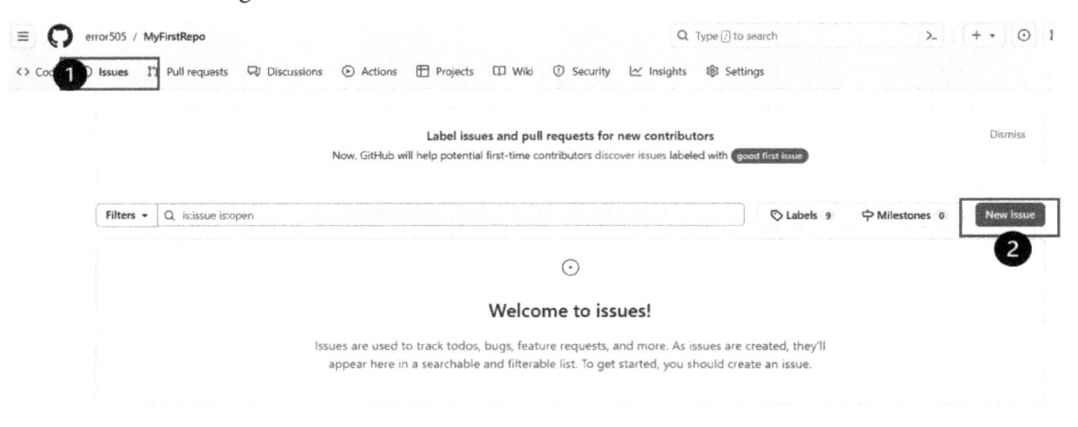

Figure 3.3 – Creating a new issue

A page for creating a new issue will open.

3. Give a title to your issue, such as `Rename the Repository to MyBiography`, and describe the problem in detail:

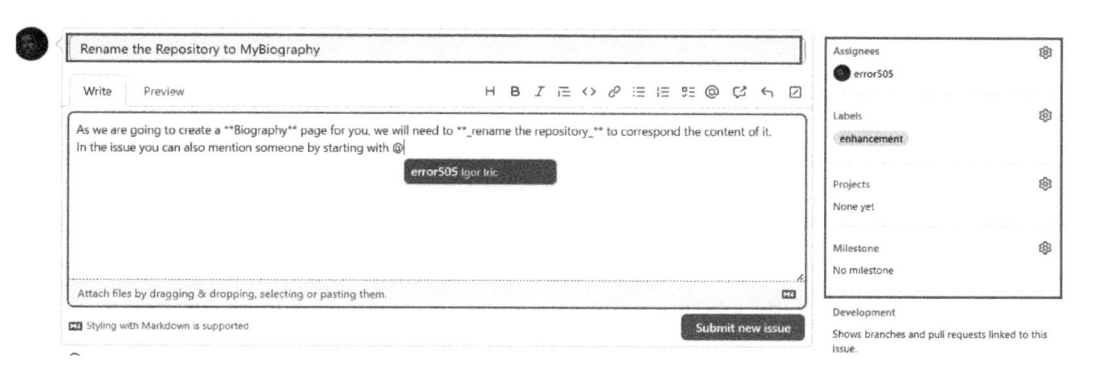

Figure 3.4 – Issue title and description

You can even add labels such as `bug` or `enhancement` to categorize it and attach it to projects, which we will create in the next section:

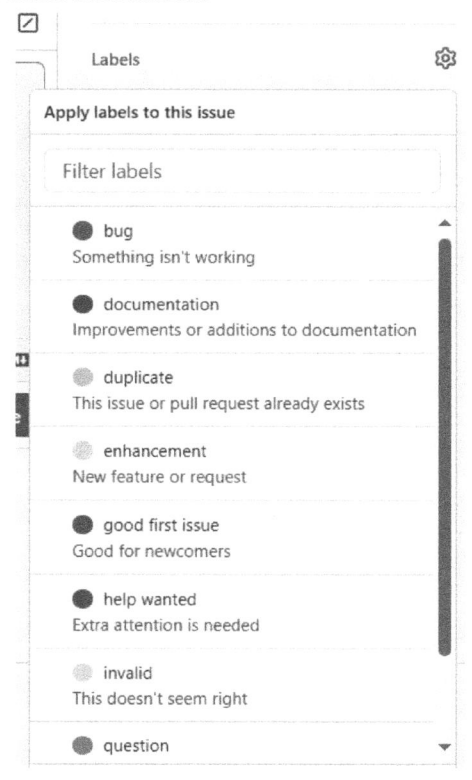

Figure 3.5 – Issue labels

Now, your issue is on the bulletin board. Your teammates can see it, comment on it, and even suggest solutions.

Discussing the issue

You can have a conversation right there on the issue page, sharing code snippets, screenshots, or any other useful information:

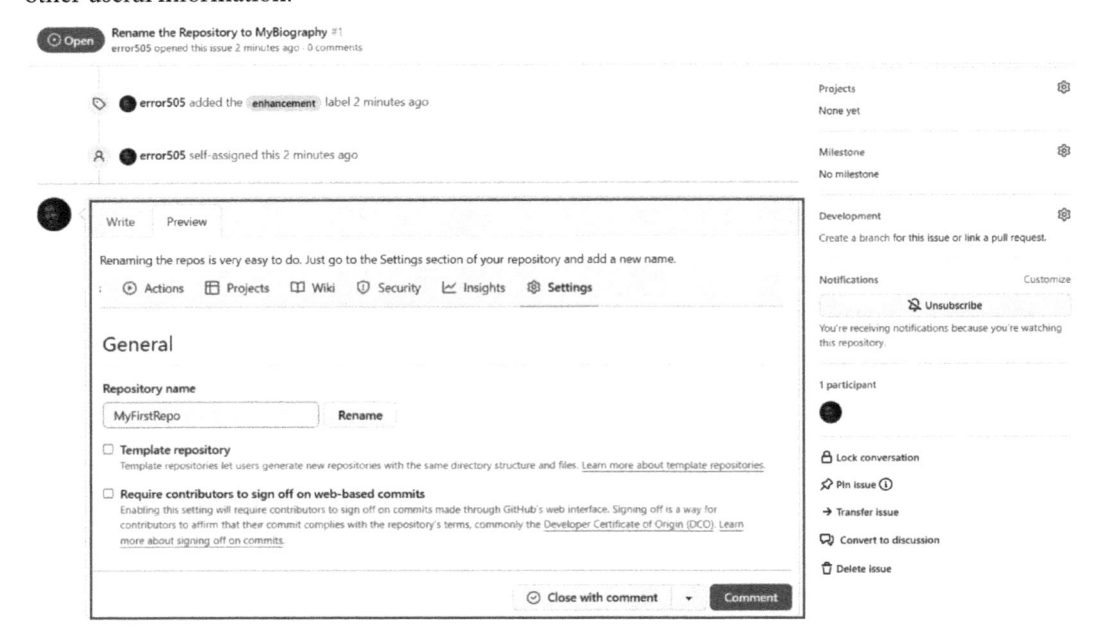

Figure 3.6 – Commenting on the issue

If one of your teammates has a good idea of how to fix the problem, you or your project manager can assign the issue to them.

Assigning the issue

You can assign the issue to a person like so:

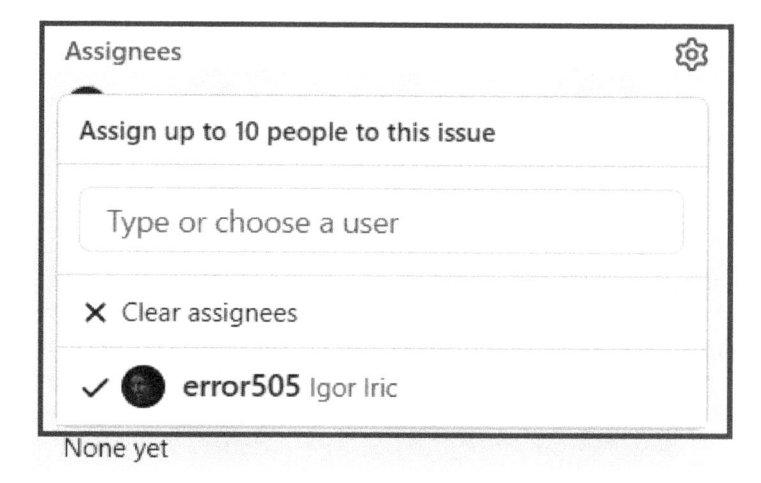

Figure 3.7 – Assigning the issue to a team member

They now have a task to work on, and everyone knows they're the one tackling this problem.

Once your teammate has fixed the problem on their local machine, they can create a pull request (which we'll discuss in the *Introduction to pull requests* section).

Linking to a pull request

In the pull request, they can mention the issue number, such as `Fixes #1`, to show that the changes in this pull request aim to resolve this issue. GitHub is smart; once the pull request is merged, it will automatically close the issue, indicating that the problem has been resolved:

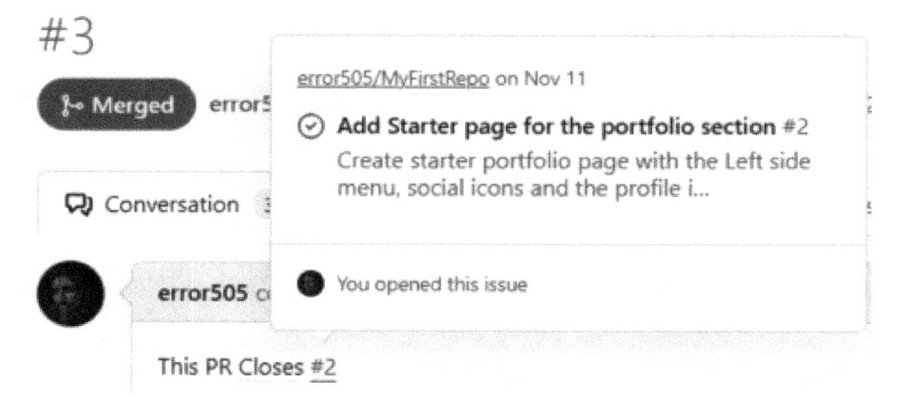

Figure 3.8 – Linking issue to pull request

Through this simple, organized system, GitHub Issues helps keep everyone on the same page about what needs attention in the project. It's like having a living, breathing to-do list that everyone can contribute to and interact with, ensuring nothing falls through the cracks as your project progresses toward perfection. With this knowledge, let's move on to the next section!

Introduction to pull requests

You're working on a one-page portfolio website with your friends. Each of you has different ideas and contributions to make the website look great, like a group of artists working on a big mural. Now, how do you make sure everyone's ideas fit well together and the mural ends up looking awesome? That's where GitHub pull requests come in! *Figure 3.9* shows how pull requests appear on GitHub:

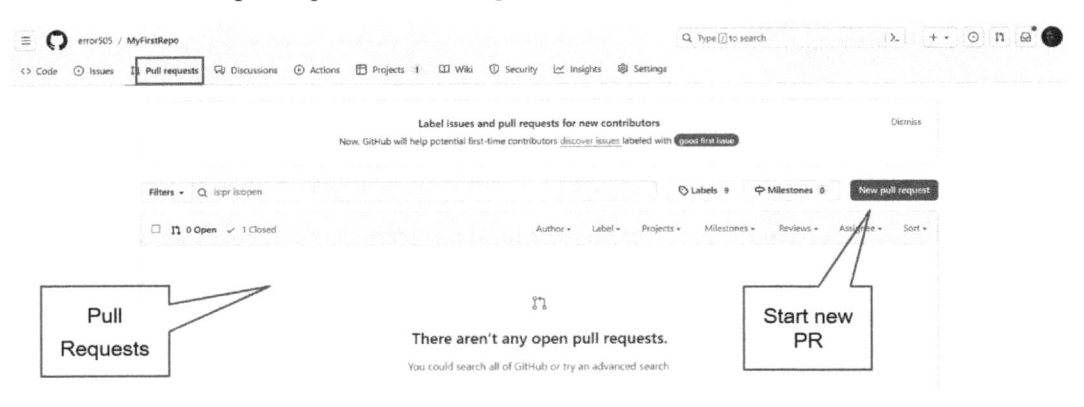

Figure 3.9 – Managing pull requests on GitHub

What are pull requests?

Basically, a **pull request** is a request to merge code changes. A pull request is like asking the team, "*Hey, can we add my blog page to the website?*"

It's a way to discuss changes: before the team agrees, they'll look at your blog page, give suggestions, or say it's perfect as is.

Using pull requests in your website project

Suppose you've created a new section for your website, such as an **About Me** page. You submit a pull request on GitHub, which is like showing your draft to the team. *Figure 3.10* demonstrates how to open a pull request:

Introduction to pull requests | 71

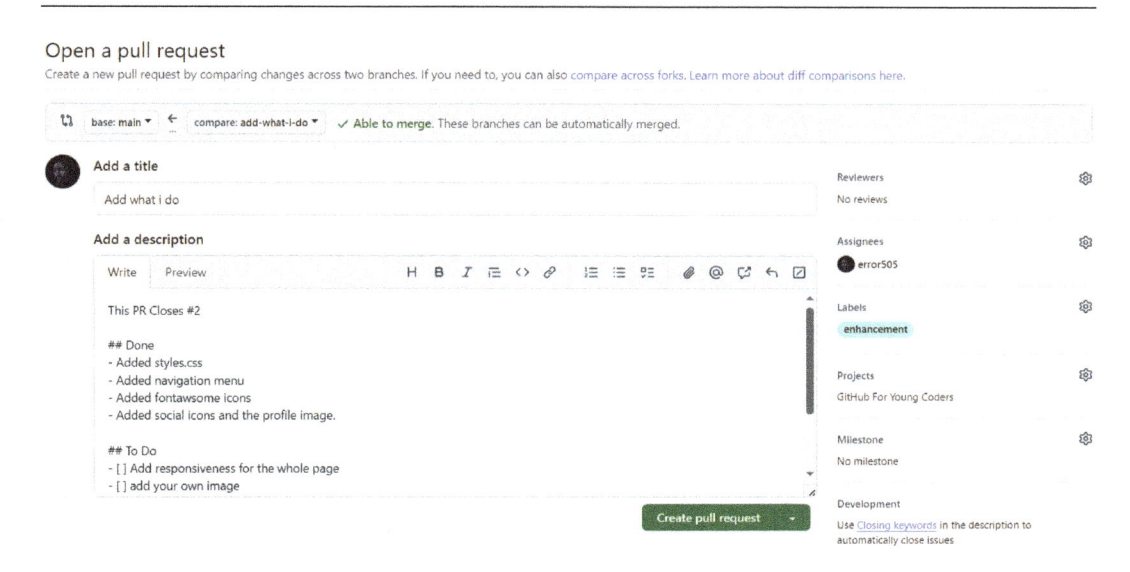

Figure 3.10 – Opening a pull request for code changes

Review and feedback

Your teammates check your **About Me** page. They can suggest changes, such as *Maybe use a different font here?*, or give a thumbs up if they love it.

It's a chance for everyone to discuss and refine the idea, ensuring the website stays cohesive. The following figure shows how to review pull requests:

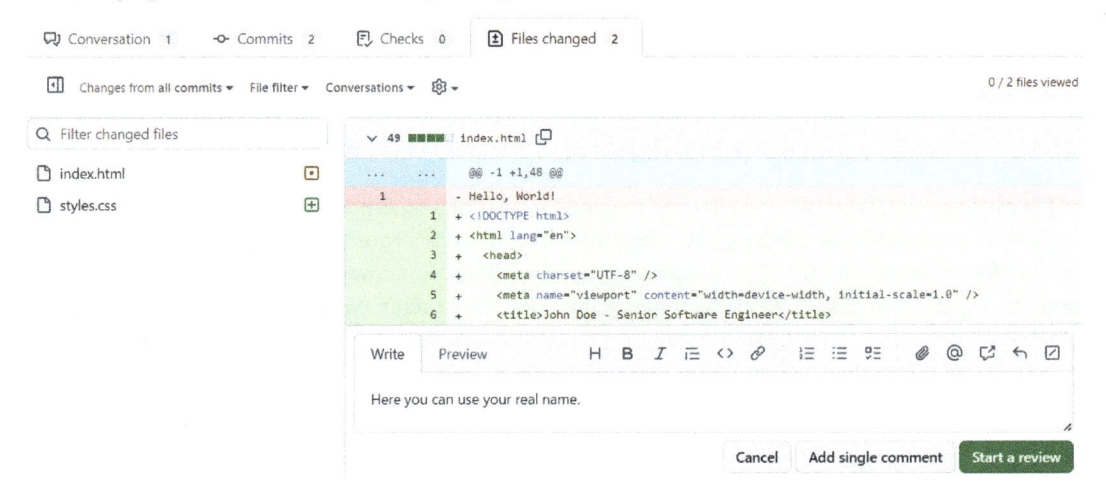

Figure 3.11 – Reviewing pull request

Merging the pull request

Once everyone agrees, your changes are merged into the main project:

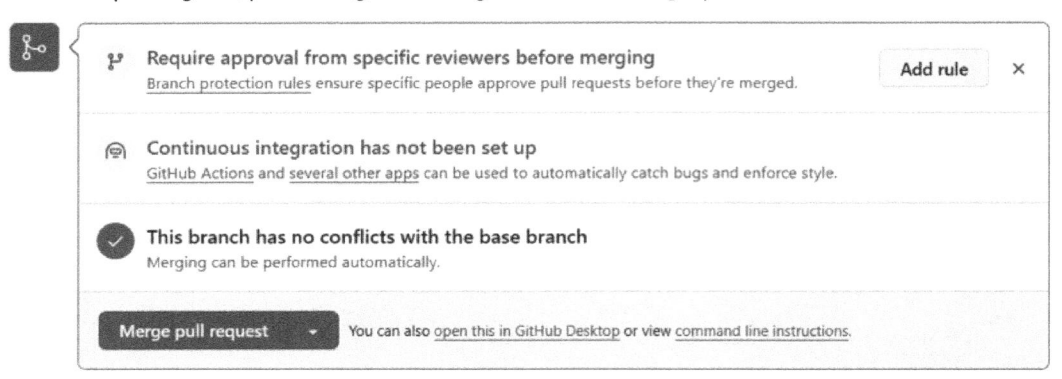

Figure 3.12 – Merging pull request

Your **About Me** section is now part of the official website!

It's like adding your piece to the mural, making it richer and more complete.

Why are pull requests important?

Pull requests ensure that changes are reviewed and fit well with the overall design.

Pull requests encourage collaboration and learning. They open conversations about different ideas and approaches. It's a great way to learn from each other and make the project better as a team. Pull requests ensure that the website stays functional, looks good, and is free from errors. Next, let's learn about optimizing your GitHub experience by managing the notifications.

What's buzzing? Managing GitHub notifications

Ever felt overwhelmed with the buzzing and beeping of your phone, especially when you're part of an active online community or a busy project team? It's like being in a room where everyone's talking at the same time, and you are trying to keep up with everything. Now, imagine this happening in your digital workspace on GitHub. Each comment on a code line, each new issue opened, or each pull request made sends a notification your way. It can quickly become annoying! So, how do you tune into the conversations that matter without getting lost in a sea of notifications?

GitHub notifications are your way to keep a tab on discussions and activities that concern you. They are like those little nudges or taps on the shoulder, telling you where your attention is needed. Let's navigate the buzzy marketplace of ideas and information on GitHub without losing our way.

Spotting notifications

When you log in to GitHub, you'll notice an inbox icon in the top-right corner of the page:

Figure 3.13 – Notifications

That's where all your notifications gather, waiting for your attention.

Tuning channels

Click on the bell icon, and you'll see a list of all your notifications. On the left side, there's a sidebar that helps you filter these notifications. You can choose to view notifications from a particular repository or those that directly mention you:

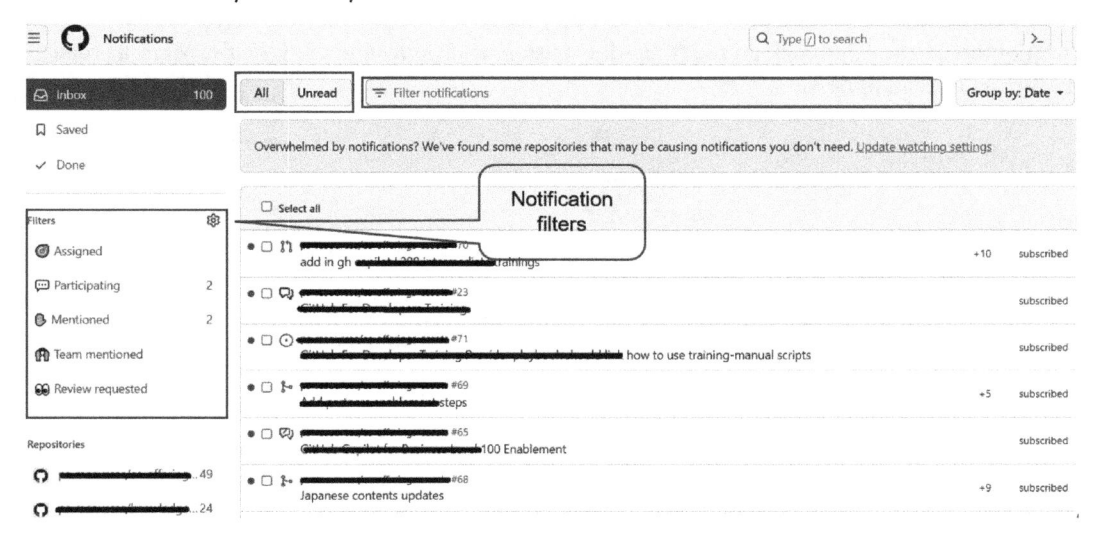

Figure 3.14 – Notification list

74 GitHub Features for Collaborating on Projects

There is a possibility to filter notifications by repository, by status or discussion type, by notification reason, author, or organization:

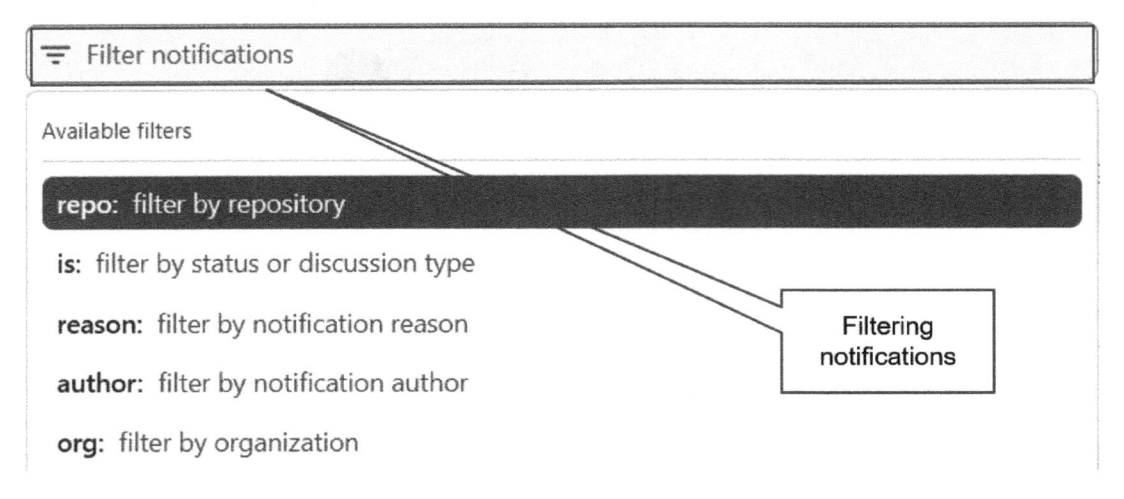

Figure 3.15 – Filtering notifications

Marking notifications

As you browse through the notifications, you can mark them as read so that they no longer grab your attention, or you can keep them unread for later if they require more of your time. Select the checkbox in front of the notification, click on the three-dots button, and there you will have the **Mark as read** and **Mark as unread** options:

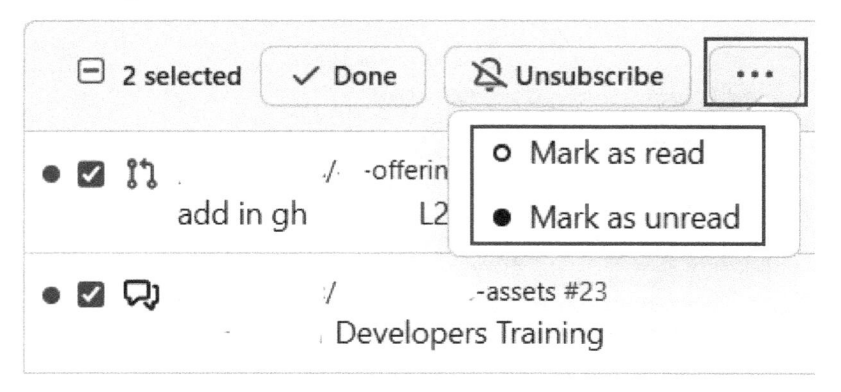

Figure 3.16 – Reading and marking notifications

Reducing notification flood with filters

GitHub also allows you to filter your notifications based on their status, such as **Unread**, **Participating**, or **Mentioned**. It's like having a smart assistant who knows exactly which papers to keep on your desk. You can even create your own filters by clicking on the cog icon:

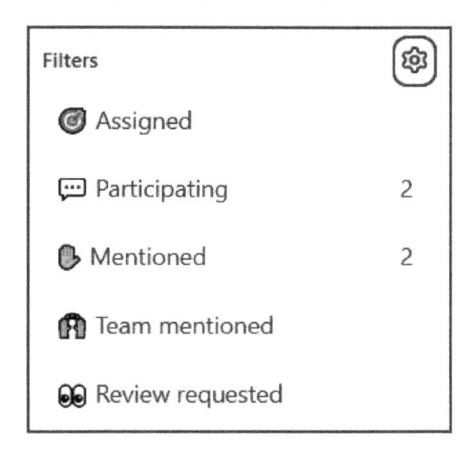

Figure 3.17 – Filters

A new window will open with a few default filters, and from there, you can now create your new filter:

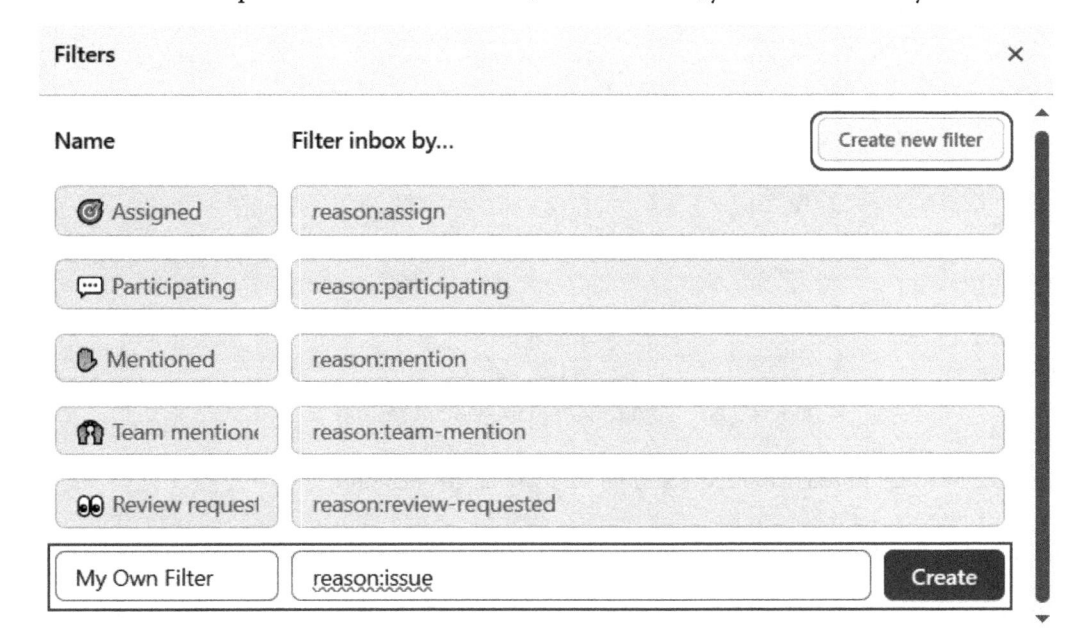

Figure 3.18 – Creating custom filters

Customizing alerts

If you find the buzz too much, you can customize your notification settings by clicking on the **Manage notifications** option on the left side:

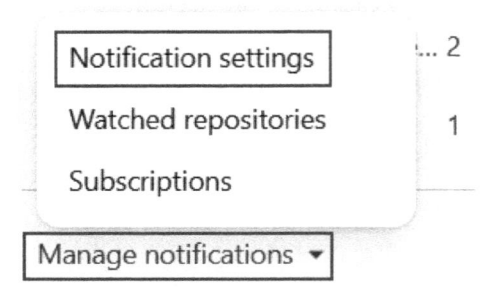

Figure 3.19 – Manage notifications option

If you click on **Notification settings**, you will be redirected to the **Notifications** page where you can choose when and how GitHub should send you notifications:

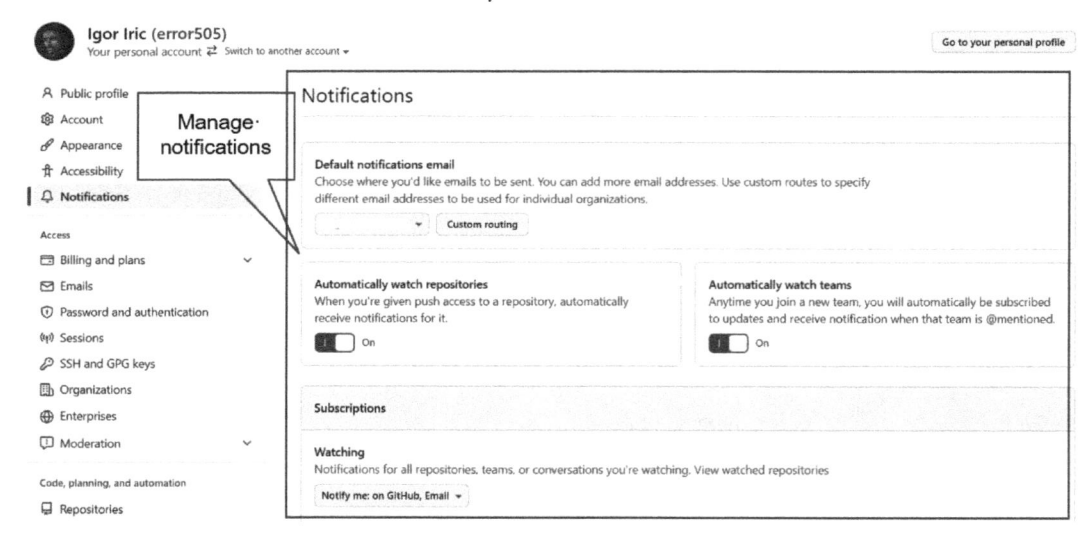

Figure 3.20 – Notifications management section

You can decide to get notified about every single activity or just highlights.

Emailing updates

If you prefer a less interruptive way, you can have GitHub send all notifications to your email. You can then check these notifications when you have a moment to spare.

Now that you've learned how to manage GitHub notifications effectively and keep your project updates tidy and relevant, let's shift our focus to another interactive feature of GitHub. We're going to look at how you can use GitHub Discussions to further enhance collaboration on your project. This section will show you how to use Discussions as a platform to share ideas, gather feedback, and communicate more openly with your team.

Exploring GitHub Discussions

Have you ever wished for a place where you and your teammates could chat about your project, share ideas, or ask questions? It's like your project's own social media platform. That's exactly what **GitHub Discussions** is for!

The cool factor in project collaboration

So, you're teaming up with friends to create a one-page portfolio website. It's not just about coding and design, right? You need to brainstorm ideas, sort out problems, and sometimes just chat about cool features. That's where GitHub Discussions comes in.

It's like a chat room for your project

Think of Discussions as a digital coffee shop where you and your friends can talk about everything related to your website, from design ideas to deciding what content to include.

Instead of juggling conversations across different apps or losing track of important suggestions, everything related to your project is in one organized place.

What types of discussions could we have?

There are a number of discussions that one could have, such as:

- **General**: The all-purpose space for anything that doesn't fit elsewhere. It's like the *miscellaneous* drawer:

    💬 **General**
    Chat about anything and everything here

    Figure 3.21 – General discussions

- **Ideas**: Got a bright idea or suggestion? This is your stage!

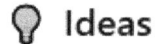

    💡 **Ideas**
    Share ideas for new features

    Figure 3.22 – Ideas discussions

- **Q&A**: If you've got questions, this is where you ask. And if you've got answers, here's where you share them:

🙏 **Q&A**
Ask the community for help

Figure 3.23 – Q&A discussions

- **Show and tell**: Proud of something you've done? This is your spotlight moment to share and shine!

🙌 **Show and tell**
Show off something you've made

Figure 3.24 – Show and tell discussions

- **Polls**: Imagine you have a question, and you want to know what the group thinks. A poll lets everyone cast their vote. It's like asking, "Should our project logo be blue or green?" and everyone gets to click on their choice:

🗳️ **Polls**
Take a vote from the community

Figure 3.25 – Polls discussions

- **Announcements**: This is where important news is shared. Think of it as a bulletin board in a coffee shop where important notices are pinned:

📣 **Announcements**
Updates from maintainers

Figure 3.26 – Announcements discussions

How do I start a discussion?

To begin a discussion, follow these steps:

1. Navigate to the **Discussions** tab.
2. Select one of the categories on the left-side menu to start the discussion there:

Exploring GitHub Discussions   79

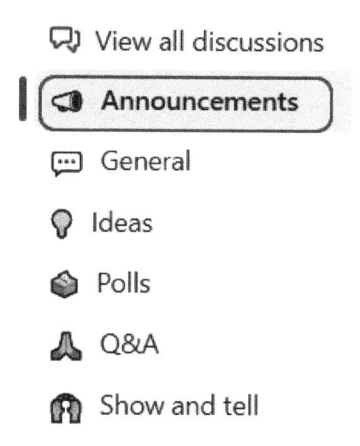

Figure 3.27 – Discussions categories

3. To start a new discussion, there's a button that says **New discussion**. Clicking it is like raising your hand to start a new topic:

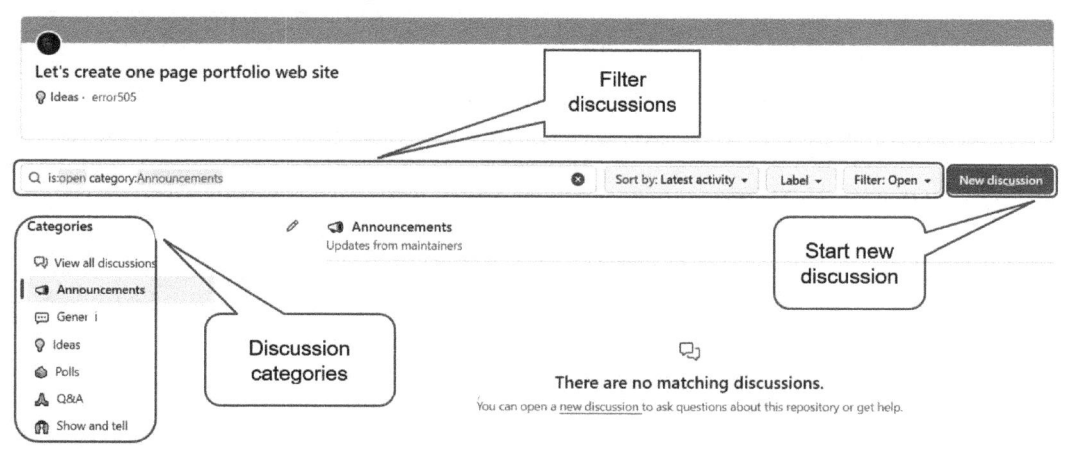

Figure 3.28 – Discussions overview

4. Just like a conversation starter, give your discussion a catchy title and share your thoughts, questions, or ideas here. It's like typing out a social media post:

Start a new discussion

General
Chat about anything and everything here

If this doesn't look right you can choose a different category.

Add a title

> One Page Portfolio web site

Add a body

> We will use this discussions for discussing new features for One Page Portfolio web site

Start discussion

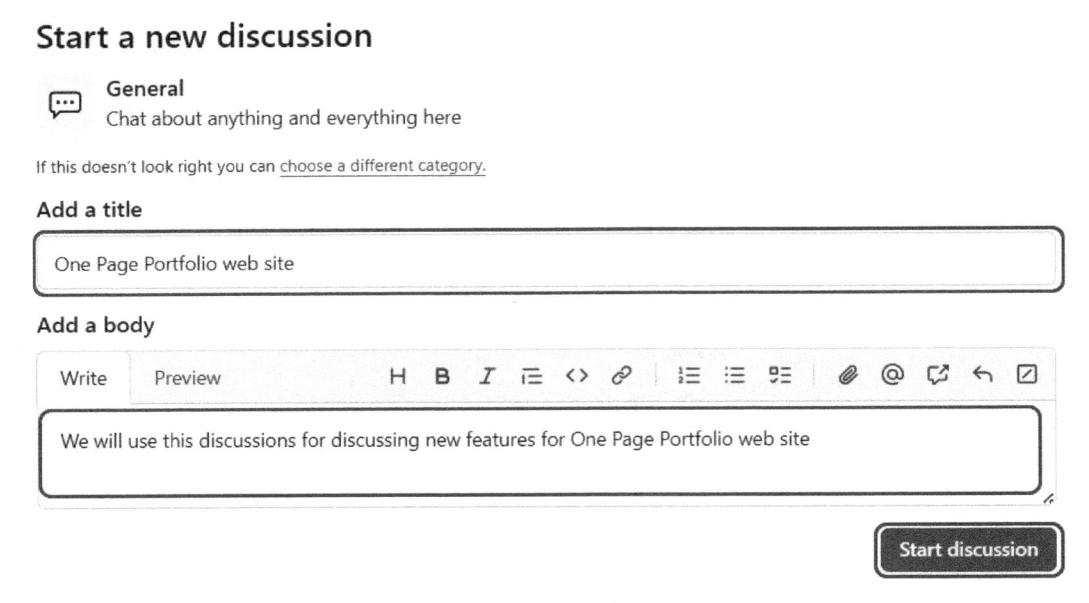

Figure 3.29 – Creating discussions

5. **Post it**: Hit the **Start discussion** button, and you're live!

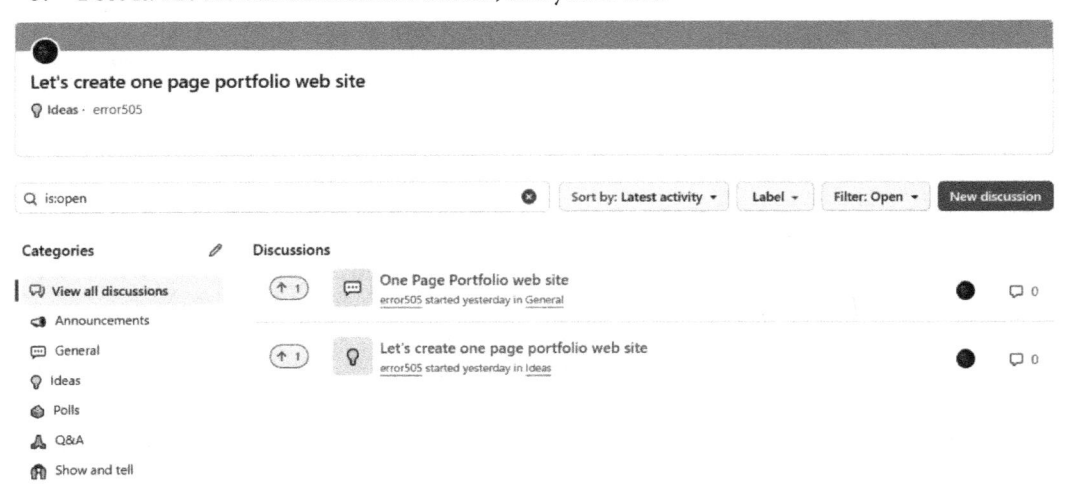

Figure 3.30 – List of discussions

Building a stronger team

These discussions create a history of your project. They're valuable for any new team members to get up to speed and for keeping track of past decisions and ideas. When everyone's involved in the conversation, you get better ideas, more commitment, and a sense of community. It's like having all your friends contribute their unique ideas to your website, making it richer and more personal.

Setting up projects on GitHub

Let's say you're planning a backyard barbecue party with your friends. You've got a checklist to buy burgers, set up the grill, invite people, and maybe plan some games. Setting up a project on GitHub for your one-page website is kind of like that but for coding.

Why do projects matter in GitHub?

Just like a party needs a plan, your website needs organization. GitHub Projects helps you and your friends (or teammates) know who's doing what and when. But it's not just about code. It's about people working together, chatting, sharing ideas, and helping each other out. Think of it as the fun chats and high-fives while grilling those burgers.

Setting up a project for your one-page portfolio website

1. **Start with a plan**: Imagine you're drawing a map of your party. In GitHub, you create a *project* for your website. This is where you'll outline all the steps to make your website awesome.
2. **Break it down into tasks**: You wouldn't make a guest grill their own burger, right? Similarly, in your project, you create tasks (such as designing the home page, writing content, and so on) and assign them to your friends. Everyone knows what they need to do.
3. **Track progress**: Just like checking off items on your party checklist, you can track which tasks are done, which are in progress, and what's still left. It's like knowing you've got the drinks sorted but still need to set up the games.
4. **Collaborate and communicate**: Use GitHub's features to chat about tasks, give feedback, and make changes. It's like discussing where to set up the grill or what music to play.
5. **Celebrate milestones**: Finished the home page? Got the contact form working? Celebrate these small wins! It's like cheering when the first batch of burgers is ready.

Setting up a Kanban project for your one-page website on GitHub

What's Kanban, anyway?

Kanban is like a bulletin board with sticky notes that show different tasks such as *Add profile picture* or *Add contact section*. Each note moves from one category to another, such as **Todo**, **In Progress**, and **Done**:

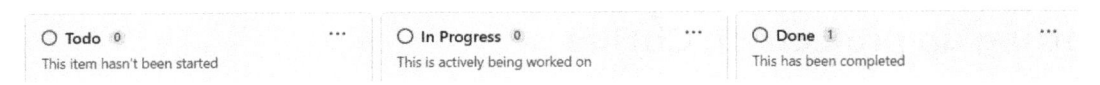

Figure 3.31 – Kanban lanes

That's Kanban – a way to see your project's progress visually.

Okay – so how to create a project on GitHub?

Creating a project on GitHub

1. In GitHub, go to your website's repository and click on **Projects**. There, you will have an overview of projects. Click on **Projects**, then **New project**:

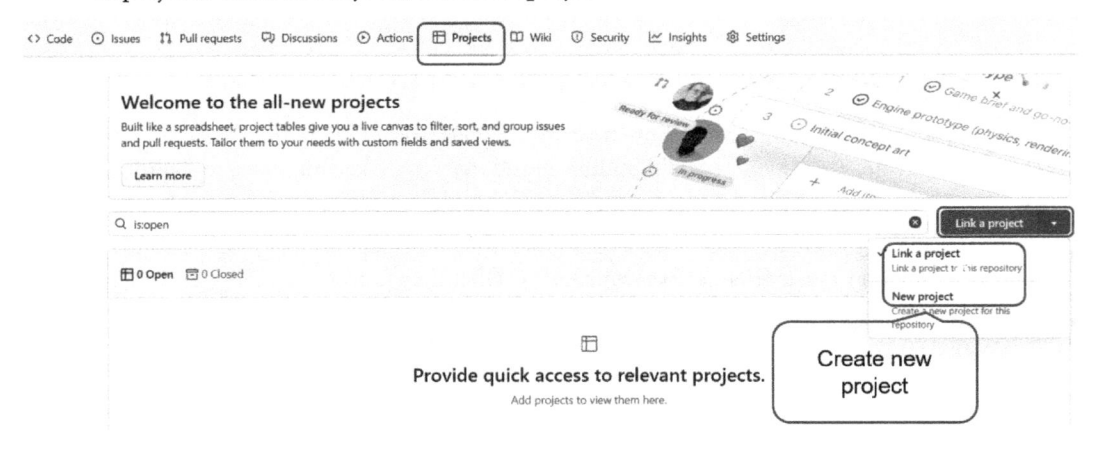

Figure 3.32 – Projects overview page

2. A new popup will open where you can choose to create a project from scratch as a table, board, or roadmap, or to start from a template:

Setting up projects on GitHub | 83

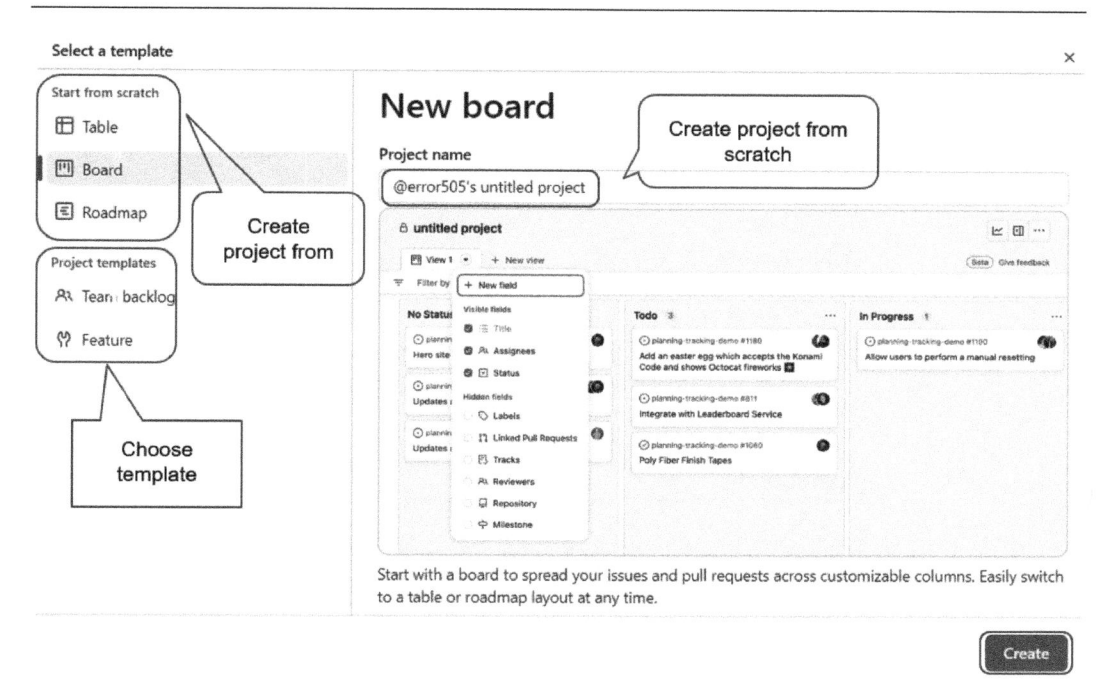

Figure 3.33 – Creating a new project

Name it something like `One-Page Website Development`. It's like naming your DIY bookshelf project. After you click on **Create a new project**, a **One-Page Website Development** project will be created, with already created columns such as **Todo**, **In Progress**, and **Done**. These are like sections on your bulletin board, showing what's planned, what you're working on, and what's completed:

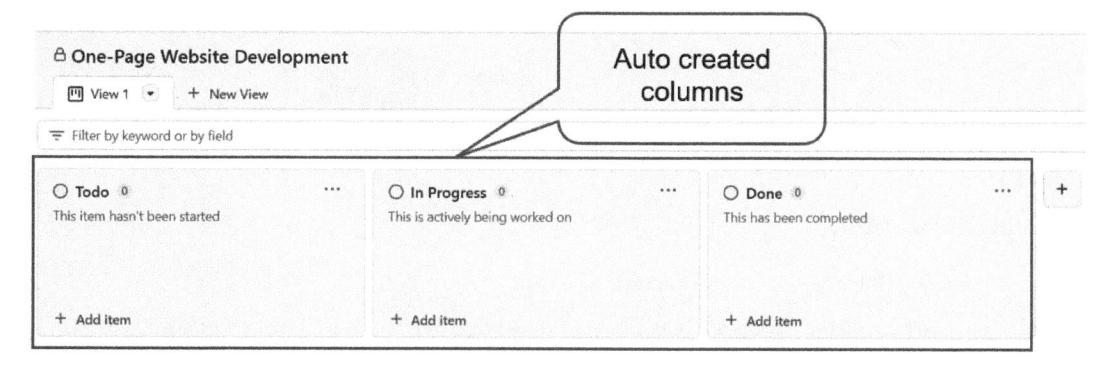

Figure 3.34 – New project created

3. **Create new columns**:

   I. Most probably, you would need to have more columns for your project. To set up new columns, just click on the plus icon and start creating your new columns.

   II. Give your column a label, color, and description:

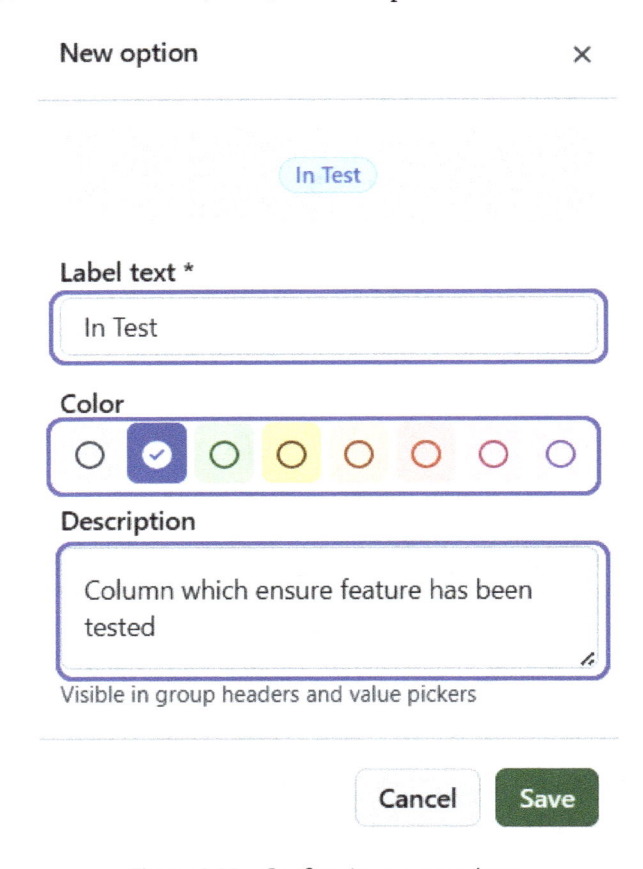

   Figure 3.35 – Configuring a new column

4. **Add tasks as cards**:

   I. Each task for your website becomes a card.

   II. Create cards for things such as *Design home page*, *Write About Me section*, and so on.

III. Place them in the **Todo** column. It's like putting sticky notes on the **Todo** part of your board. Just click on **Add Item** and start adding:

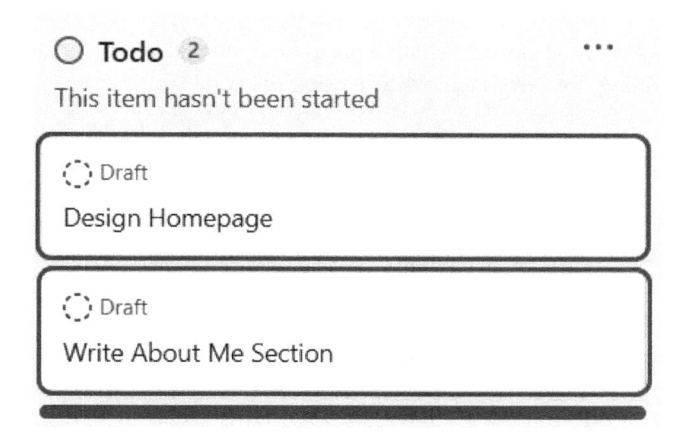

Figure 3.36 – Creating tasks

Add more details to your new items by clicking on one of them. A new page will open where you can add additional details such as a description, assign the issue to a team member, archive it, or delete it:

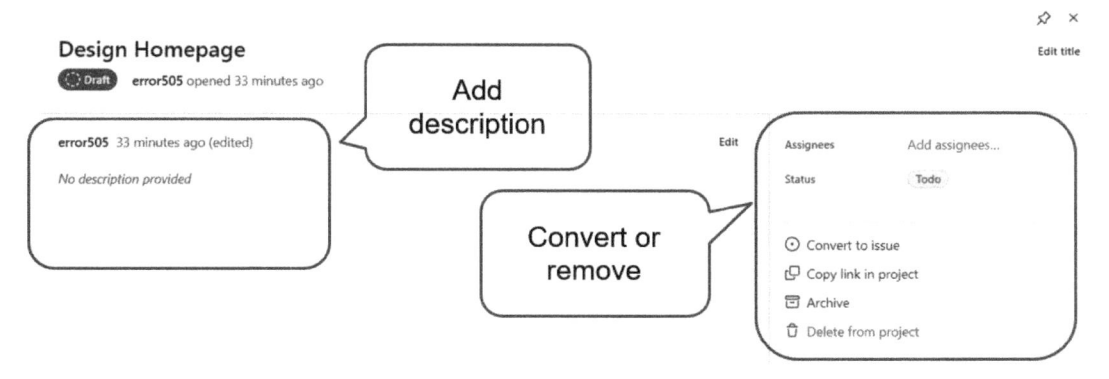

Figure 3.37 – Task view

5. **Move cards along**:

   I. When you start working on a task, move its card to **In Progress**.

   II. Once done, move it to **Done**.

   It's like moving a sticky note from **Todo** to **Done** when a part of your bookshelf is complete.

6. **Collaborate and update**: Your team can add comments, suggestions, or updates to each card. It's like your friends writing notes on sticky notes about the website's progress.

Why use a Kanban board?

It's satisfying and motivating to see cards move from **Todo** to **Done**, like watching the website take shape.

Collaboration is easy as everyone can see what needs to be done and what's already completed. No confusion about who's doing the design and who's writing the code.

It's a very flexible way to add new tasks or rearrange them as needed. It's like adjusting your steps while building the website if you find a better way.

Setting up a Kanban project on GitHub for your one-page website is like organizing and visualizing the steps of a fun DIY project. It keeps everything clear, collaborative, and flexible, making the process of building your website enjoyable!

How can I change my project settings?

In the **Settings** section of your projects, you have many options, such as changing your project name and description, adding a README file, managing access, and, additionally, a possibility to add custom fields:

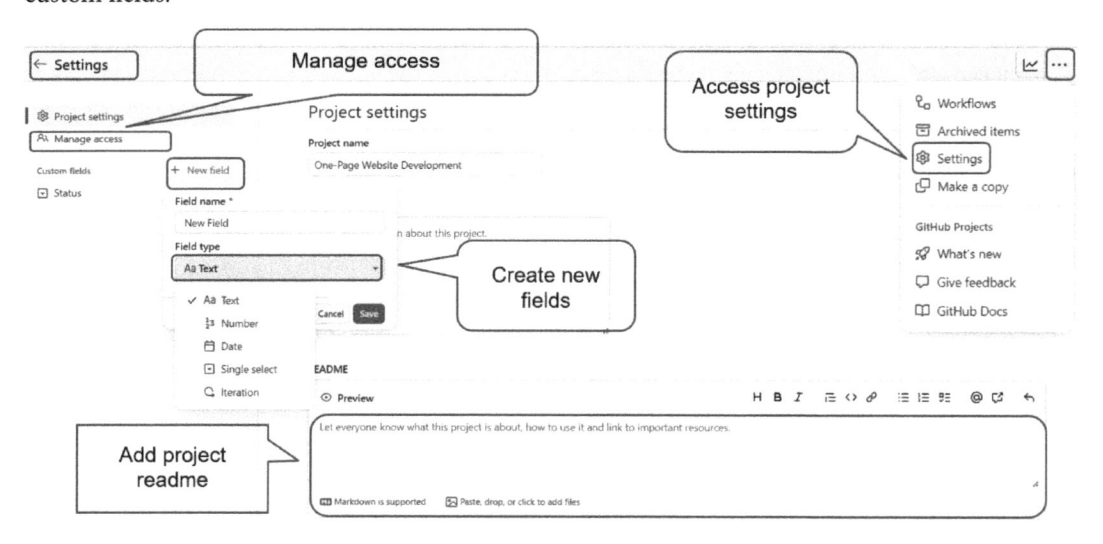

Figure 3.38 – Project settings

Now that we've covered how to set up your project on GitHub, use Kanban for organization, and dived into the insights to monitor your project's progress, let's move forward. Next, we're going to learn about using GitHub wikis. This will help you create a comprehensive guide for your one-page website that anyone can reference. It's like building a manual for your website, making it easier for others to understand and contribute to your project.

Understanding wikis on GitHub

What's a **wiki** on GitHub? It's a place on your GitHub repository where you can write and store important information about your project. Think of it as a digital notebook where you and your team can write down everything about building your one-page website.

How can wikis help?

Just like having all the instructions for building a website in one place, a wiki keeps all the information about your project in one spot. This can include design guidelines, coding standards, or even meeting notes.

Anyone on your team can add or edit pages. If someone finds a better way to code a part of the website or update the design, they can quickly update the wiki.

When someone new joins your project, they can read the wiki to get up to speed. It's like giving them the instruction manual to your website so that they know exactly what to do.

Creating a wiki for your one-page website on GitHub

To create a wiki for your one-page website, follow these steps:

1. Go to your website's repository on GitHub and find the **Wiki** tab, usually next to **Projects**. Click on **Create the first page** to start writing your first wiki page:

Figure 3.39 – Wiki overview

2. Write a brief introduction to your website project as the introduction page of the website. Include what the website is about and the goal of this wiki:

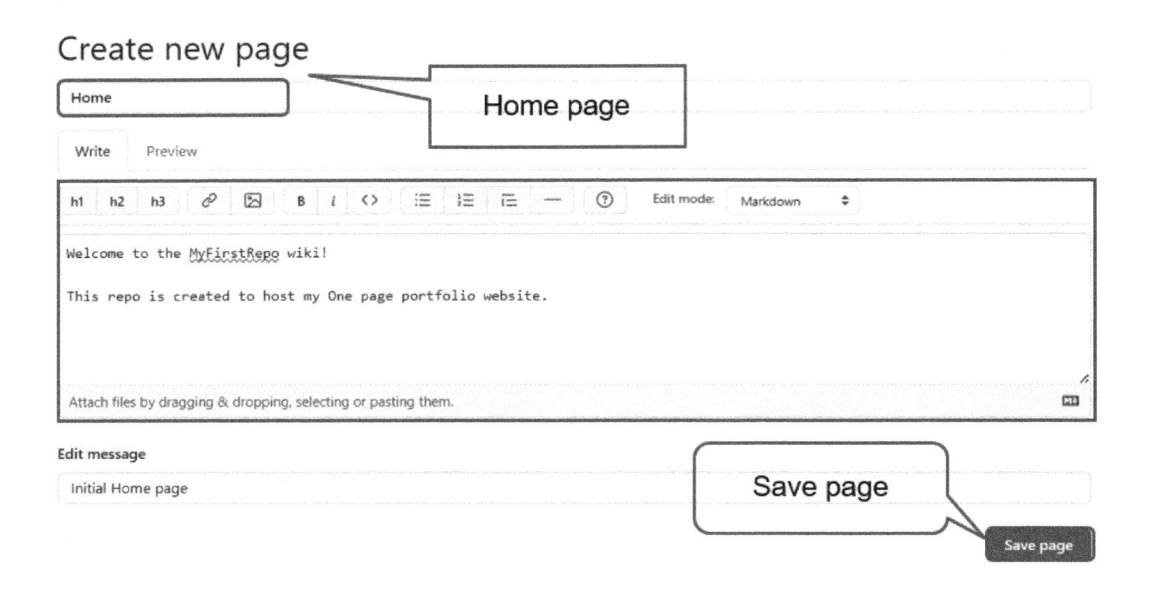

GitHub Features for Collaborating on Projects

Figure 3.40 – Creating a new wiki

3. Click **New page** to add more content:

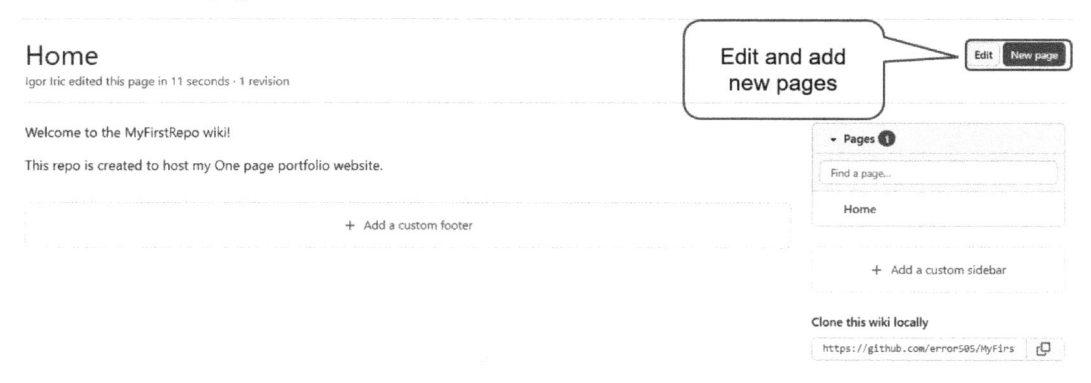

Figure 3.41 – Creating new wiki pages

4. Create pages for different aspects, such as **Design Guidelines**, **Coding Standards**, or **Project Timeline**. It's like adding sections to your website – one for contact, one for portfolio, and so on.

Next, let's look at some best practices for using wikis:

- **Keep it organized**: Just like how you'd organize your notebook with sections and a table of contents, do the same for your wiki. Create different pages for different aspects of your website project, such as **Design Guidelines**, **Coding Standards**, and so on. Use a clear and descriptive title for each page. Don't create a page for every tiny detail. Use a similar format for all pages.

- **Regular updates**: Keep your wiki current. If something changes in your project, update the wiki to reflect it. It's like updating the instruction manual when you find a better way to build a part of an airplane.
- **Encourage collaboration**: Make sure everyone feels comfortable adding to and editing the wiki. The more people contribute, the richer the content will be.
- **Link to other resources**: If you have designs stored elsewhere or code snippets in your repository, link to them in your wiki. It's like having references in your instruction manual to the parts of the airplane.

Using a wiki on GitHub for your one-page website is like having a comprehensive, collaborative instruction manual for your project. It centralizes knowledge, makes collaboration easy, and keeps everyone on the same page. This way, building your website becomes a well-organized team effort, just like assembling a model airplane with a clear set of instructions. Up next, we will learn about GitHub repository insights.

What are GitHub repository insights?

Imagine you and your friends are working together on a group project, such as baking a cake. You'd want to know who's great at making icing, who's a pro at baking, and how the whole cake-making process is going, right? That's where **GitHub repository insights** come in for your one-page website project. They're like a dashboard showing how everyone's contributing to the cake, or in this case, your website. They give you charts and data about your website project. It's like a summary report of how cake baking (website building) is progressing.

How to use Insights

You can easily access **Insights** by clicking on the ⌁ **Insights** tab in your website's GitHub repository:

Figure 3.42 – Insights tab

Contributors

In **Contributors**, you can see a chart of commits. Each commit is like a small contribution someone made, such as adding a new line of code to the website.

You'll see who's contributing and how often. Maybe Alice added most of the home page content, or Bob fixed a bunch of typos:

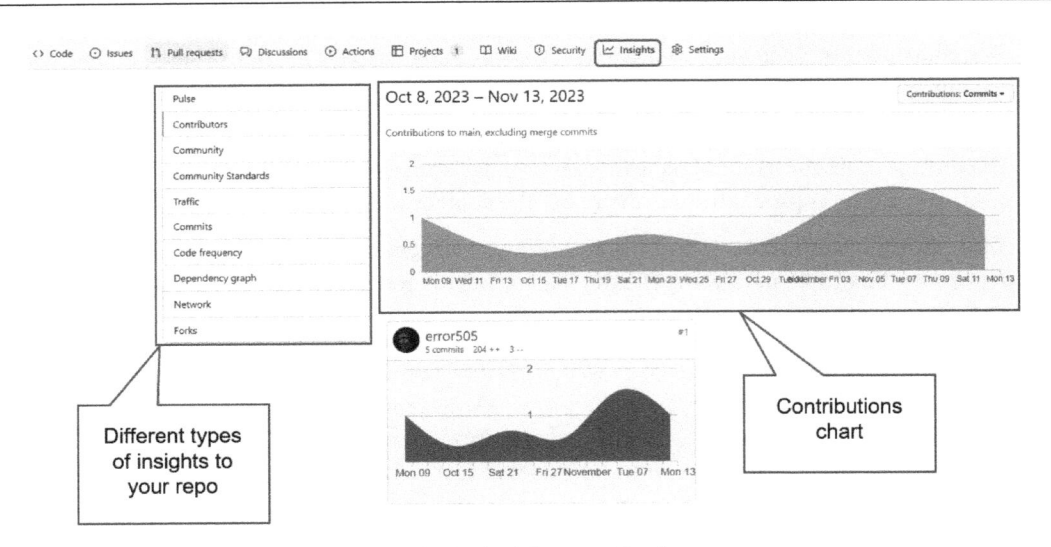

Figure 3.43 – Insights contributions

Traffic

Traffic shows how many people are visiting your project. It's like tracking how many friends peeked into the website to see your portfolio.

You can see what's attracting more viewers. Maybe your **About Me** page is really popular!

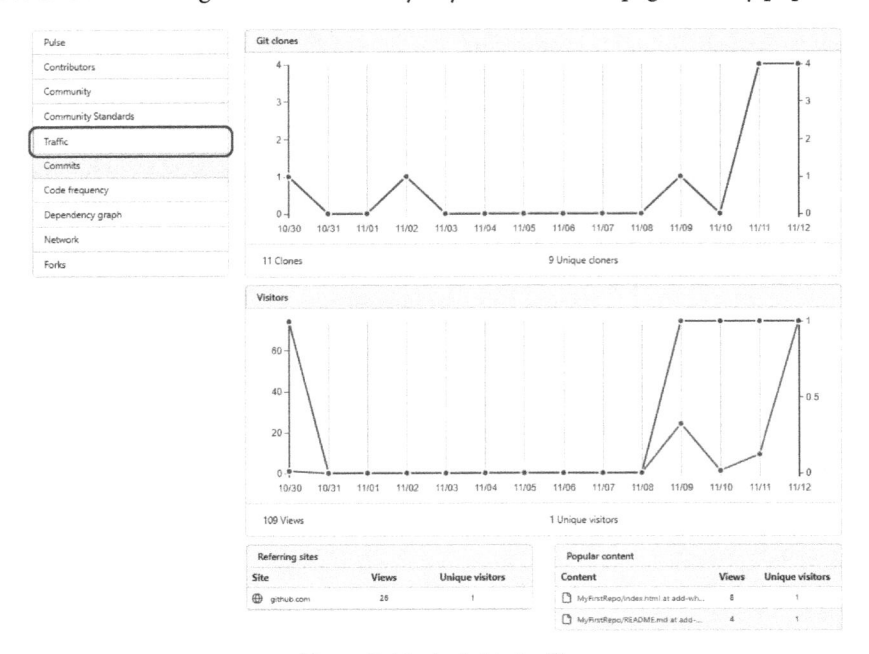

Figure 3.44 – Insights traffic

Pulse

Pulse gives you a quick look at what happened over a period, such as a week. It's like a weekly summary of your website progress: who did what; what got finished.

It shows recent changes, open and closed issues (tasks or suggestions), and more:

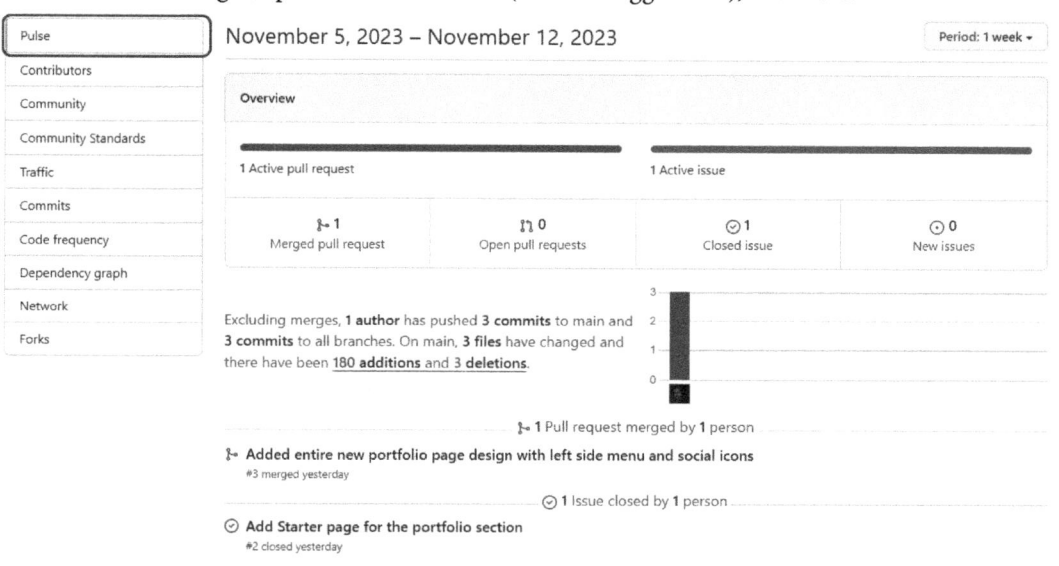

Figure 3.45 – Insights Pulse

Code frequency

The **code frequency** chart shows additions and deletions in your code over time. Think of it as showing how many lines of code were added or removed from your website each week:

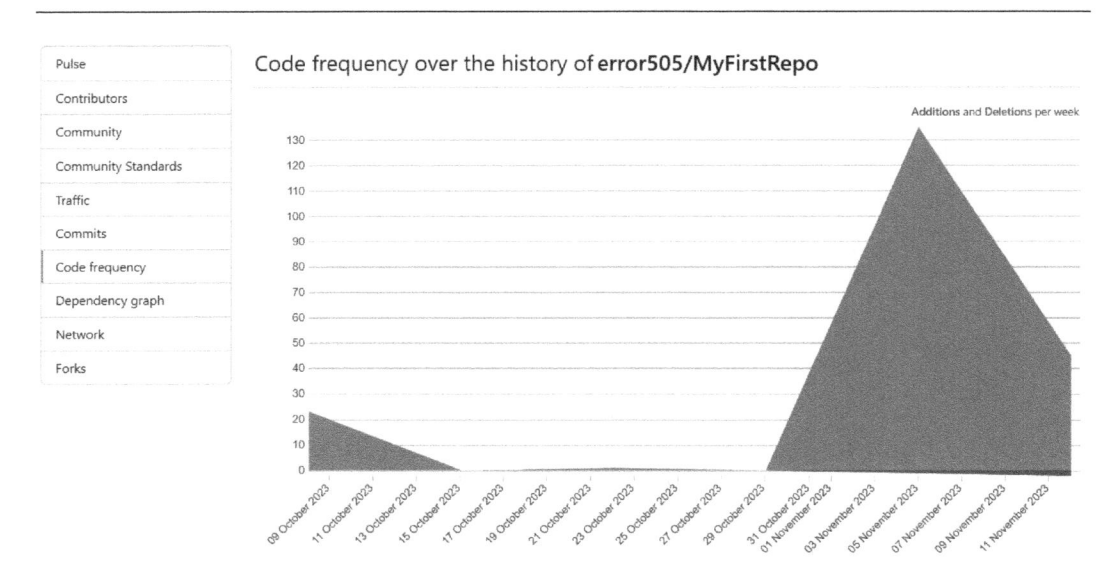

Figure 3.46 – Insights Code frequency

Why are insights helpful?

You can see how your website is evolving. It's like watching your cake rise in the oven and seeing it come together.

Insights show who's contributing more and in which areas. Maybe Alice is great at design while Bob excels at writing content.

By understanding each other's strengths and work patterns, you can collaborate better. It's like knowing who's best at whisking eggs or icing cakes.

GitHub repository insights for your one-page website offer a clear, visual way to understand how your project is coming along and how everyone's pitching in. It's like having a handy report on your group cake-baking project, helping you see the big picture and fine-tune your teamwork!

Summary

In this chapter, we learned how to use GitHub for effective collaboration on projects, including inviting friends, managing issues, sharing ideas through pull requests, handling notifications, engaging in discussions, setting up projects, and utilizing insights and charts. These skills are crucial for working together on coding projects, making collaboration smoother and more efficient.

In the next chapter, we'll be covering GitHub Flow, creating branches, understanding GitHub Flow in practice, cloning, adding, and committing files locally, and pushing changes to GitHub. This will help us manage different versions of our work more effectively, building on our collaborative skills.

Quiz

Test your knowledge with the following questions:

1. What is the purpose of inviting friends to your GitHub project?

    A. To make the project private

    B. To collaborate and contribute to the project

    C. To view the project without editing

    D. To compete in coding challenges

    Answer: B. To collaborate and contribute to the project

2. Issues in GitHub are used to track and discuss problems within a project.

    A. True

    B. False

    Answer: A. True

3. In GitHub, "_____" are used to show your part of a project to the group for feedback.

    Answer: pull requests

4. Why is it important to discuss and find solutions to issues in GitHub?

    Answer: It helps in identifying, tracking, and resolving problems in a project, ensuring smooth collaboration and progress.

5. What can you use to share and discuss ideas with your team on GitHub?

    A. Emojis

    B. Issues and pull requests

    C. Direct messages

    D. Profile pictures

    Answer: B. Issues and pull requests

6. You can use GitHub to work on coding projects alone, without any collaboration.

    A. True

    B. False

    Answer: A. True

7. Which feature in GitHub is like throwing a party and inviting your buddies to join in?

    A. Creating a repository

    B. Setting up a dashboard

    C. Inviting collaborators

    D. Writing a README file

    Answer: C. Inviting collaborators

8. To see how your project is doing, you can set up a _____ in GitHub.

    Answer: Dashboard

9. Pull requests in GitHub are only for showing completed work, not for receiving feedback.

    A. True

    B. False

    Answer: B. False

10. What is one benefit of using discussions in GitHub for your project?

    Answer: Discussions allow team members to communicate effectively, share ideas, and make decisions collaboratively, enhancing the project's development.

Challenge – Teaming up for a stellar mission!

Welcome back, space explorers! *Chapter 3* was all about teamwork and collaboration in the vast universe of GitHub. Now, it's time to apply those skills to improve your *Space Explorer* game. Ready to turn your solo space mission into a galactic team adventure?

Figure 3.47 – Space Explorer game screen

Your mission

Expand your *Space Explorer* game by teaming up with friends, managing issues, and getting feedback.

Steps to success

Here are your steps to success:

1. **Invite your crew**:

    I. Open your `Space Explorer` game repository on GitHub.

    II. Just like inviting friends to a party, add collaborators to your project. Navigate to **Settings | Manage access | Invite a collaborator**. Enter your friends' GitHub usernames and invite them aboard your spaceship!

2. **Identify and create issues**:

    I. Play your game and look for any glitches or areas for improvement. Maybe an obstacle is too hard, or the stars aren't shiny enough?

    II. Create issues in your GitHub repository for each of these points. It's like pinning a note in your spaceship about what needs fixing or upgrading.

3. **Pull requests for new features**:

    I. We've added flying obstacles and stars to the game! Time for you to integrate them.

    II. Create a new branch and add these features. Then, open a pull request and invite your team to review your code. It's like showing a sketch of your spaceship's new wings and asking for feedback before you paint them.

4. **Manage notifications**: With your crew onboard, your GitHub might get busy. Set up your notification preferences to keep track of updates without getting lost in space.

5. **Start a discussion**: Use GitHub Discussions to chat about your next big feature or how to improve the game. It's like having a team meeting in the control room of your spaceship.

6. **Plan your next adventure**: Use the **Projects** feature to organize your upcoming game enhancements. Create a project board and add tasks such as *Add new levels* or *Design alien spaceships*.

7. **Review the galaxy (insights and charts)**: Check out the *insights* and *charts* in your repository. See how your project is progressing and who's contributing the most. Who's the most active astronaut in your crew?

4
Branching in GitHub and Git

Ready for *Chapter 4*? It's like leveling up in your coding adventure. This chapter is all about making your own space in a project, adding cool new features, and sharing your work with your team.

The first section is where you will learn how to create your own little world within a project where you can try out new ideas without affecting the main project. You'll learn how to make these branches and why they're super helpful.

Then, you'll learn how to make changes, commit parts of your project, and then save them, like checkpoints in a video game. You'll see how to keep track of all the cool stuff you add or change.

We will also talk about pushing your awesome changes, where you'll learn how to share them with your team on GitHub.

We will cover the following topics in this chapter:

- Branching with GitHub Flow and creating a branch on GitHub
- Understanding GitHub Flow in practice – two developers, two features, one project
- Creating a local copy of the repository (cloning)
- Adding and committing files to a local repository
- Understanding Git's working area, staging area, and history and pushing your changes to GitHub

Technical requirements

For this chapter, which focuses on branching in GitHub and Git, you will need to do the following:

- Install Git on your computer to handle version control tasks
- Create a GitHub account to perform actions such as creating branches and submitting pull requests

Please refer to this book's GitHub repository to find the code examples for this chapter: `https://github.com/PacktPublishing/GitHub-for-Next-Generation-Coders/tree/main/Chapter%204`. Make sure you clone the repository so that you have a local copy to work on. This is essential for practicing branching and merging. Also, organize your work by creating a folder in GitHub named `ch4`. This will help you manage different chapters separately.

This setup will allow you to effectively learn and apply branching strategies with GitHub, making your coding projects more manageable and collaborative.

Branching with GitHub Flow and creating a branch on GitHub

Let's dive further into GitHub, where we'll learn how branching works and why branches are important. You and your friends are building a one-page portfolio website. You want to add new features without messing up what's already there. That's where GitHub Flow comes in.

GitHub Flow, in the context of branching, is a simple yet effective way to manage changes in a project, such as when building a one-page portfolio website.

Setting up your own space – creating a new feature branch

To start using GitHub Flow, we need to create a new feature branch.

So, what's a branch? Think of it as a copy of the website where you can experiment without affecting the main website (the *main* branch).

You decide to add a **Skills** section to the website. So, you create a new branch named `add-skills-section`. Now, you have a separate space to work on this feature.

Understanding the main branch

The main branch is the final, public version of your website. Everything here is in good order, tested, and presentable.

Why not work directly on the main branch? Doing so is risky, like cooking a new dish for the first time in the middle of a busy kitchen. Mistakes can affect everyone.

Deciding to add a new feature

Let's say you want to add a **Contact Me** form to your website. It's a new form style that you've never developed before.

You create a branch that comes off the main branch. This is your workspace. Name it something clear, such as `add-contact-form`.

Why use a feature branch?

If something goes wrong in your branch, the main website is unaffected. It's like experimenting with your new contact form on a separate website. You can focus on your **Contact Me** form without worrying about messing up the main website. It's your area to experiment, make mistakes, and learn. Once created, make sure to switch your working directory to your new branch.

Creating a feature branch in GitHub Flow is like setting up a workstation in a shared project. It gives you the freedom to experiment and develop your part of the project (such as your **Contact Me** form) without stepping on anyone else's toes. Once your feature is ready and tested, it can be seamlessly integrated into the main project, enriching everything with your contribution.

Making changes and committing

Figure 4.1 shows the process of creating a new feature branch in git. It visually represents how a feature branch is created from the main branch, allowing you to work on new features independently without affecting the main project. This helps in keeping the main branch stable while new features are being developed and tested. Once the new feature is ready and tested, it can be merged back into the main branch. This workflow ensures that the development process is organized and that the main project remains stable.

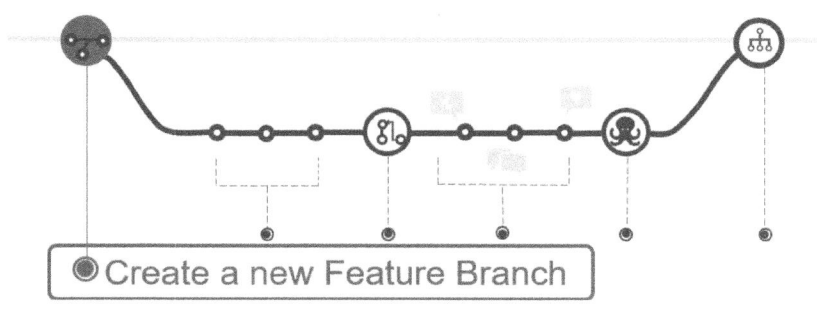

Figure 4.1 – GitHub Flow – creating a new feature branch

Once you've created your new feature branches, you can start making some changes while adding new files to your feature branch.

You start adding elements for the **Contact Me** form. This could involve writing HTML for the layout, CSS for the style, or JavaScript for interactive features. Maybe you'll lay out the basic structure, then add some style, and finally fine-tune the details.

In this personal workspace, you can add, edit, or delete files. Feel free to experiment here because your changes won't impact the main project until you decide to merge them.

Once you're happy with your changes, you *commit* them. This is like saying, "*This part of my section is done for now.*"

What's a commit?

Think of a commit as a checkpoint. Each commit represents a set of changes or additions you made.

Using Git, you'll add the files you've changed to the staging area (with `git add`), and then you'll create a commit with `git commit`. In this commit, you'll write a message describing what you did – for example, `Added basic layout for the blog section`. It's best to commit often because doing so allows you to see what changes you made and when. If something breaks or doesn't work as expected, it's easier to revert to a previous commit if you have smaller, more frequent commits. Each commit is a snapshot of your work at a specific point in time. If something goes wrong later, you can always go back to a previous commit.

Every commit is recorded in the branch history. This means you can see the evolution of your **Contact Me** form over time, understanding what changes were made and when:

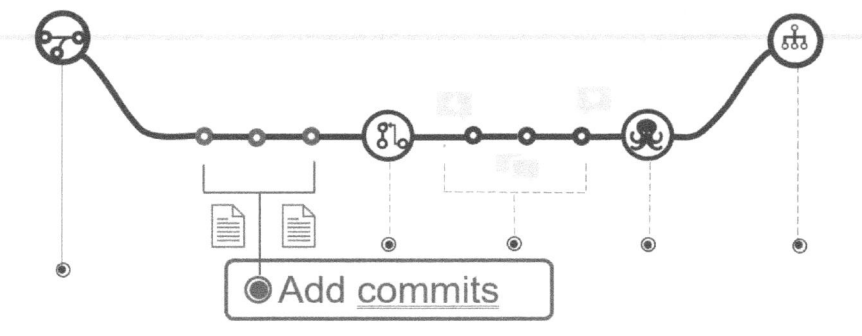

Figure 4.2 – GitHub Flow – adding commits to a feature branch

Pushing changes

Once you've made a series of commits that represent a significant portion of your work (such as completing the layout of the **Contact Me** form), you *push* these commits to GitHub. This updates your branch on GitHub with all the changes you've made locally.

After pushing, your commits are visible to your teammates. They can see exactly what changes you've made, commit by commit.

Show and tell – creating a pull request

Once you've made your changes and you're ready to show your **Skills** section to your friends, you can create a pull request. This is like putting your part of the mural on display for feedback. It's like saying, *"Hey, I've made some changes, can someone check them out?"*

Why a **pull request** (**PR**)? This lets your team see what you've done, give suggestions, or say it's ready to be part of the main website. This step is great for getting feedback from others on your team:

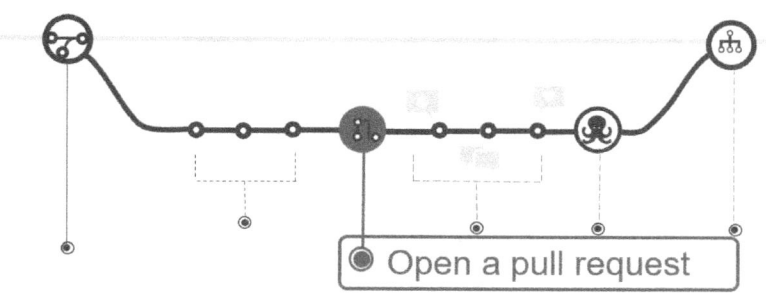

Figure 4.3 – GitHub flow – Opening a PR

Opening a PR is like a mini-celebration of your work, combined with a brainstorming session. It's where your effort becomes a team effort, bringing everyone's talents together to make your one-page portfolio website the best it can be!

Teamwork – discussing and reviewing the PR

Your friends look at your new **Skills** section. Say someone suggests *Add more color here* or *Align this text better*. They might give you tips on how to add some tests or which naming conventions to follow. It's all about working together and learning from each other. Based on this feedback, you tweak your section. Every change you make is updated in the PR:

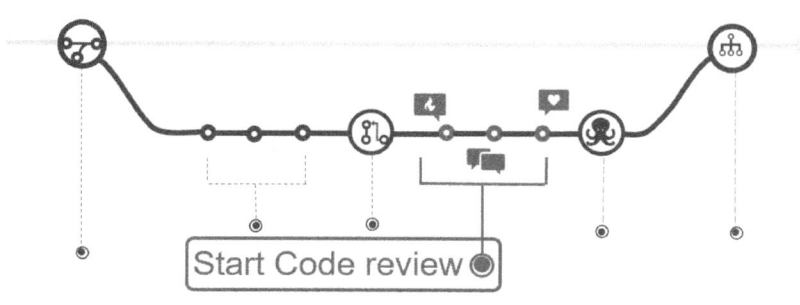

Figure 4.4 – GitHub Flow – starting the code review

Approving the PR

Once everyone is happy with the proposed changes and improvements and likes the **Skills** section, it's time to finalize this and make it officially part of the website:

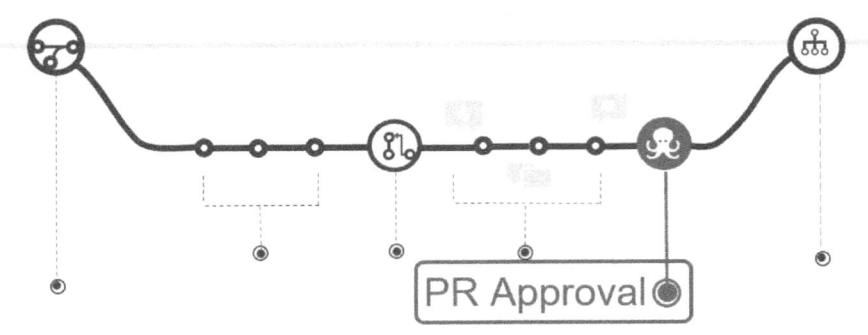

Figure 4.5 – GitHub Flow – PR approval

Making it official – merging the PR

Once your changes have been reviewed and approved, you can merge the PR. This means your changes will be added to the main project. It's like saying, *"Okay, these changes are good to go!"*

Now, your **Skills** section is ready to go live on the website:

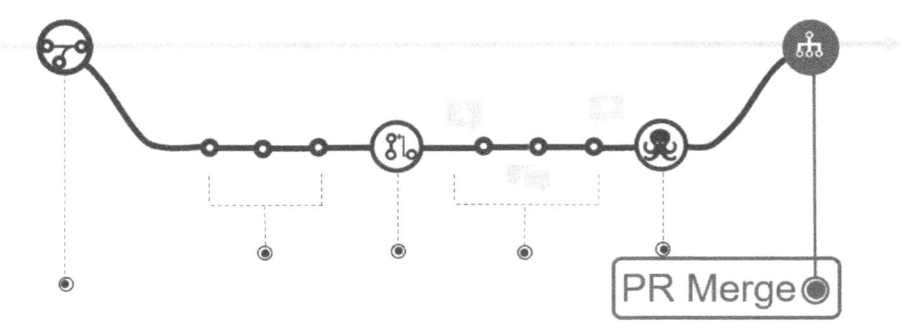

Figure 4.6 – GitHub Flow – merging the PR to the main branch

Cleanup time

After merging, you can delete the branch you worked on. It's a bit like cleaning up your workspace after completing a project. But don't worry – the history of your changes is still saved.

And that's it! By following these simple steps, you've learned how to use GitHub to build something amazing with your friends. Using GitHub Flow, you and your friends can work on different parts of your one-page portfolio website simultaneously and efficiently. Each branch is a playground for new ideas, and PRs are where those ideas get polished and readied for the world to see. It's a smooth, collaborative process that helps keep your project organized and moving forward!

Understanding GitHub Flow in practice – two developers, two features, one project

Let's explore how GitHub Flow works through an example with two developers, Alex and Jamie, working on a project to improve a website. They are each responsible for different features, and they use GitHub Flow to manage their contributions seamlessly.

The main branch is always a starting point for creating a feature branch. This is the stable version of the project, like the foundation of a building. Alex and Jamie will both start their work from here.

Let's try to explain this step by step.

Alex needs to add a contact form. They create a branch called **feature/contact-form** from the main branch. This branch is their workspace.

While Alex was working on the new feature branch they created previously, Jamie started on a gallery feature in a branch called **feature/gallery**:

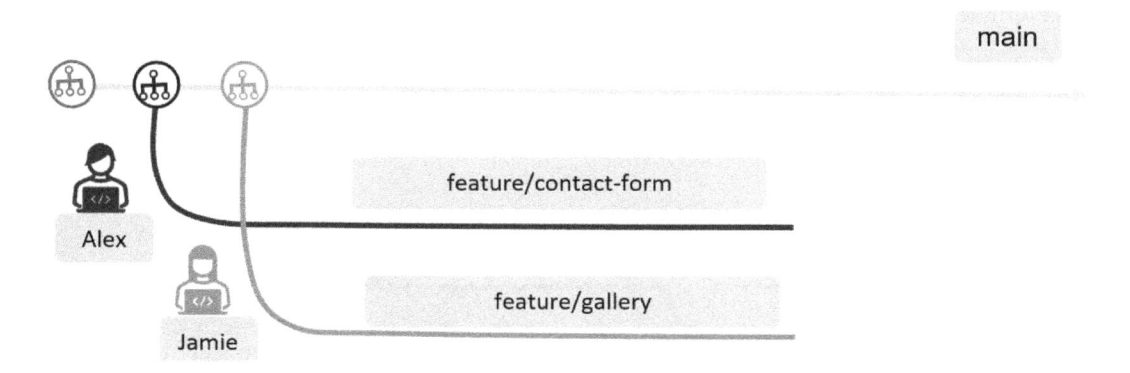

Figure 4.7 – GitHub Flow – branch creation

Both Alex and Jamie are coding their features, regularly committing their changes to their feature branches. These commits act like checkpoints in their work:

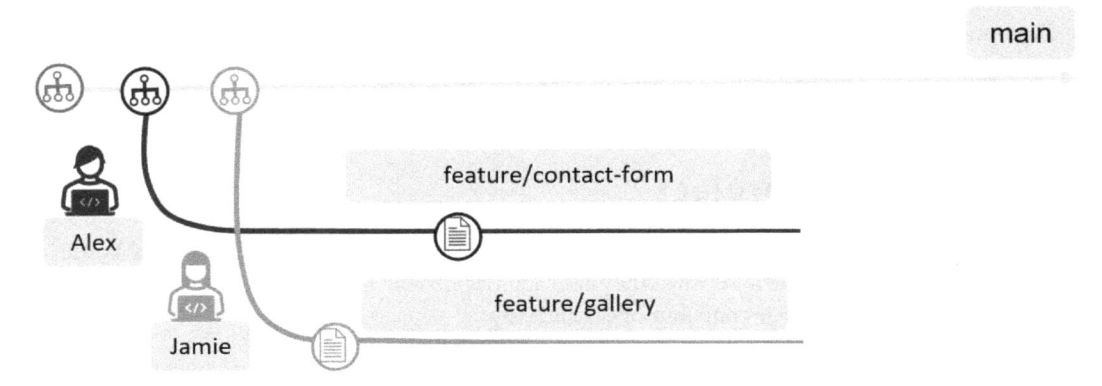

Figure 4.8 – GitHub Flow – adding and committing files

Once the contact form works well, Alex pushes their branch to GitHub. They create a PR on GitHub, asking to merge **feature/contact-form** into the main branch. This is like saying, "*Hey team, I think this is ready to be part of our main project.*"

The team reviews Alex's work. Once it's approved, the PR is merged into the main branch. The **feature/contact-form** branch is deleted to clean up the branches. The main branch now contains the new contact form feature:

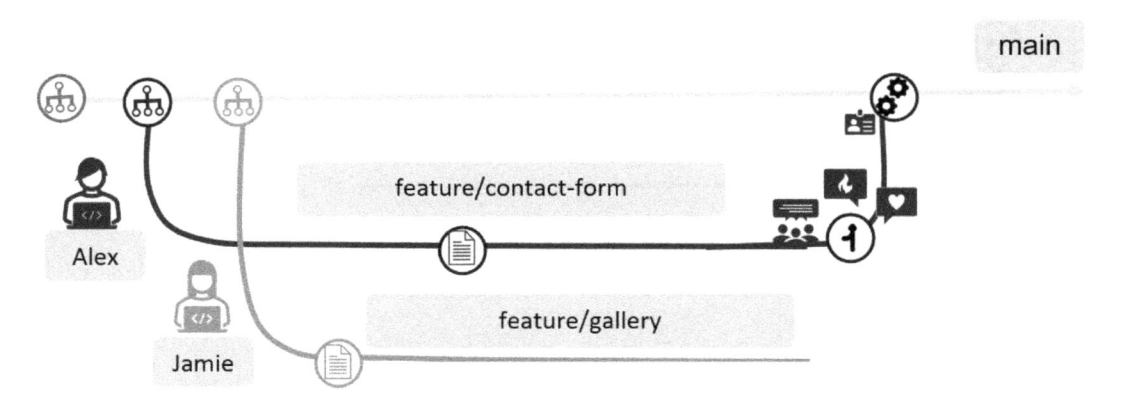

Figure 4.9 – GitHub Flow – pushing the branch, creating a PR on GitHub, and merging the PR

After Alex merges their contact form, Jamie pulls the latest main branch to update **feature/gallery**. This step ensures Jamie's branch is up to date, reducing the risk of conflicts. Jamie keeps working on the gallery, making commits to their branch:

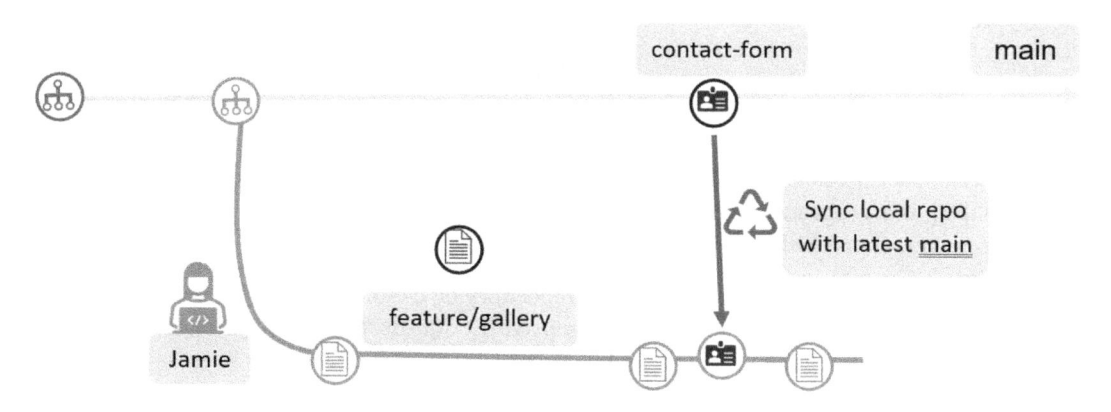

Figure 4.10 – GitHub Flow – syncing the local repository with the newest main branch

Once satisfied, Jamie pushes their branch to GitHub and opens a PR for merging into the main branch. The team reviews Jamie's gallery feature. They might suggest changes or improvements. After any final adjustments and team approval, Jamie's gallery feature is merged into the main branch:

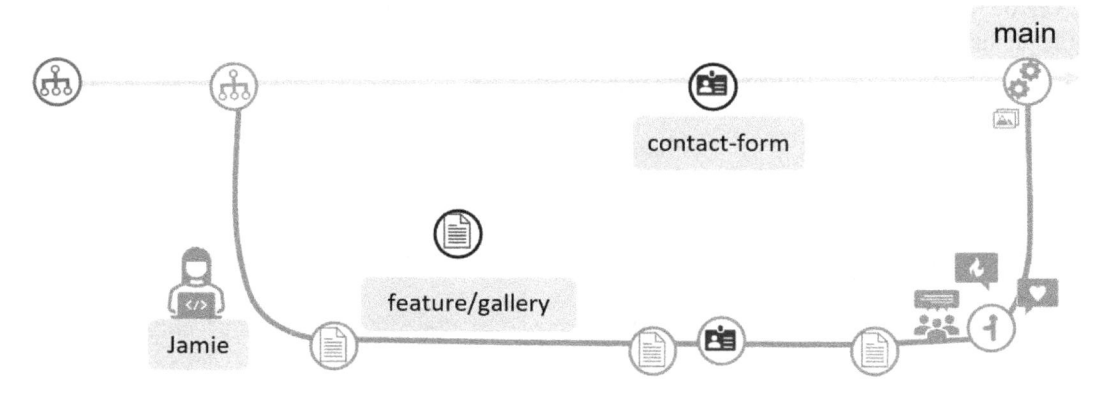

Figure 4.11 – GitHub Flow – creating a PR on GitHub and merging it

And what's the result of this type of flow? A collaboratively enhanced project.

The main branch now contains both Alex's contact form and Jamie's gallery as they have been seamlessly integrated. By regularly updating their feature branches with the latest main branch, Alex and Jamie avoided merge conflicts, making the integration smoother. The use of PRs allowed for team input and review, ensuring high-quality additions to the project.

Through GitHub Flow, Alex and Jamie were able to work independently on their features while ensuring their work aligned with the evolving project. This approach encourages collaboration, maintains project stability, and ensures new features are integrated smoothly:

Figure 4.12 – GitHub Flow – actual main with all feature branches

You've now seen how GitHub Flow works in practice with two developers working on different features and keeping the project organized. This process helps maintain a smooth workflow and ensures that new features are integrated without disrupting the main project.

Next, let's learn how to create a local copy of your repository by cloning it. This step is essential for working on your projects locally on your computer. We'll walk through the process of cloning a repository so you can start making changes and contributing to your project right away.

Creating a local copy of the repository (cloning)

If you still have not cloned your repository, `My First Repo`, from GitHub, now would be the time to do so. In *Chapter 2*, we explained how to clone the repository from GitHub, but let's quickly recap it here.

You will be working on a one-page portfolio website and you need to add a left-hand side menu, a profile picture, social icons, and some CSS styling. Let's go through the steps, starting from cloning the project to adding and committing your changes.

Cloning the project

Follow these steps:

1. **Navigate to the Code tab on GitHub**: Find your project's repository on GitHub:

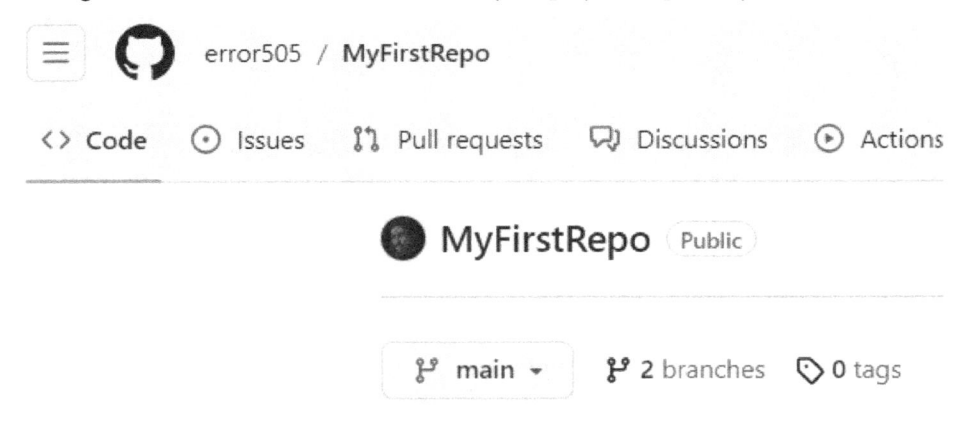

Figure 4.13 – GitHub repository

2. **Clone the repository**: Click the **Clone** or **Download** button and copy the URL provided:

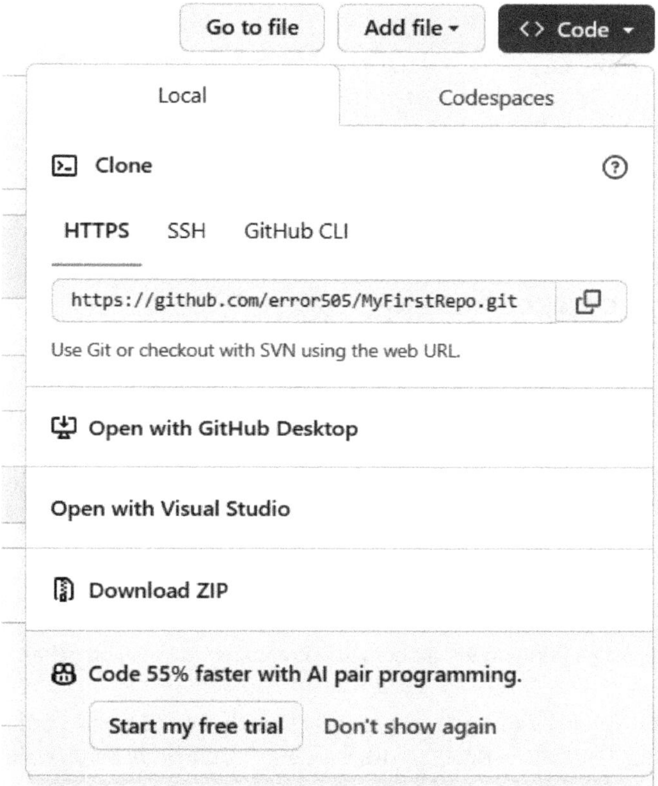

Figure 4.14 – GitHub repository clone URL

3. **Open Git Bash or use your Terminal**: On your computer, open your Git Bash application or your terminal.

4. **Run git clone**: Type `git clone <CLONE-URL>`, replacing `<CLONE-URL>` with the URL you copied. This command downloads a full copy of the repository to your local machine:

```
PS C:\Users\igor.iric\Desktop> git clone https://github.com/error505/MyFirstRepo.git
Cloning into 'MyFirstRepo'...
remote: Enumerating objects: 17, done.
remote: Counting objects: 100% (17/17), done.
remote: Compressing objects: 100% (15/15), done.
remote: Total 17 (delta 4), reused 0 (delta 0), pack-reused 0
Receiving objects: 100% (17/17), 6.37 KiB | 2.13 MiB/s, done.
Resolving deltas: 100% (4/4), done.
```

Figure 4.15 – Running the git clone command

5. **Change directory**: Enter cd <REPOSITORY-NAME>, substituting <REPOSITORY-NAME> with the name of the repository you just cloned. This will move you into the repository's directory:

```
PS C:\Users\igor.iric\Desktop> cd .\MyFirstRepo\
PS C:\Users\igor.iric\Desktop\MyFirstRepo>
```

Figure 4.16 – Entering your new repository folder

Now that we have cloned the project, let's switch to the correct branch.

Switching to the correct branch

Follow these steps:

1. **List all branches**: Type git branch -a to see all available branches, both local and remote:

```
PS C:\Users\igor.iric\Desktop\MyFirstRepo> git branch -a
* main
  remotes/origin/HEAD -> origin/main
  remotes/origin/add-what-i-do
  remotes/origin/main
```

Figure 4.17 – Running the git branch -a command to list all branches

2. **Switch branch**: If you need to work on a specific branch, use git switch <BRANCH-NAME>, replacing <BRANCH-NAME> with the name of the branch you want to switch to. If you need to create a new branch, then use git switch -c <BRANCH-NAME>:

```
PS C:\Users\igor.iric\Desktop\MyFirstRepo> git switch -c feature/leftside-menu
Switched to a new branch 'feature/leftside-menu'
```

Figure 4.18 – Running the git switch command to create a new branch and switch to it

Now that you've successfully cloned a project from GitHub and switched to the correct branch, you are ready to start adding your new files, various sections of the website, and different bits of styling. In the next section, you'll start building your one-page portfolio website.

Adding and committing files to a local repository

Now that you've cloned your repository and switched to the correct branch, you can start building your one-page portfolio website. In this hands-on section, you will create a left-hand side menu, add a profile picture and some social icons, and style everything with CSS. You'll want these changes to be recorded in your local Git repository.

Adding your new files to Git

For this hands-on section, we will add some HTML code to the `index.html` file to replace the welcome text.

Use your favorite code editor to make changes in the `index.html` file. You can use Visual Studio Code for this, a free code editor created by Microsoft. You can download it from `https://code.visualstudio.com/download` and install it locally. In *Chapter 8*, *Helpful Tools and Git Commands*, we will talk more about Visual Studio Code and its cool features.

Now, you can replace the content of the file with the following code:

```html
<> index.html > ...
 2   <html lang="en">
 3     <head>
11       <link rel="stylesheet" href="styles.css" />
12     </head>
13
14     <body>
15       <div class="sidebar">
16         <img src="https://via.placeholder.com/150" alt="John Doe" class="profile-image" />
17         <h2>John Doe</h2>
18         <p>Senior Software Engineer</p>
19         <!-- Social Media Icons -->
20         <div class="social-icons">
21           <a href="#"><i class="fab fa-twitter"></i></a>
22           <a href="#"><i class="fab fa-linkedin"></i></a>
23           <a href="#"><i class="fab fa-github"></i></a>
24           <a href="#"><i class="fab fa-instagram"></i></a>
25         </div>
26         <!-- Navigation Menu -->
27         <nav class="navigation">
28           <ul>
29             <li><a href="#about" class="about-button"><i class="fas fa-user"></i><span>About Me</span></a></li>
30             <li><a href="#portfolio" class="portfolio-button"><i class="fas fa-briefcase"></i><span>Portfolio</span></a></li>
31             <li><a href="#services" class="service-button"><i class="fas fa-dollar-sign"></i><span>Services & Pricing</span></a></l
32             <li><a href="#blog" class="blog-button"><i class="fas fa-blog"></i><span>Blog</span></a></li>
33             <li><a href="#contact" class="contact-button"><i class="fas fa-envelope"></i><span>Contact</span></a></li>
34           </ul>
35         </nav>
36       </div>
37
38       <div class="main-content">
39         <h1>What I Do</h1>
40         <p>
41           I have more than 10 years' experience building software for clients...
42         </p>
43         <!-- Add other sections similarly -->
44       </div>
```

Figure 4.19 – Code for the website

Replace John Doe with your name. Once you've done this, you can add some CSS to a new file called `styles.css`.

Here's the code snippet you will use for your `styles.css` file:

```css
body {
    font-family: Arial, sans-serif;
    margin: 0;
    padding: 0;
}

.sidebar {
    background-color: #1a855f;
    width: 300px;
    height: 100vh;
    float: left;
    padding: 20px;
    box-shadow: 2px 0px 5px rgba(0, 0, 0, 0.1);
}

.profile-image {
    width: 200px;
    border-radius: 50%;
    margin: 0 auto;
}

.social-icons {
    margin-top: 15px;
    display: flex;
}

.social-icons a {
    display: inline-flex; /* Ensures the circular background spans the icon */
    align-items: center;
    justify-content: center;
    width: 40px; /* Set the width and height to create a circular background */
    height: 40px;
    border-radius: 50%; /* Make the background circular */
    background-color: #333; /* Background color for the circle */
    margin: 0 5px;
    color: white;
    text-decoration: none;
```

Figure 4.20 – CSS style for the website

Once you've saved your `index.html` and `styles.css` files, you can run `index.html` in your browser. You should see a page similar to the following:

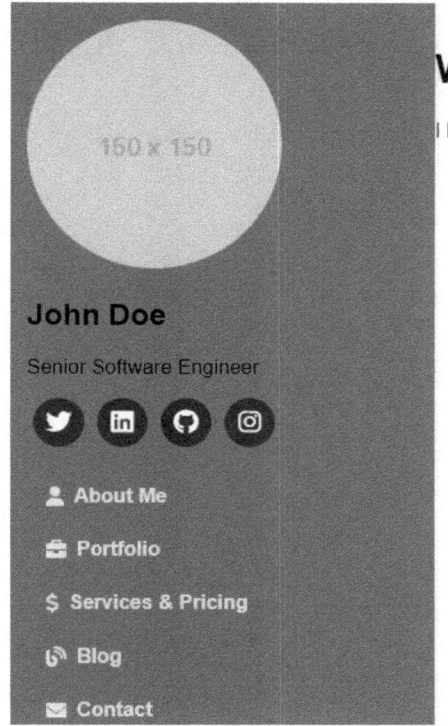

Figure 4.21 – One-page portfolio website – left-hand side menu

Understanding "Check Untracked Files" in Git

What is **Untracked Files**? When you're working with Git, it keeps an eye on the files in your project directory. But when you add new files or make changes that Git hasn't seen before, it labels them as *untracked*.

Why "Untracked"?

Untracked means new files or changed files in your GitHub repository. Git sees these files but hasn't been told to keep track of them yet. They're like strangers to Git's current understanding of your project. Since Git hasn't recorded these files in its history, they aren't part of your repository's timeline.

Check Untracked Files

This command is like asking Git, "*Hey, what's going on with the files in my project?*"

Type `git status`. You'll see the files you've created or changed listed as *untracked* (Git knows they're there, but they're not part of your project's history yet).

Git responds with a list showing the status of files. Untracked files are usually highlighted or marked clearly:

```
PS C:\Users\igor.iric\Desktop\MyFirstRepo> git status
On branch feature/leftside-menu
Changes not staged for commit:
  (use "git add <file>..." to update what will be committed)
  (use "git restore <file>..." to discard changes in working directory)
        modified:   index.html
        modified:   styles.css

no changes added to commit (use "git add" and/or "git commit -a")
PS C:\Users\igor.iric\Desktop\MyFirstRepo>
```

Figure 4.22 – Running the git status command

Why Check Untracked Files?

It helps you see what's new or changed in your project that Git hasn't started tracking yet. You might not want to track every file (such as temporary files or personal notes). Checking untracked files lets you decide which ones to add to Git.

Checking untracked files would show you the new elements you added (such as the left-hand side menu, profile picture, and social icons) and any CSS files you've modified. It's like taking inventory of what's new or changed before deciding what to include in your project's official history.

Adding your new file changes to the Git staging area

What does *adding to Git* mean? Your Git repository is like your digital scrapbook for your project. When you *add* files to Git, you're telling Git, "*Hey, I want these new pieces (your menu, picture, icons, and so on) to be part of my project's scrapbook.*" You're marking them for inclusion in your next snapshot.

How do you add files?

If you just want to add one file, such as `index.html`, you would type `git add index.html`:

```
PS C:\Users\igor.iric\Desktop\MyFirstRepo> git add index.html
PS C:\Users\igor.iric\Desktop\MyFirstRepo>
```

Figure 4.23 – Running git add index.html to add the index

If you want to add everything new or changed, type `git add .`:

```
PS C:\Users\igor.iric\Desktop\MyFirstRepo> git add .
PS C:\Users\igor.iric\Desktop\MyFirstRepo>
```

Figure 4.24 – Adding all files

Here, . tells Git to include all new and modified files to the staging area. Once you've added your files, they move to the **staging area**. Think of the staging area as a pre-commit holding zone. It's like you're saying, *"These are the files I want to be remembered in my next snapshot."*

After adding your files, your changes are ready to be committed. Committing is like gluing everything down in your scrapbook, making your layout permanent. Adding your new files to Git is like selecting which new items you want to permanently include in your project's history. It's a crucial step in organizing and tracking the evolution of your portfolio website. You can type `git status` again to see all the files that have been added to the staging area:

```
PS C:\Users\igor.iric\Desktop\MyFirstRepo> git status
On branch feature/leftside-menu
Changes to be committed:
  (use "git restore --staged <file>..." to unstage)
        modified:   index.html
        modified:   styles.css
```

Figure 4.25 – Running git status after all files have been added to the staging area

Now that we know how to add and stage files in your local Git repository, we'll learn about Git's different areas, such as the working area, staging area, and the repository itself. We'll also cover how to push your changes to GitHub so that everyone can see the updates you've made. This way, you'll get a complete picture of how your work progresses from your local machine to being accessible to collaborators.

Understanding Git's working area, staging area, and history and pushing your changes to GitHub

When you're using Git for your projects, such as your one-page portfolio website, it helps to picture Git as having three main areas. Think of these like spaces in your workshop where different stages of your project happen.

Working area (or working directory) – what is it?

This is where you do all your work – that is, create, edit, delete, and organize files. It's like your coding workspace, where everything is laid out in front of you. When you're adding text to your portfolio, choosing fonts, or adjusting your layout, you're in the working area:

Figure 4.26 – Git stages – the working area

Staging area (or index) – what is it?

This is a preparation area for your changes before they become part of your project's history. Think of it as a waiting room where you gather and review your changes.

Let's say you've added a new section to your portfolio and fixed a typo. You're happy with these changes and want them to be part of your next snapshot. You move these changes to the staging area with `git add`:

Figure 4.27 – Git stages – the staging area

History (or repository)

This is where all your snapshots (commits) are stored. It acts as a record of every change you've made, safely stored in your project's history.

After staging your changes, you commit them. This process takes a snapshot of your staging area and stores it in your history. It's like taking a photo of your workbench after you've completed an important part of your project and putting it in an album. To create a history of your changes, you can use the `git commit -m "your commit message"` command.

> **Tip**
> Keep in mind that you should always write short commit messages that are around 50 characters in length.

Carefully describe the changes you introduced with your commit so that your team can better understand what the commit is about. The last step here is to push your changes to the remote GitHub repository with the `git push` command.

Pushing your changes to GitHub

You have worked on your one-page portfolio website and you've just finished your new left-side menu and provided it with links and social icons. You've made your changes locally in your feature branch, `feature/leftside-menu`. Now, you want these changes to be available on GitHub so that others can see or contribute. This is where `git push` comes into play.

Normally, you would just type `git push`, which tells Git you want to transfer your commits from your local repository to a remote repository (such as GitHub).

In this case, you will get an error message saying that you cannot push to the remote repository as the branch you created locally still doesn't exist in your GitHub repository:

```
PS C:\Users\igor.iric\Desktop\MyFirstRepo> git push
fatal: The current branch feature/leftside-menu has no upstream branch.
To push the current branch and set the remote as upstream, use

    git push --set-upstream origin feature/leftside-menu

To have this happen automatically for branches without a tracking
upstream, see 'push.autoSetupRemote' in 'git help config'.
```

Figure 4.28 – Running the git push command

As you can see, the message also gives you instructions on how to push your local feature branch to the remote repository and set it upstream using `--set-upstream`. This is like setting a default path for your future pushes. You're telling Git, *"Hey, next time I push changes from this branch, I want them to go to the same place without me having to specify it again."*

The next part of the command is `origin`. This is the default name Git gives to the remote repository you cloned from. In most cases, `origin` is just shorthand for the URL of your GitHub repository and `feature/leftside-menu` is your branch name. You're specifying that you want to push the commits from this branch.

What happens when you run this command?

By using `--set-upstream`, you're setting up a link between your local branch, `feature/leftside-menu`, and a branch on GitHub (`origin`). If this branch doesn't exist on GitHub yet, this command will create it for you.

Git takes all the commits you've made in your `feature/leftside-menu` branch and uploads them to GitHub, placing them in a branch of the same name.

Once you've set the upstream once, you can just use `git push` while in your `feature/leftside-menu` branch and Git will know where to send your changes. So, let's run this command in our terminal and see what happens:

```
PS C:\Users\igor.iric\Desktop\MyFirstRepo> git push --set-upstream origin feature/leftside-menu
Enumerating objects: 7, done.
Counting objects: 100% (7/7), done.
Delta compression using up to 8 threads
Compressing objects: 100% (4/4), done.
Writing objects: 100% (4/4), 1.80 KiB | 1.80 MiB/s, done.
Total 4 (delta 0), reused 2 (delta 0), pack-reused 0
remote:
remote: Create a pull request for 'feature/leftside-menu' on GitHub by visiting:
remote:      https://github.com/error505/MyFirstRepo/pull/new/feature/leftside-menu
remote:
To https://github.com/error505/MyFirstRepo.git
 * [new branch]      feature/leftside-menu -> feature/leftside-menu
branch 'feature/leftside-menu' set up to track 'origin/feature/leftside-menu'.
```

Figure 4.29 – Git push to remote

By pushing your branch to GitHub, you make it possible for team members to see your new left-hand side menu, give feedback, or add their own tweaks.

As you can see, our new feature branch has been created in GitHub and all our changes to the files have been pushed with the branch. GitHub already provides us with a message to visit `<YOUR-REPOSITORY-URL>/MyFirstRepo/pull/new/feature/leftside-menu`. This is where we should create a PR.

Now, you can create a PR to merge it into the main project, integrating your stylish left-hand side menu into the main portfolio website.

Why are these areas important?

These areas help you organize your work. You can experiment in the working area, decide what changes you like in the staging area, and save your best work in your history. This separation gives you control over how your project evolves. You can make changes, try new things, and only commit to the ones you're sure about. And if something goes wrong, you can always go back to a previous state in your history.

Summary

This chapter provided a masterclass in teamwork and creativity in coding. You now know how to make branches in a project. It's like having a personal lab where you can experiment without messing up the main project. You also learned how to make checkpoints for your work. Every time you make a cool change or add something new, you save it as a checkpoint. Finally, you learned how to send your changes to the GitHub gallery, where your team can see what you've been up to.

In the next chapter, you'll learn how to showcase your work and ask for feedback. You'll learn the secrets to making your PRs stand out and connecting them to project tasks and how to review a PR. It's like being part of a judges' panel, where you help decide what makes it into the project.

Quiz

Answer the following questions:

1. What is *branching* in GitHub?

    A. Cutting off parts of your project

    B. Creating a copy of your project to work on separately

    C. Merging two projects into one

    D. Deleting old project files

    Answer: B. Creating a copy of your project to work on separately.

2. True or false: Branching is only used by advanced coders.

    A. True

    B. False

    Answer: B. False.

3. When you want to add a new feature to your project without affecting the main project, you create a new _____ in GitHub.

    Answer: branch.

4. Why is branching useful when collaborating on a project?

   Answer: Branching allows multiple people to work on different parts of the project simultaneously without interfering with each other's work.

5. What can you do with a branch in a GitHub project?

   A. Share it with friends

   B. Merge it back into the main project

   C. Delete it after use

   D. All of the above

   Answer: D. All of the above.

6. True or false: Every branch in GitHub must be merged back into the main project.

   A. True

   B. False

   Answer: B. False.

7. What is a common use of a branch in GitHub?

   A. To fix a bug

   B. To try out new ideas

   C. To add new features

   D. All of the above

   Answer: D. All of the above.

8. Branching in GitHub allows you to make changes in a "safe space" without affecting the _____ branch.

   Answer: main.

9. True or false: Once a branch is created in GitHub, it becomes the new main project.

   A. True

   B. False

   Answer: B. False.

10. What should you do after you've finished work on a branch?

    Answer: Once you've finished work on a branch, you should consider merging it back into the main project or sharing it for feedback.

Challenge – Navigating the stars in Space Explorer!

Hello, young space adventurers! In this chapter, you learned about some cool GitHub features for collaborating on projects. Here, you'll apply these skills to improve your *Space Explorer* game with flying obstacles and stars. Are you ready to navigate this starry challenge?

Your mission?

Figure 4.30 – Space Explorer game screen

Integrate new features – collision detection and scoring – into your *Space Explorer* game using GitHub collaboration tools.

Collision detection: `https://github.com/PacktPublishing/GitHub-for-Next-Generation-Coders/tree/main/Space%20Explorer%20Game/Game%20Obsticles%20-%20Colision%20detection`

Scoring: `https://github.com/PacktPublishing/GitHub-for-Next-Generation-Coders/tree/main/Space%20Explorer%20Game/Game%20Obsticles%20-%20Scoring`

Steps to success

Follow these steps:

1. **Create a feature branch**: In your GitHub repository for the *Space Explorer* game, create a new branch named `feature-stars-and-obstacles-collision`. This branch is your personal spacecraft where you'll add dazzling new features to the game without affecting the main code base.
2. **Pull the new features**: I'll provide you with the code for the flying obstacles and star collision. Use the `git pull` command to bring these new features into your local copy of the `feature-stars-and-obstacles` branch.
3. **Explore and integrate**: Merge the new code into your game. Tweak and play around with it. How about changing the colors of the stars or the speed of the obstacles? Test your game to ensure everything works seamlessly. Remember, a good space pilot always checks their equipment!
4. **Commit your journey**: Once you're happy with the integration, `commit` your changes. Use a message like `Added stars and obstacles to Space Explorer`. This is like logging your space journey in the captain's logbook.

5
Collaborating on Code through Pull Requests

This chapter will focus on collaborating through **pull requests** (**PRs**) on GitHub. You'll learn how to propose changes and review them, making collaborative coding projects smoother and more efficient. By the end of this chapter, you'll be adept at using PRs for feedback and improvements, ensuring clear communication and integration of changes.

This chapter covers the following main topics:

- Getting familiar with PRs and their importance
- Reviewing a PR
- Enhancing your website with GitHub's easy editing features

Technical requirements

For this chapter, you'll need Git and a GitHub account. Being familiar with basic Git commands is also recommended. For the practical exercises in this chapter, go to `https://github.com` and create a folder named `ch5` in your GitHub account to organize your work. Clone the code for *Chapter 5* from `https://github.com/PacktPublishing/GitHub-for-Next-Generation-Coders/tree/main/Chapter%205` to follow along. Alternatively, you can run `git clone https://github.com/PacktPublishing/GitHub-for-Next-Generation-Coders.git` and access the `Chapter 5` folder from there.

Getting familiar with PRs and their importance

You and your friends decide to create a one-page biography website to share your stories. Each of you has a section on the page where you can tell your own story. You've thought of adding a new section to your part of the page so that you can share what you do. But before doing that, you'd like to get some feedback from your friends and maybe have them help you pick an experience to display. How do you go about this in a well-organized, collaborative manner? How do you do that in a structured, clear, and collaborative way? This is where PRs come into play.

When you submit a PR on GitHub, you're not just proposing code changes – you're inviting your teammates to discuss your ideas and review the code together, ensuring it's good to go:

Figure 5.1 – A team working on a one-page portfolio

Creating a new PR

When you're ready to add new features or make changes to your one-page portfolio website, the first step is to create a new PR. This process allows you to propose changes to the project without affecting the main, stable version of the website. Let's look at how to safely experiment and prepare updates for review.

Branching out

In GitHub, each project has a main branch, which is like the main version of your website. When you want to make changes, you create a new branch, which is like making a copy of your website where you can experiment without affecting the main one. You make your changes in this new branch.

In Git, you can use the `git switch -c name-of-your-branch` command to create a new branch,

On GitHub, on the other hand, you can do this easily on the overview (**Code**) page of your repository. Click on the **Branch: main** drop-down menu and type the name of your new branch – for example, `feature/add-what-i-do` – and hit *Enter*:

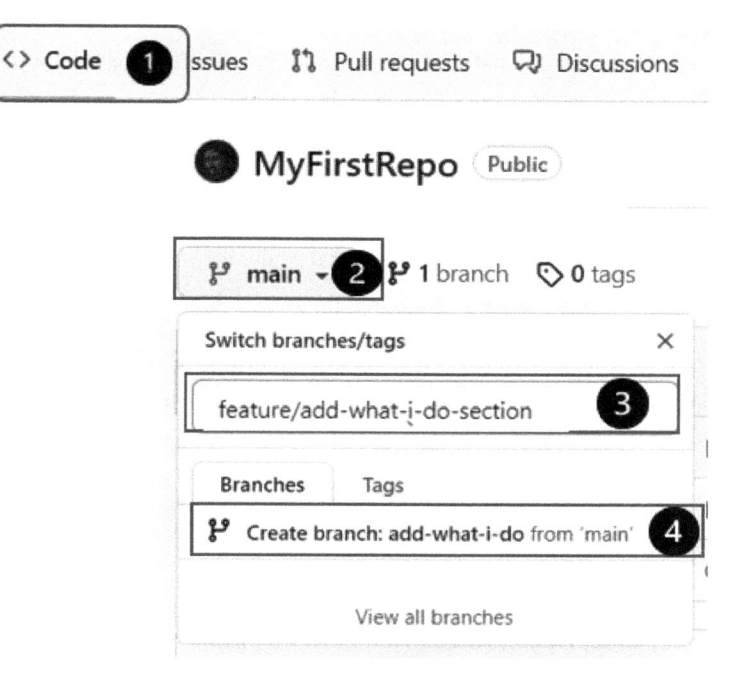

Figure 5.2 – Creating a new branch

You will see that your new branch has been created. You will be automatically switched to the newly created branch:

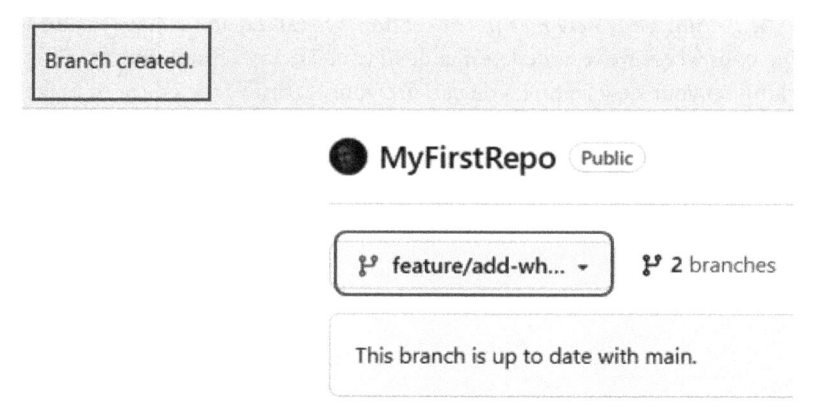

Figure 5.3 – Branch created successfully

In your local repository in Git Bash or your terminal, you can type `git pull` to get the branch you just created on GitHub:

```
PS C:\Users\igor.iric\Desktop\MyFirstRepo> git pull
From https://github.com/error505/MyFirstRepo
 * [new branch]      feature/add-what-i-do-section -> origin/feature/add-what-i-do-section
Already up to date.
PS C:\Users\igor.iric\Desktop\MyFirstRepo> git branch -a
  feature/leftside-menu
* main
  remotes/origin/HEAD -> origin/main
  remotes/origin/feature/add-what-i-do-section
  remotes/origin/main
```

Figure 5.4 – git pull – getting the new branch

Now, you can switch to your `add-what-i-do-section` feature branch and start coding this new section. Use the `git switch feature/add-what-i-do-section` command to switch to your feature branch. We will add new HTML code for the section and add proper stylings. Let's make those changes:

```
PS C:\Users\igor.iric\Desktop\MyFirstRepo> git switch feature/add-what-i-do-section
Switched to a new branch 'feature/add-what-i-do-section'
M       index.html
M       styles.css
branch 'feature/add-what-i-do-section' set up to track 'origin/feature/add-what-i-do-section'.
PS C:\Users\igor.iric\Desktop\MyFirstRepo>
```

Figure 5.5 – Running git switch

Making changes

In *Chapter 4*, in the *Adding your new files to Git* section, we talked about how to add new files or change the existing ones where we've added new code to your `index.html` and `styles.css` files. To continue working on your new feature, you can use your favorite code editor to make changes to them. Again, we can use Visual Studio Code to edit our files.

To add a new section to your one-page portfolio website. you can add an HTML code section called `<section class="what-i-do-header">` inside `<div class="main-content">`:

```html
<html lang="en">
  <head>
    <meta charset="UTF-8" />
    <meta name="viewport" content="width=device-width, initial-scale=1.0" />
    <title>John Doe - Senior Software Engineer</title>
    <link
    rel="stylesheet"
    href="https://cdnjs.cloudflare.com/ajax/libs/font-awesome/6.0.0-beta3/css/all.min.css"
    />
    <link rel="stylesheet" href="styles.css" />
  </head>

  <body>
    <aside class="sidebar">
      <img src="https://via.placeholder.com/150" alt="John Doe" class="profile-image" />
      <h2>John Doe</h2>
      <p>Senior Software Engineer</p>
      <!-- Social Media Icons -->
      <div class="social-icons">
        <a href="#"><i class="fab fa-twitter"></i></a>
        <a href="#"><i class="fab fa-linkedin"></i></a>
        <a href="#"><i class="fab fa-github"></i></a>
        <a href="#"><i class="fab fa-instagram"></i></a>
```

Figure 5.6 – HTML code

Replace John Doe with your name and add some CSS to a new file called `styles.css`. Here's the code snippet you'll need to use for the `styles.css` file:

```css
:root {
  --main-theme-color: #1a855f; /* Default theme color */
}

body {
  font-family: Arial, sans-serif;
  margin: 0;
  padding: 0;
}

.sidebar {
  background-color: #1a855f;
  width: 300px;
  height: 100vh;
  float: left;
  padding: 20px;
  box-shadow: 2px 0px 5px rgba(0, 0, 0, 0.1);
}

.profile-image {
  width: 200px;
  border-radius: 50%;
  margin: 0 auto;
}
```

Figure 5.7 – CSS style

Finally, after saving the `index.html` and `styles.css` files, you can run the website in your browser and view the new changes you've made:

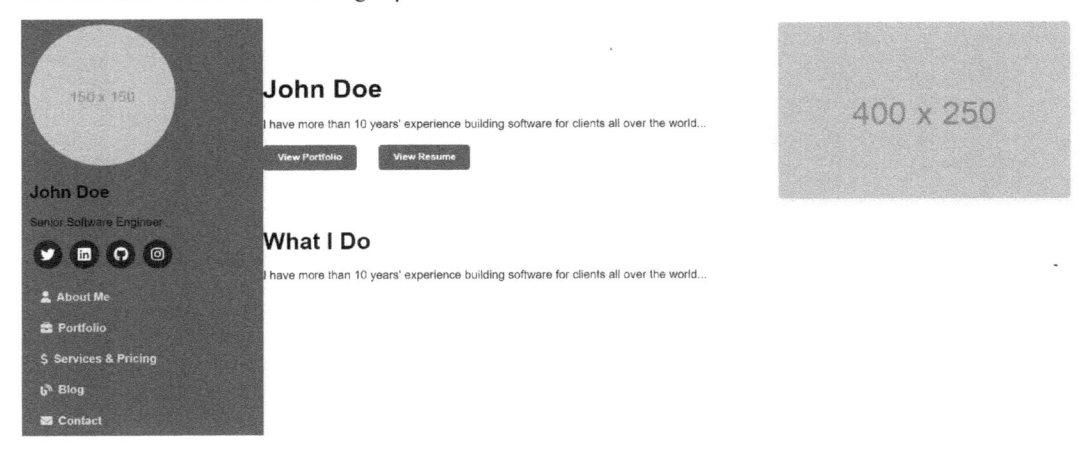

Figure 5.8 – Portfolio profile page with a What I Do section

Sharing your draft

Once you're ready to share your idea with your friends, you'll need to commit your changes. This is like saving your draft.

You can use the following commands to add and commit your changes to your branch:

- `git add .` (to stage all your changes)
- `git commit -m "Added a new what I do section"` (to commit your changes)

Now that you've learned how to share your draft changes, we'll look at how to effectively use the GitHub PR interface to propose a draft, discuss why a well-crafted PR is crucial, and go through some tips on how to create the perfect PR on GitHub. The next section will help you ensure your contributions are clear, useful to the project, and appreciated by your teammates.

GitHub PR interface – proposing your draft

At this point, you need to push your branch to GitHub and create a PR. This is like sharing your draft with your friends and asking for their feedback. Let's look at the steps:

1. Run the `git push -u origin feature/add-what-i-do-section` command to push your new files and changes to the `feature/add-what-i-do` branch on GitHub. Here, the `-u` option sets the upstream (or tracking) reference for the branch you're pushing:

Getting familiar with PRs and their importance 129

```
● PS C:\Users\igor.iric\Desktop\MyFirstRepo> git push -u origin feature/add-what-i-do-section
  Enumerating objects: 7, done.
  Counting objects: 100% (7/7), done.
  Delta compression using up to 8 threads
  Compressing objects: 100% (4/4), done.
  Writing objects: 100% (4/4), 1.15 KiB | 1.15 MiB/s, done.
  Total 4 (delta 2), reused 0 (delta 0), pack-reused 0
  remote: Resolving deltas: 100% (2/2), completed with 2 local objects.
  To https://github.com/error505/MyFirstRepo.git
     8699309..1430afa  feature/add-what-i-do-section -> feature/add-what-i-do-section
  branch 'feature/add-what-i-do-section' set up to track 'origin/feature/add-what-i-do-section'.
○ PS C:\Users\igor.iric\Desktop\MyFirstRepo> []
```

Figure 5.9 – git push changes

2. Now go to GitHub, locate your repository, and click **New Pull Request**. Alternatively, you can use the **Compare & pull request** button, which automatically detects if there are new changes in some of the branches and allows you to create the PR out of the box:

Figure 5.10 – Starting a new PR

3. When you click **Compare & pull request**, a new page will open where you can start your PR:

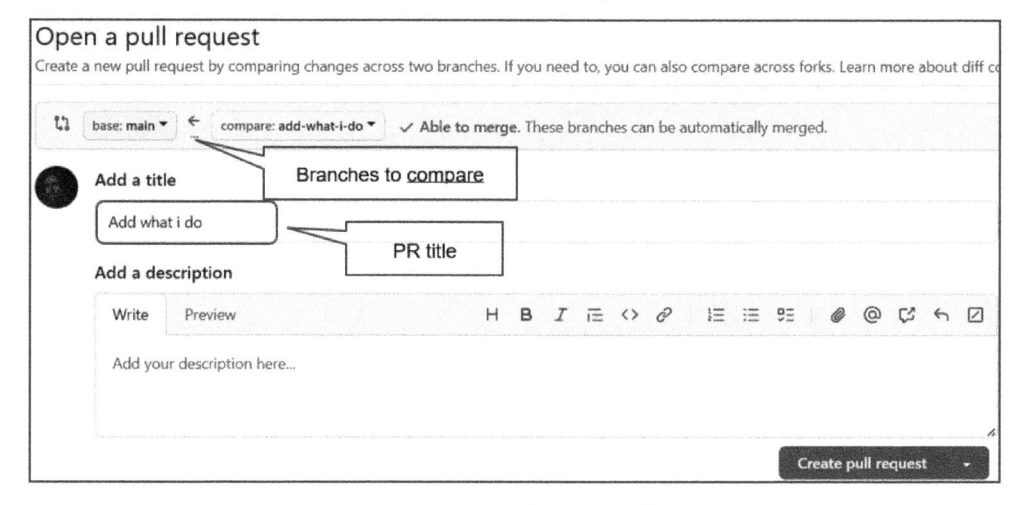

Figure 5.11 – Adding a PR title

Why is a good PR important?

Using a PR is your chance to show off the work you've done and make sure it fits perfectly with the rest of the website you're creating.

A well-crafted PR ensures the following:

- **Your team understands your work**: They'll know exactly what changes you've made and why.
- **Smooth integration**: It's easier to combine your work with others' without issues.
- **Better collaboration**: Your team can give you helpful feedback, making the final project stronger.

Let's look at how to create a perfect PR on GitHub:

1. **Start with a clear title**: Your title should be like a headline in a newspaper – short and descriptive. An example of such as title is *Improved social icon design for better visibility*:

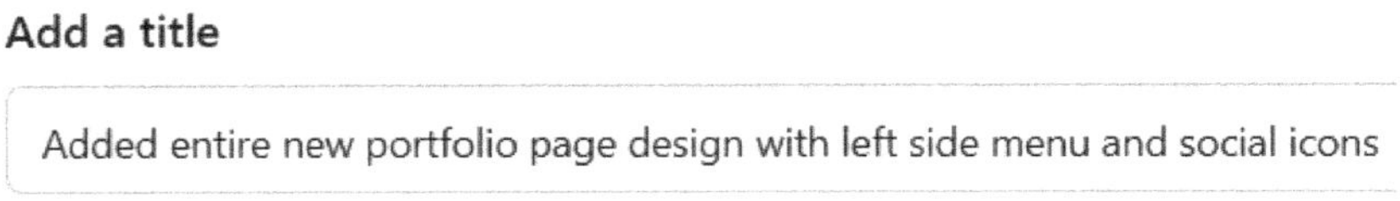

    Add a title

    Added entire new portfolio page design with left side menu and social icons

    Figure 5.12 – PR title

2. **Write a detailed description**: This is like telling the story of what you did. Include the following details:

    - **What changes you made**: Think of this as explaining the modifications you made to the social icons
    - **Why you made them**: Maybe you read that a certain shape helps improve the visibility of the page

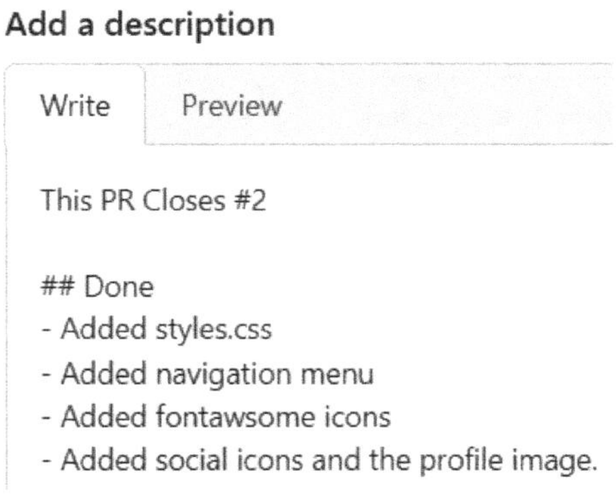

    Add a description

    | Write | Preview |

    This PR Closes #2

    ## Done
    - Added styles.css
    - Added navigation menu
    - Added fontawsome icons
    - Added social icons and the profile image.

    Figure 5.13 – PR detailed description

3. **Link to related issues**: If your work is solving a specific problem that was discussed earlier (such as an issue with the previous social icons design), make sure to link that discussion. This is like saying, *"This is the fix for the problem we talked about last week."*

Add a description

Write	Preview

This PR Closes #2

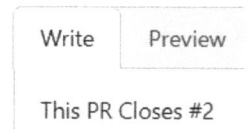

Figure 5.14 – Linking the PR to a related issue

4. **Outline what's left to do**: Sometimes, you might not be completely done. It's important to note what's still pending, such as if you still need to test the icons in different resolutions and devices:

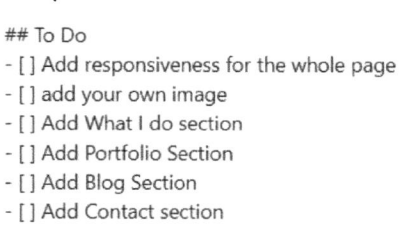

```
## To Do
 - [ ] Add responsiveness for the whole page
 - [ ] add your own image
 - [ ] Add What I do section
 - [ ] Add Portfolio Section
 - [ ] Add Blog Section
 - [ ] Add Contact section
```

Figure 5.15 – The To Do section in a PR

5. **Add labels for clarity**: Labels are like sticky notes that help categorize your PR. You can add labels such as `bug fix`, `enhancement`, or `needs review`:

Labels

Apply labels to this issue

Filter labels

✓ enhancement ✕
New feature or request

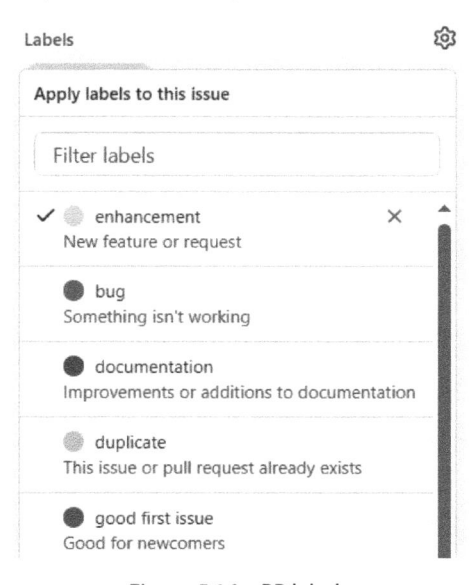

● bug
Something isn't working

● documentation
Improvements or additions to documentation

● duplicate
This issue or pull request already exists

● good first issue
Good for newcomers

Figure 5.16 – PR labels

6. **Request reviewers**: Ask specific teammates to look at your work. It's like asking your friends who are good at CSS to check your responsive design:

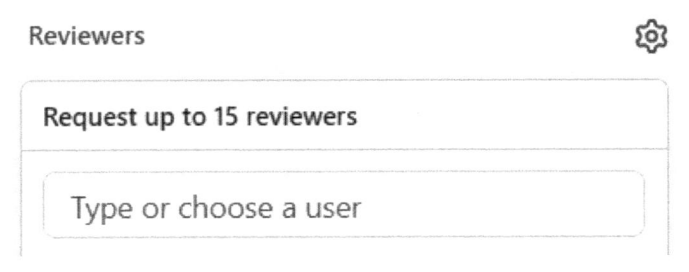

Figure 5.17 – PR reviewers

7. **Be responsive**: After submitting, stay active. If your teammates ask questions or suggest changes, respond promptly.
8. **Keep it up to date**: If other parts of the project change (such as if someone modifies the body of the page), update your PR so that it matches these changes.
9. Finally, click on the **Create pull request** button to create your first PR:

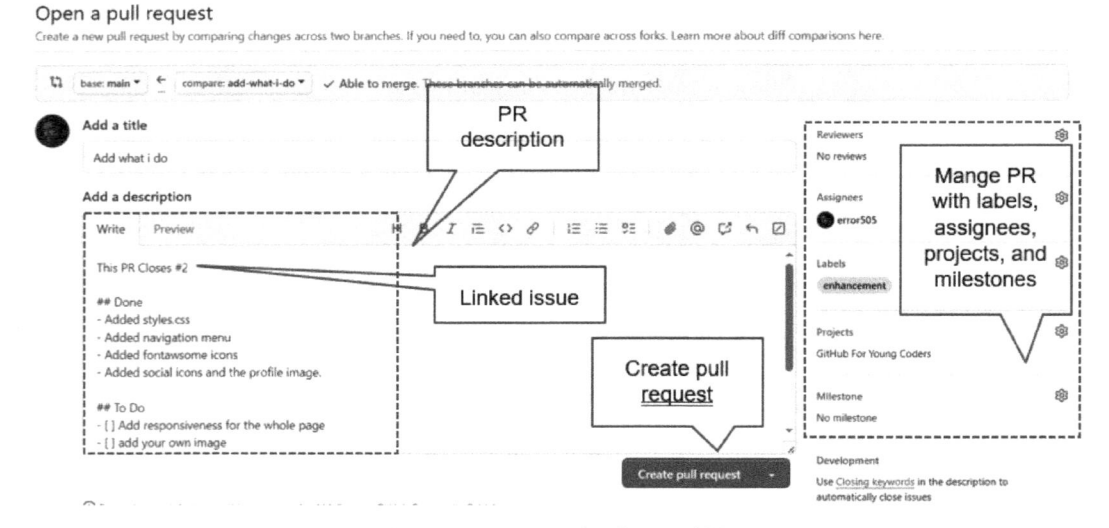

Figure 5.18 – Example of a good PR

Before you create a PR, you can compare files on the main branch and the branch you want to merge with the PR by scrolling down. You will see how many commits you had, how many files have been changed, and how many contributors worked on this branch:

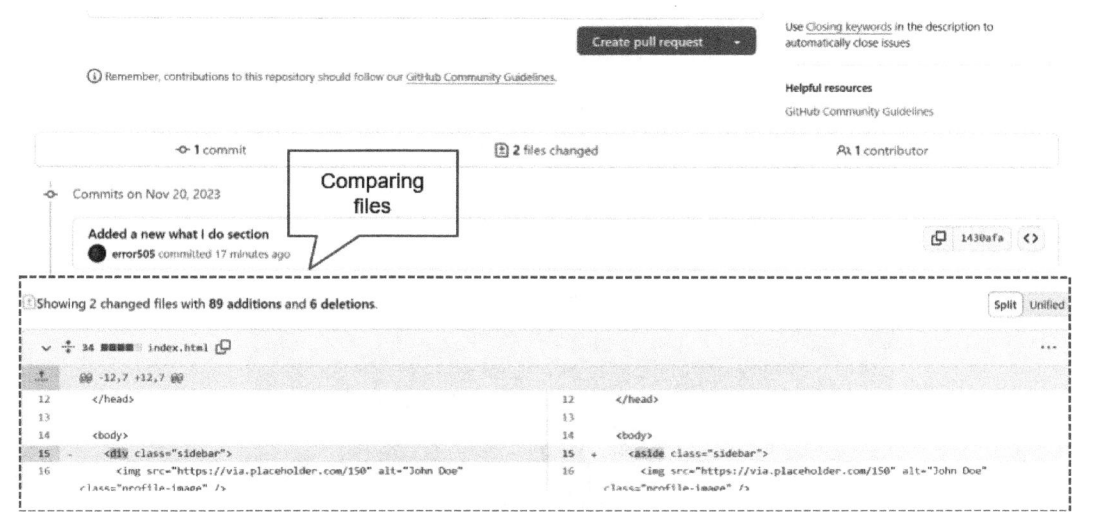

Figure 5.19 – Comparing files

Having covered how to create a perfect PR on GitHub, let's shift our focus. In the next section, we'll explore how to review a PR effectively. This includes reviewing changes line by line and ensuring the quality of the contributions. After the review, we'll go through how to merge those changes and the best practices for managing your branches post-merge, including why and how to delete branches. This process helps maintain a clean project environment and ensures your repository stays organized.

Reviewing a PR

Your teammates can now see your proposed changes on GitHub. They can leave comments, suggest modifications, or even add more code. This is the code review process. It's like showing your work to friends and getting their feedback and ideas. Let's look at how this can be done on GitHub while focusing on the **Files changed** tab:

Figure 5.20 – PR review

What's the Files changed tab?

When you or your friend submits a PR on GitHub, it means they're suggesting some changes to the website. The **Files changed** tab is where you can see exactly what has been changed:

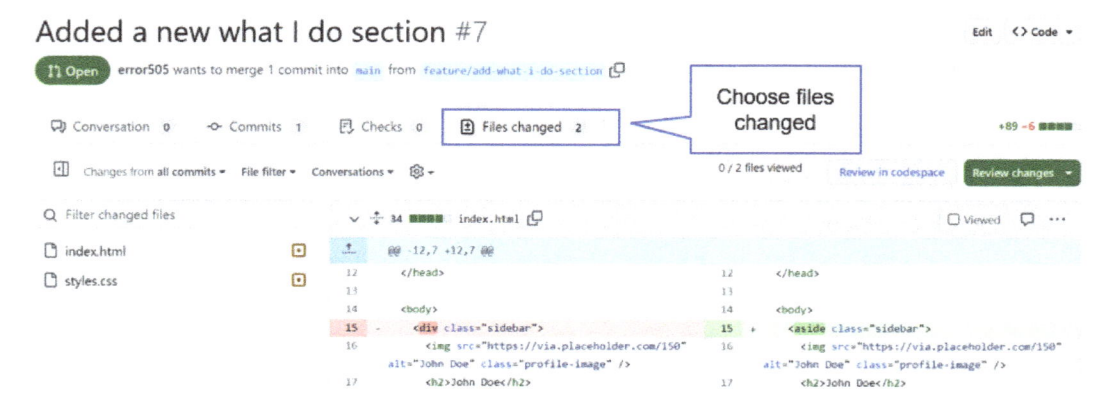

Figure 5.21 – The GitHub Files changed tab

Think of it like a highlighted version of a document, where the edits are clearly marked.

Reviewing changes line by line

Here's how you can review changes in detail, line by line:

1. Open the PR that has just been made and go to it.
2. You'll find a tab labelled **Files changed**.
3. Click on this tab; you'll see the files that have been changed or added. GitHub shows these changes in two colors: **red** for what's been removed and **green** for what's been added. It's like tracking changes in a Word document – you can easily spot what's new and what's been taken out.
4. If you have a question or a suggestion about a specific part of the change, you can comment right there. Hover your mouse over the line of code; a blue plus sign will appear. Upon clicking it, a comment box will open.

   Here, you can type your thoughts or suggestions. Maybe you think a different color would look better for the background, or you have a question about a new section they added:

Figure 5.22 – Commenting on the PR

5. Once you've added all your comments, you can submit your review. You'll have options such as **Approve** (if you think everything looks great), **Request changes** (if you think some things need to be altered), or to just leave comments without any specific approval or request:

136   Collaborating on Code through Pull Requests

○ **Comment**
Submit general feedback without explicit approval.

● **Approve** ──────────── Approve review.
Submit feedback and approve merging these changes.

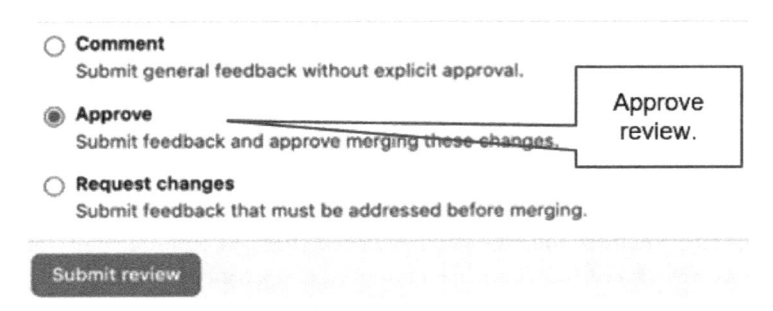

○ **Request changes**
Submit feedback that must be addressed before merging.

[Submit review]

Figure 5.23 – Submitting a PR review

What happens next?

Your friend will see your comments and can respond to them. They might make more changes based on your feedback. This back-and-forth will continue until you both agree the changes are good.

Ensuring quality

PRs also ensure that the code meets the project's standards. Is the code clean and well-commented? Does it follow the project's coding conventions? Are there tests to ensure the code works as expected? This step is like ensuring your new web page has been tested on all devices and it can serve as many users as possible.

This can be automated by using GitHub workflows and actions. The workflow will start automatically when you create a PR to ensure that the code has been checked for any vulnerabilities and that unit tests are running correctly. Everything happens on the **Checks** tab of your PR.

Right now, we don't have any actions in place to check our code. We will talk about this in more detail in *Chapter 9*:

Added a new what I do section #7

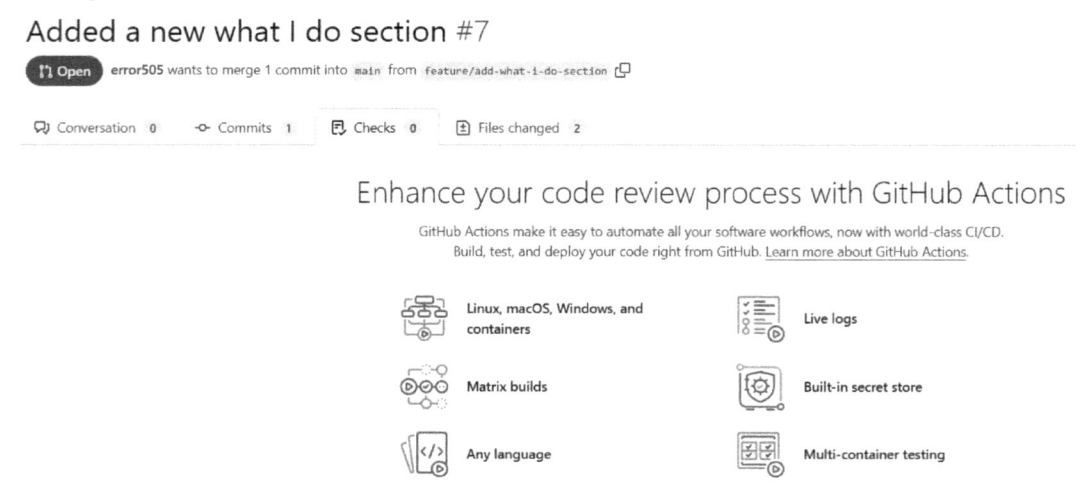

[Open] error505 wants to merge 1 commit into `main` from `feature/add-what-i-do-section`

Conversation 0 · Commits 1 · Checks 0 · Files changed 2

Enhance your code review process with GitHub Actions

GitHub Actions make it easy to automate all your software workflows, now with world-class CI/CD. Build, test, and deploy your code right from GitHub. Learn more about GitHub Actions.

- Linux, macOS, Windows, and containers
- Matrix builds
- Any language
- Live logs
- Built-in secret store
- Multi-container testing

Figure 5.24 – PR checks

Once we have actions in place, the build process should look like this:

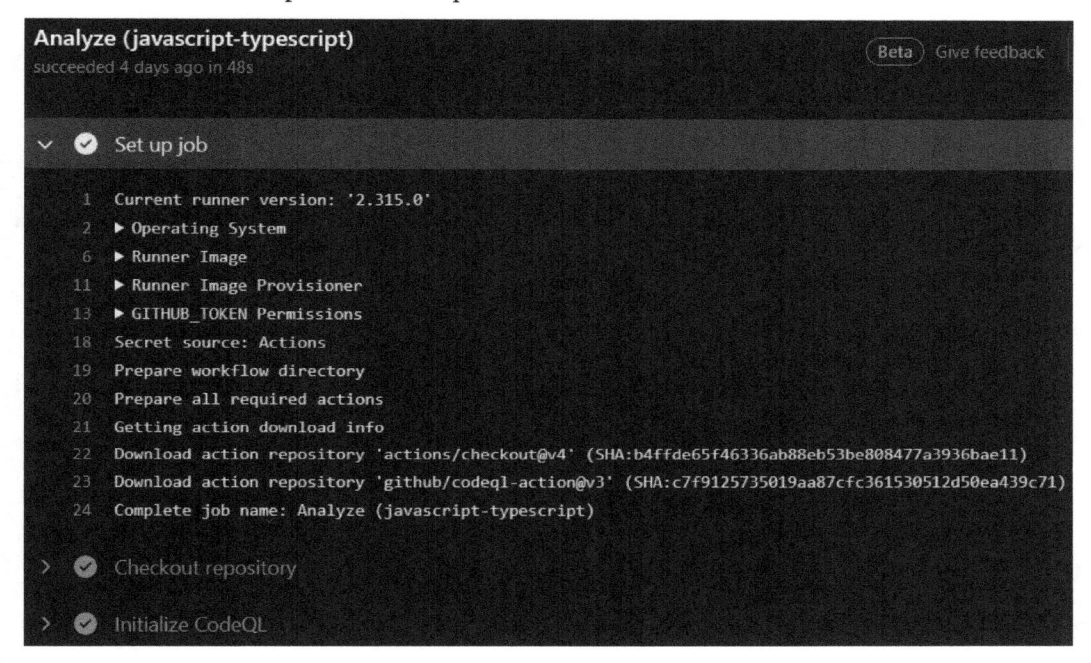

Figure 5.25 – PR actions

Learning and growing

By discussing and providing feedback on PRs, everyone learns and grows. You learn better ways to code, understand different approaches to solving problems, and immerse yourself in the collective wisdom of your team.

Based on this feedback, you might make more changes to your branch. These updates will automatically be part of the same PR:

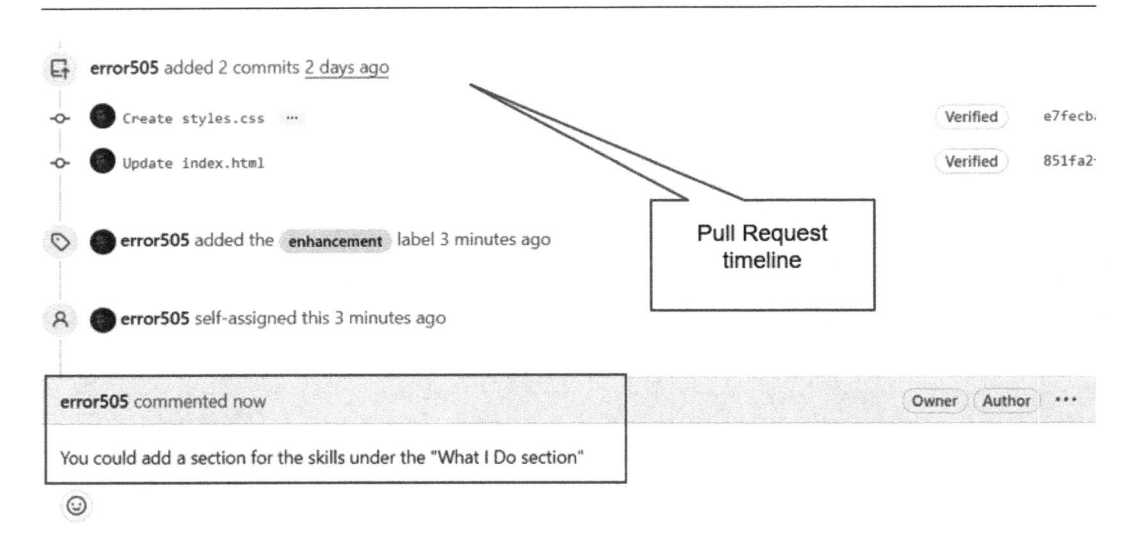

138   Collaborating on Code through Pull Requests

Figure 5.26 – Reviewing PR comments

Merging the changes

PRs embody the essence of teamwork in software development. They ensure that everyone is on the same page, the code is robust and clean, and the journey of building something great is a shared adventure.

Once everyone's happy with the changes, a teammate or you (if you have the permissions) can merge your branch with the main branch. This means your changes are now part of the main website.

Now, you are ready to finish the PR by clicking **Merge pull request**:

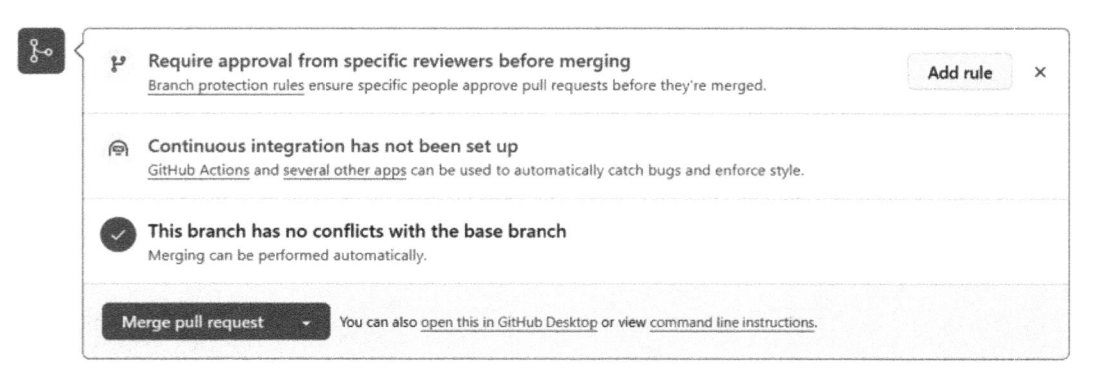

Figure 5.27 – Merge pull request

Then, click **Confirm merge**:

Figure 5.28 – Confirm merge

Deleting a GitHub branch

Think of your GitHub repository as a workspace in a community workshop. Every branch you create is like a separate workbench. When you're done with a project (such as your **What I Do** section) and have moved it to the main display area (merged it into the main branch), you don't need that workbench anymore. Cleaning up by deleting the branch keeps your workshop tidy, making it easier for you and others to see what's currently being worked on and what's finished.

The PR is now closed, marking the end of this collaboration. At this point, you can click **Delete branch** to remove the branch from the repository. This will remove the `feature/add-what-i-do-section` branch from your GitHub repository:

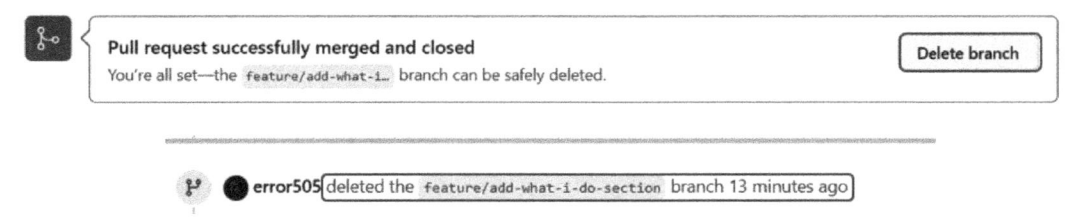

Figure 5.29 – Deleting the merged branch

Why is it important to delete a branch?

It's good practice in software development workflow. Once a feature has been integrated, its branch has served its purpose, and confusion about which branches are active or needed for ongoing work will be prevented. Deleting branches after their PRs are merged ensures you maintain a clean, understandable, and well-organized GitHub repository. This practice is especially important in team projects where clarity and organization can save time and avoid confusion.

Deleting a local branch after your PR is approved

Now, you want to clean up your local Git environment by deleting the feature branch since it's no longer needed.

Switch back to the main branch by typing the `git switch main` command in Git Bash or your terminal. This command moves you out of your feature branch and back into the main branch of your project:

```
PS C:\Users\igor.iric\Desktop\MyFirstRepo> git switch main
Switched to branch 'main'
Your branch is behind 'origin/main' by 2 commits, and can be fast-forwarded.
  (use "git pull" to update your local branch)
```

Figure 5.30 – Git terminal – Switching to the main branch

Why do you need to switch back to the main branch?

You can't delete a branch you're currently in. So, you need to switch to a different branch (main, in this case) to delete the feature/add-what-i-do-section branch.

Updating your local main branch

Now, you can update your local main branch using the `git pull` command as there is a message that your branch is behind the origin/main by two commits. This command (`git pull`) updates your local main branch with the latest changes from the remote main branch on GitHub, including the changes from your recently merged PR.

It's good practice to keep your local main branch updated with the remote main branch so that you have the latest code, ensuring everything is synchronized:

```
PS C:\Users\igor.iric\Desktop\MyFirstRepo> git pull
From https://github.com/error505/MyFirstRepo
 - [deleted]         (none)     -> origin/feature/add-what-i-do-section
remote: Enumerating objects: 1, done.
remote: Counting objects: 100% (1/1), done.
remote: Total 1 (delta 0), reused 0 (delta 0), pack-reused 0
Unpacking objects: 100% (1/1), 641 bytes | 213.00 KiB/s, done.
   8699309..cdadca9  main       -> origin/main
Updating 8699309..cdadca9
Fast-forward
 index.html | 34 ++++++++++++++++++++++++++++++------
 styles.css | 61 ++++++++++++++++++++++++++++++++++++++++++++++++++++++++++++
```

Figure 5.31 – Git terminal – pulling the latest changes to the main branch

Pruning unneeded branches

At this point, you must clean up references to remote branches that no longer exist on GitHub with the `git pull --prune` command.

Why should you do this?

After merging your PR, the feature branch on GitHub might be deleted (a common practice). The `--prune` option ensures your local Git knows that this remote branch is gone:

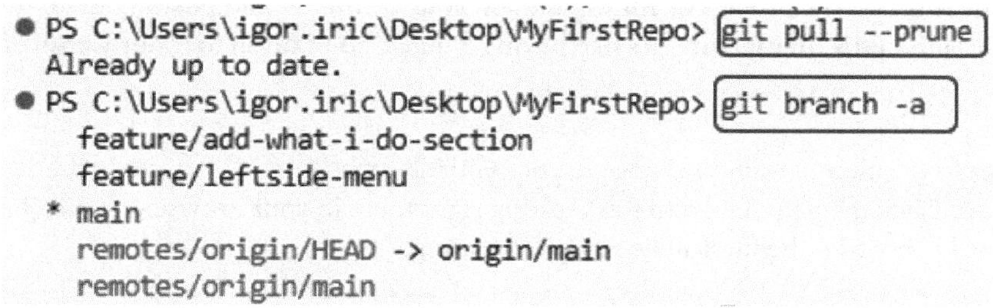

```
PS C:\Users\igor.iric\Desktop\MyFirstRepo> git pull --prune
Already up to date.
PS C:\Users\igor.iric\Desktop\MyFirstRepo> git branch -a
  feature/add-what-i-do-section
  feature/leftside-menu
* main
  remotes/origin/HEAD -> origin/main
  remotes/origin/main
```

Figure 5.32 – Git terminal – pruning and checking for branches

Deleting the local feature branch

Now, it's time to delete the branch from your local repository. You can do this by running the `git branch -d feature/add-what-i-do-section` command, which will delete the branch from your local repository:

```
PS C:\Users\igor.iric\Desktop\MyFirstRepo> git branch -a
  feature/add-what-i-do-section
  feature/leftside-menu
* main
  remotes/origin/HEAD -> origin/main
  remotes/origin/main
PS C:\Users\igor.iric\Desktop\MyFirstRepo> git branch -d feature/add-what-i-do-section
Deleted branch feature/add-what-i-do-section (was 1430afa).
PS C:\Users\igor.iric\Desktop\MyFirstRepo> git branch -a
  feature/leftside-menu
* main
  remotes/origin/HEAD -> origin/main
  remotes/origin/main
```

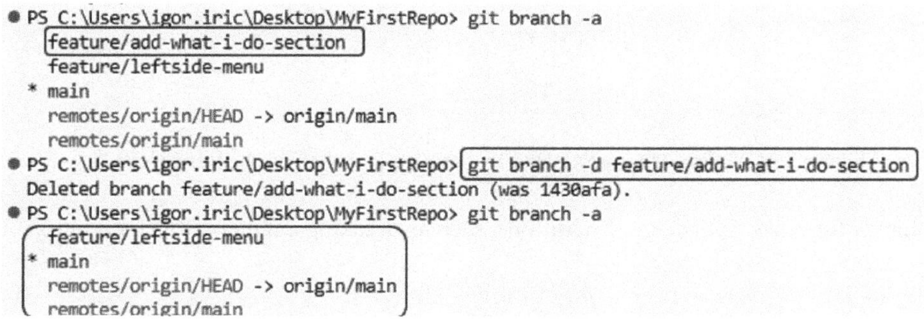

Figure 5.33 – Git terminal – deleting the feature branch from your local

Now that your changes have been merged and you've updated your main branch, the feature branch is no longer needed. Deleting it keeps your local repository clean and organized. This command is like cleaning up your workspace after completing a project section. In the next section, we'll cover enhancing your website with GitHub's easy editing features. You'll learn about direct editing on GitHub for quick changes and using *github.dev* as your full-fledged editor. This will help you make improvements to your website efficiently and effectively.

Enhancing your website with GitHub's easy editing features

Let's dive a bit more into how you can edit your one-page portfolio website using GitHub's handy editing features. Besides the traditional way of editing files on your computer and then pushing the changes to GitHub, you've got some quicker, user-friendly options right on GitHub!

Direct editing on GitHub – quick and easy

Instead of going through the process of editing files on your computer and pushing them to GitHub, you can use GitHub's **Edit files** feature. It's like having a quick-edit button for your website's files.

So, how does it work?

Simply navigate to the file you want to change in your GitHub repository – for example, `index.html` or `style.css`. Click the pencil icon to start editing right there in your browser, or even better, just press *E* on your keyboard – the file will be ready for editing:

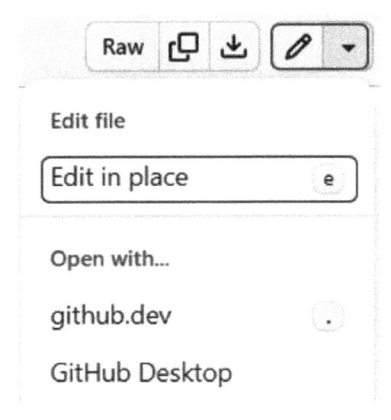

Figure 5.34 – Editing files on GitHub by pressing E or clicking the pencil icon

It's super handy for quick fixes or small additions, such as tweaking a sentence or changing a color:

Figure 5.35 – Editing files on GitHub

Using github.dev – your full-fledged editor

For more extensive editing, GitHub offers `github.dev`, which gives you the full functionality of Visual Studio Code (a popular coding editor) right in your browser.

github.dev is a lightweight, in-browser code editor that allows you to edit files in your repository. It doesn't provide any compute resources, so you can't run or debug your code directly within this environment. No setup or configuration is needed, and it's free to use.

Activating github.dev

You can activate `github.dev` by pressing . (the period key) while you're in your repository or clicking on the arrow near the pencil icon and clicking **github.dev** under **Open with…**, as shown here:

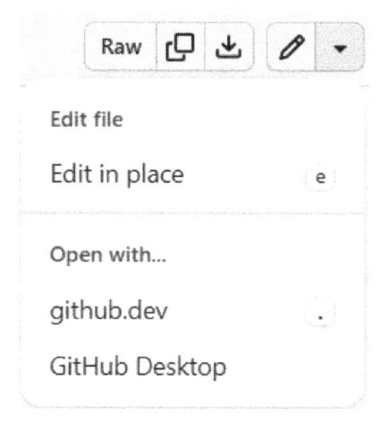

Figure 5.36 – Editing files on GitHub by pressing "." or clicking on the pencil icon

A powerful editor will open where you can work on multiple files, run commands, and save your work directly to your branch.

Why use github.dev?

This is perfect when you're away from your usual coding setup or when you want to work directly on GitHub without the back-and-forth of local editing. It's like having your full craft workshop available anywhere, anytime:

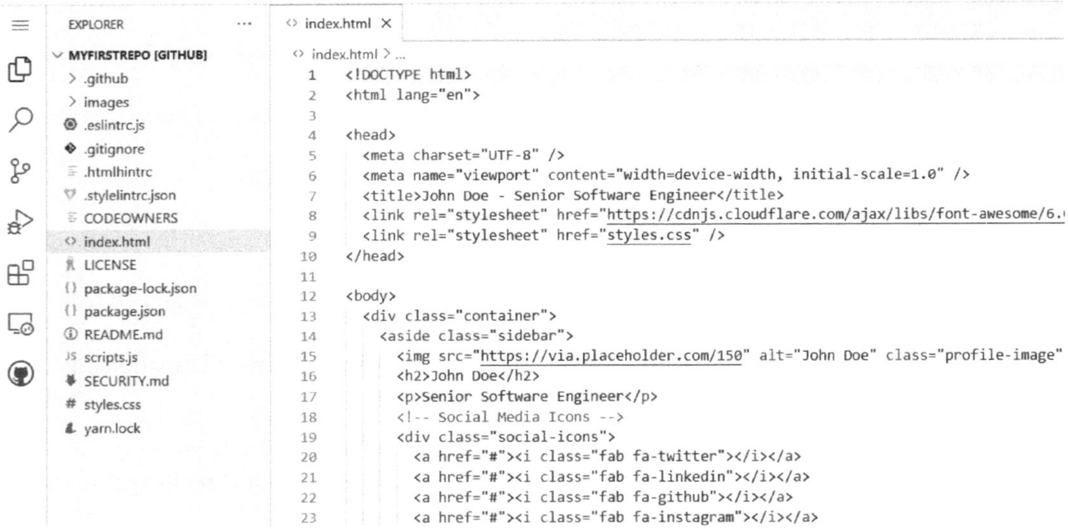

Figure 5.37 – Editing files on GitHub with github.dev

The best of both worlds

Depending on what you're doing, you can choose the best editing method. Quick change? Direct editing is perfect. More complex work? `github.dev` is your friend.

These GitHub features make it faster and simpler to collaborate on your website. It's like having a set of tools that are perfect for quick fixes or a full day of development.

Summary

In this chapter, you learned how to effectively collaborate through PRs on GitHub. This involves proposing changes to a project, reviewing these changes with teammates, and integrating feedback to improve the overall quality of the project. These skills are essential for teamwork in coding projects, ensuring clear communication and smooth integration of new features or fixes.

In the next chapter, we'll explore how to handle conflicts that arise when integrating changes. We'll learn how to use GitHub's UI for simple conflicts, tackle multiple conflicting issues, and resolve conflicts locally, ensuring seamless collaboration and project progress.

Quiz

Answer the following questions:

1. What is a PR in GitHub?

    A. A request to pull code from a repository

    B. A proposal for changes to a project

    C. A request to delete code

    D. A message to GitHub support

    Answer: B. A proposal for changes to a project

2. True or false: PRs are only used for adding new features.

    A. True

    B. False

    Answer: B. False

3. Crafting _____ in GitHub helps you propose changes in a way that's easy for your teammates to review.

    Answer: PRs

4. Why is it important to review others' work on GitHub?

    Answer: Reviewing others' work helps identify potential issues, ensures quality, and promotes collaboration, making the project stronger

5. What is the purpose of giving feedback on PRs?

    A. To make the project more complicated

    B. To improve the quality and effectiveness of the code

    C. To slow down the project

    D. To decide who is the best coder

    Answer: B. To improve the quality and effectiveness of the code

6. True or false: You can create PRs for both small and large changes.

   A. True

   B. False

   Answer: A. True

7. Which feature in GitHub is used for quick changes?

   A. Issues

   B. Wiki

   C. Codespaces

   D. Direct editing

   Answer: D. Direct editing

8. For more complex work, _____ is a helpful GitHub feature.

   Answer: `github.dev`

9. True or false: PRs can only be made by the project owner.

   A. True

   B. False

   Answer: B. False

10. What is one best practice for creating PRs in GitHub?

    Answer: Creating clear, concise, and effective PRs that are easy for teammates to understand and integrate.

Challenge – Cosmic collaboration in Space Explorer!

This chapter took you through the essentials of crafting and collaborating on PRs in GitHub. It's time to bring those skills to the forefront with our *Space Explorer* game. Let's add some stellar backgrounds and refine the obstacle interaction!

Your mission?

Improve the *Space Explorer* game by adding a new background and refining the obstacle interaction, using PRs for collaboration and review:

Figure 5.38 – Space Explorer game

Do the following:

1. **Branch out for a new background**: On GitHub, create a new branch from your main project named `feature-new-background`. This branch is your canvas to paint the cosmic backdrop of the game.
2. **Integrate the new background**: I've prepared some cool space backgrounds for you. Use the `git pull` command to get it into your local `feature-new-background` branch. Integrate it into the game. Feel the universe come alive!
3. **Improve obstacle interaction**: On another branch, `enhance-obstacle-interaction`, work on refining how the spaceship interacts with obstacles. Could you add a cool sound effect when the spaceship hits an obstacle? I have already prepared some sounds for you that you could use.
4. **Commit and open PRs**: Commit your changes with descriptive messages such as `Added cosmic background` and `Enhanced spaceship-obstacle interaction`.
5. Open PRs for each branch back to the main project. Describe your changes and ask for feedback.
6. **Review and collaborate**: Invite friends or fellow coders to review your PRs. Are your new background and obstacle interactions out of this world?

7. Discuss, revise if needed, and then, when everything looks stellar, merge your PRs into the main game.

8. **Clean up your space**: After successfully merging, delete the feature branches from your GitHub repository. Keeping things tidy is part of being a great coder!

Bonus exploration

Experiment with different backgrounds or obstacle effects. Add a personal touch and share your ideas in the PR description. Why did you choose that particular background or effect?

6
Resolving Merge Conflicts – on GitHub and Locally

This chapter will serve as a guide through the sometimes challenging landscape of merge conflicts, illustrating how to navigate and resolve these issues both on GitHub and locally. By understanding and addressing merge conflicts, you'll ensure smoother collaboration in coding projects.

In this chapter, we're going to cover the following main topics:

- Understanding merge conflicts and how they occur
- Addressing merge conflicts

You'll dive into practical, hands-on methods to resolve these conflicts, ensuring your website reflects the best of both worlds.

By the end of this chapter, you'll be equipped to handle tricky technical situations like a pro, ensuring smooth sailing in your collaborative coding projects!

Technical requirements

You'll need a basic setup including Git installed on your computer and a GitHub account. Familiarity with command-line operations and GitHub's web interface will be beneficial. For practice exercises and code examples, visit `https://github.com/PacktPublishing/GitHub-for-Next-Generation-Coders/tree/main/Chapter%206` and create a folder named `ch6` in your GitHub repository. You can use the `git clone https://github.com/PacktPublishing/GitHub-for-Next-Generation-Coders.git` Git command and from there access the `ch6` folder.

Understanding merge conflicts and how they occur

You and your team colleague are both updating a one-page portfolio website. You're working on the **About Me** section, while your colleague is adding new fonts and styles to the entire page. This is where **merge conflicts** come into play:

Figure 6.1 – Merge conflicts

Understanding merge conflicts and how they occur | 151

What is a merge conflict?

Think of a merge conflict like two people trying to paint different pictures on the same canvas at the same time. You both have your ideas, but they clash in the same space:

Figure 6.2 – What is a merge conflict?

How does a merge conflict happen?

Both you and your colleague pull the latest version of the website from the `main` branch to start your work. You add a detailed biography and some images to the **About Me** section, and your colleague changes the font style and color scheme of the entire page, including the **About Me** section. *Figure 6.3* demonstrates this clash:

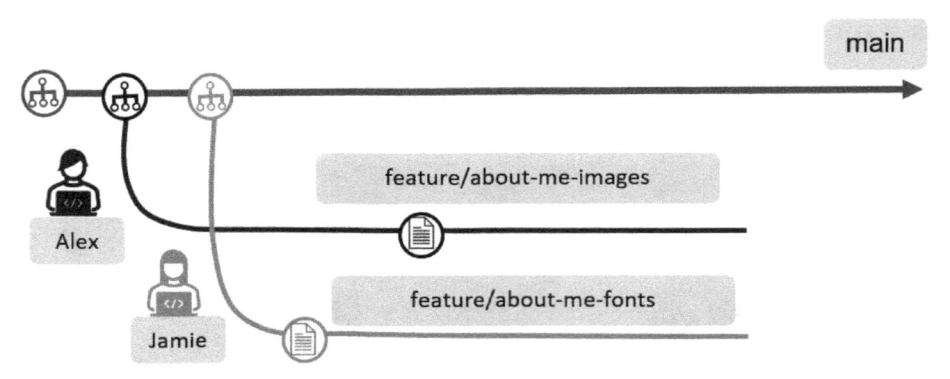

Figure 6.3 – Merge conflict feature branch

You both finish your work around the same time and commit your changes to your respective branches. Your colleague finishes first and merges their changes to the `main` branch. The `main` branch now has your colleague's new fonts and color scheme.

Now, when you try to merge your **About Me** changes, there's a problem. The **About Me** section you worked on doesn't have your colleague's new fonts and colors. Git gets confused because the same part of the website looks different in your branch compared to the `main` branch, as shown in *Figure 6.4*. This is the definition of a merge conflict:

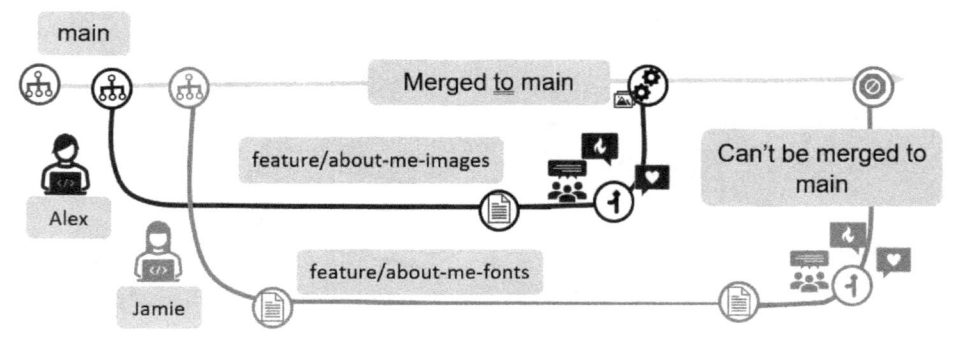

Figure 6.4 – Merge conflict in PR

When you try to make a PR for your changes on GitHub, but there is a conflict in your branch, GitHub will give you a hint that there is a merge conflict in the branch you are trying to merge and that you should resolve it first. You can see this in *Figure 6.5*:

Figure 6.5 – Conflicted PR branch error

As you are now aware of how merge conflicts could happen to you, let's try to demonstrate them with a real example.

Addressing merge conflicts

As you and your colleague are building a one-page portfolio website, you both start updating the **Skills** section, but you accidentally make different changes to the same part of the web page. You are making changes to the title of a skill, while your colleague also updates the title, icon, and description:

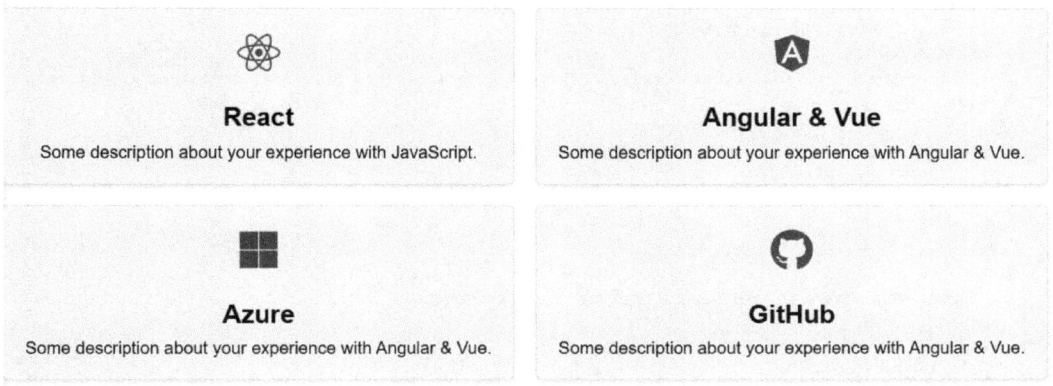

Figure 6.6 – What I Do section of the portfolio web page

Merge conflict in action

Both you and your colleague pull the latest version of the website from the `main` branch to start your work. You update the title of a skill from *React* to *Web Development*, and at the same time, your colleague changes the title to *Google Cloud Provider*, changes the icon for the same skill, and modifies its description.

You both commit and push your changes to GitHub. You finished your work first and you pushed your changes to the `main` branch. After you, your colleague tries to push their changes. But now, GitHub alerts them of a merge conflict. Why? Because the section your colleague edited has been changed by you since they last pulled from the `main` branch.

Here is the sample HTML code on which you will work with your colleagues:

```html
<!-- What I Do Section -->
<section class="what-i-do">
  <div>
    <h1>What I Do</h1>
    <p>
      I have more than 15 years' experience building software for clients
      all over the world...
    </p>
  </div>
  <div class="services-grid">
    <div class="service">
      <div class="service-icon">
        <i class="fab fa-microsoft"></i>
      </div>
      <div class="service-description">
        <h3>Azure</h3>
        <p>Microsoft Certified: Cybersecurity Architect Expert & Azure Solutions Architect Expert</p>
      </div>
    </div>
    <div class="service">
      <div class="service-icon">
        <i class="fas fa-code"></i>
      </div>
      <div class="service-description">
        <h3>C# .net</h3>
        <p>More than years of experience working with C# .net</p>
      </div>
    </div>
    <div class="service">
      <div class="service-icon">
        <i class="fab fa-github"></i>
      </div>
```

Figure 6.7 – Sample code

And here are the changes to the `styles.css` file:

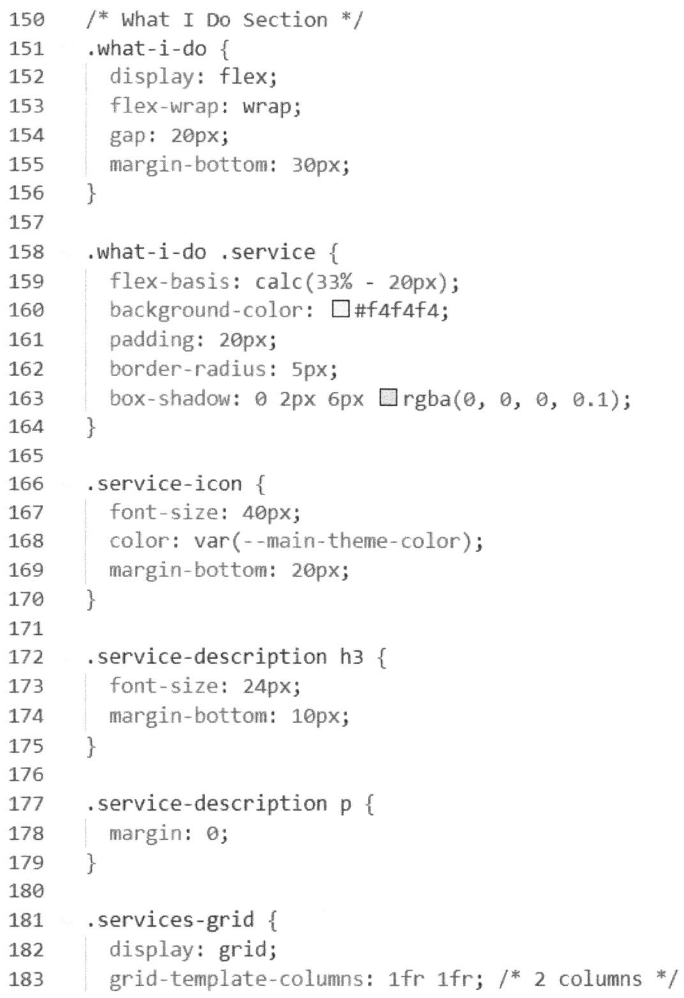

```css
150    /* What I Do Section */
151    .what-i-do {
152      display: flex;
153      flex-wrap: wrap;
154      gap: 20px;
155      margin-bottom: 30px;
156    }
157
158    .what-i-do .service {
159      flex-basis: calc(33% - 20px);
160      background-color: ☐#f4f4f4;
161      padding: 20px;
162      border-radius: 5px;
163      box-shadow: 0 2px 6px ☐rgba(0, 0, 0, 0.1);
164    }
165
166    .service-icon {
167      font-size: 40px;
168      color: var(--main-theme-color);
169      margin-bottom: 20px;
170    }
171
172    .service-description h3 {
173      font-size: 24px;
174      margin-bottom: 10px;
175    }
176
177    .service-description p {
178      margin: 0;
179    }
180
181    .services-grid {
182      display: grid;
183      grid-template-columns: 1fr 1fr; /* 2 columns */
```

Figure 6.8 – styles.css code

So, you both edited the React service, but because you have already merged your changes into the main branch, your colleague will face their first merge conflict and will not be able to merge the PR before the conflict has been resolved.

Resolving merge conflicts using GitHub's UI

GitHub shows your colleague exactly where the conflict is – in the **Skills** section. It's like the website is saying, "*Hey – two people have changed this same part, and I'm not sure which version to keep*":

Added web development skill in what I do section #10

error505 wants to merge 1 commit into `main` from `feature/webdevelopment-skill`

error505 commented now

This PR closes #8

Done:
- New skill (Web development)
- Changed Icon
- Updated description

Added web development skill in what I do section — 13633d2

Resolve conflicts in the GitHub UI

Add more commits by pushing to the **feature/webdevelopment-skill** branch on error505/MyFirstRepo.

This branch has conflicts that must be resolved
Use the web editor or the command line to resolve conflicts.
Conflicting files
  index.html

File where the conflict occured

[Resolve conflicts]

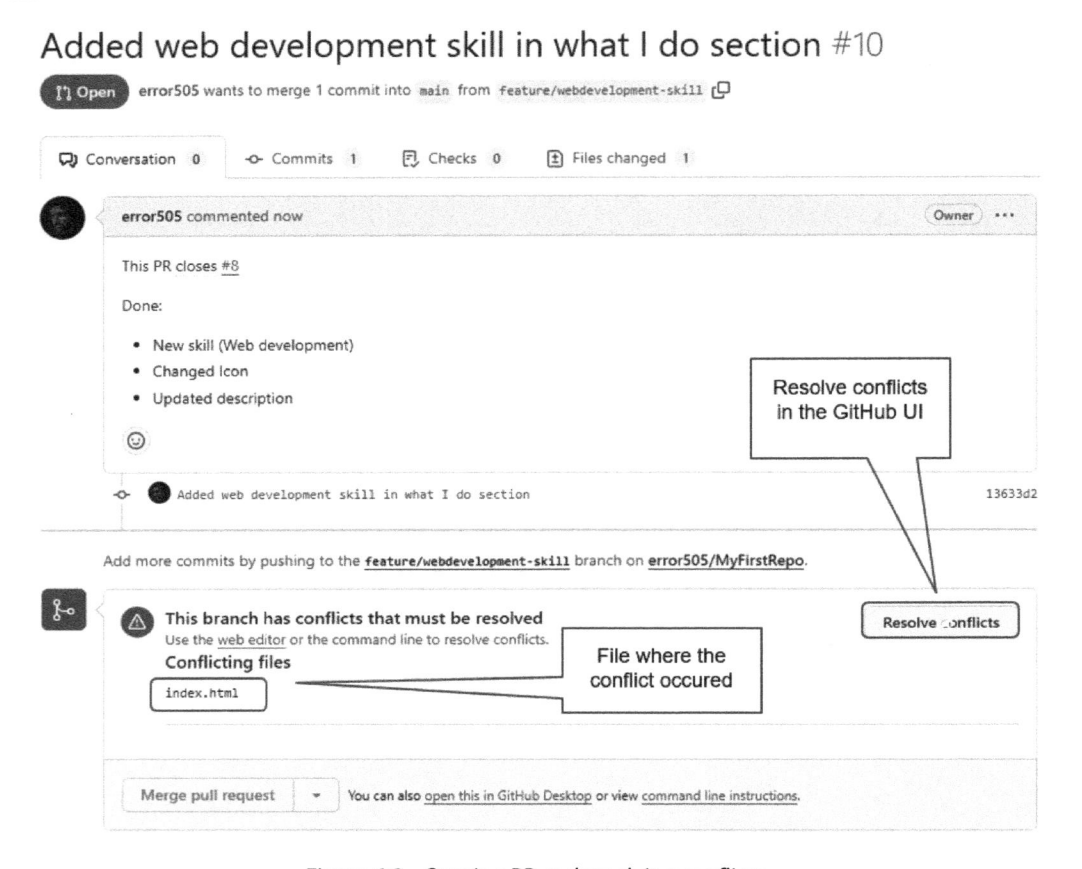

Merge pull request ▼ You can also open this in GitHub Desktop or view command line instructions.

Figure 6.9 – Starting PR and resolving conflicts

Now, when your colleague clicks on the **Resolve conflicts** button, the file in GitHub's editor will open, where the conflicting changes are clearly marked. Then, your colleague will see their code changes marked with >, =, and < signs, as shown in *Figure 6.10*:

```
73   <<<<<<< feature/webdevelopment-skill
74           <h3>Web Development</h3>
75           <p>I have more than 5 years of experience with building web sites.</p>
76   =======
77           <h3>React TS</h3>
78           <p>Many years of experience in building React websites</p>
79   >>>>>>> main
```

Figure 6.10 – PR conflicts marked on GitHub

- The `>>>>>>> main` section shows the current `main` branch changes with the changes you made previously

- The ======= section divides the two conflicting changes
- The <<<<<<< feature/webdevelopment-skill section shows the changes your colleague made

Your colleague Jamie now must decide how to integrate both changes. They might choose to keep the new title and incorporate their icon and description updates.

You can navigate the PR conflicts directly on the PR sections by clicking on the **Prev** and **Next** buttons, and it will move from one conflict to another. There, you will also have the count of how many conflicts there are in the current file that must be merged marked in red:

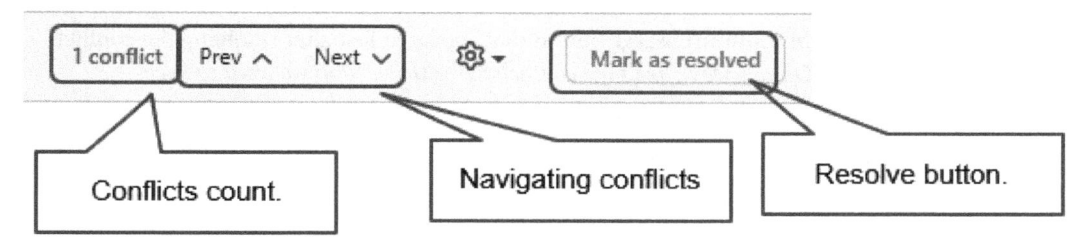

Figure 6.11 – Conflicts navigation menu

On the **Conflicts** menu, you can choose how you would like to configure **Indent mode** between spaces and tabs, **Indent size 2**, **4**, **8**, and **Line wrap mode** between **No wrap** and **Soft wrap**. This has been demonstrated in *Figure 6.12*:

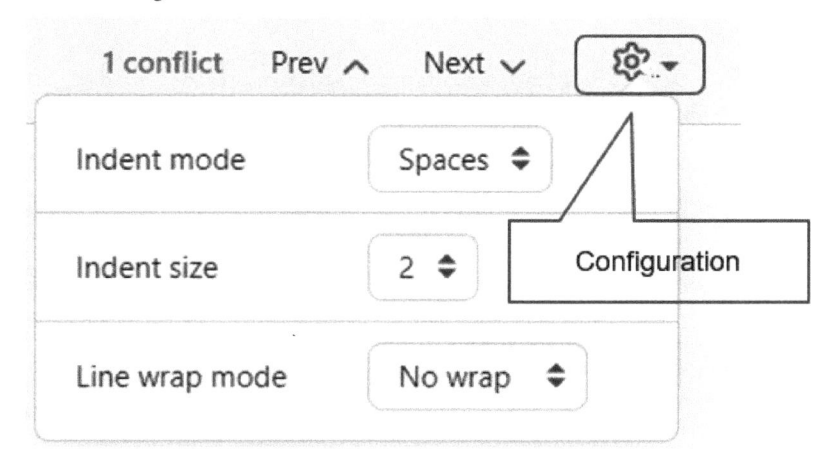

Figure 6.12 – Conflicts configuration menu

Jamie removes conflicting markers and the code from the file to resolve the merge conflict and clicks on the **Mark as resolved** button from the **Conflicts** menu, combining the changes in a way that makes sense:

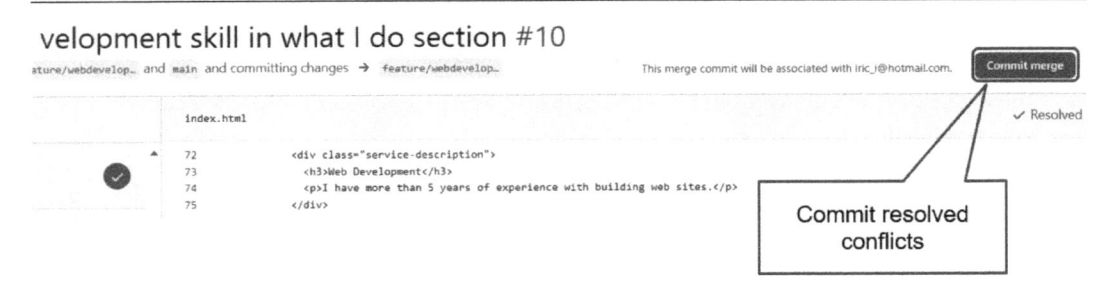

Figure 6.13 – Merge conflict resolved

After making their edits and resolving the conflict, Jamie will have the possibility to commit the updated file by clicking on the **Commit merge** button that appeared just after resolving the conflict. This commit tells GitHub, "*Okay – we've fixed the conflict; here's the version we want to keep.*"

With the conflict resolved, you and Jamie might chat to make sure you agree on the final version of the **Skills** section. With the conflict resolved, the portfolio website now accurately reflects both contributions to the **Skills** section.

The **Pull Request** page for the branch will now detect that no more conflicts are to be fixed, and you will have the possibility to finally merge the PR. The PR can now be completed by merging the feature branch into the `main` branch:

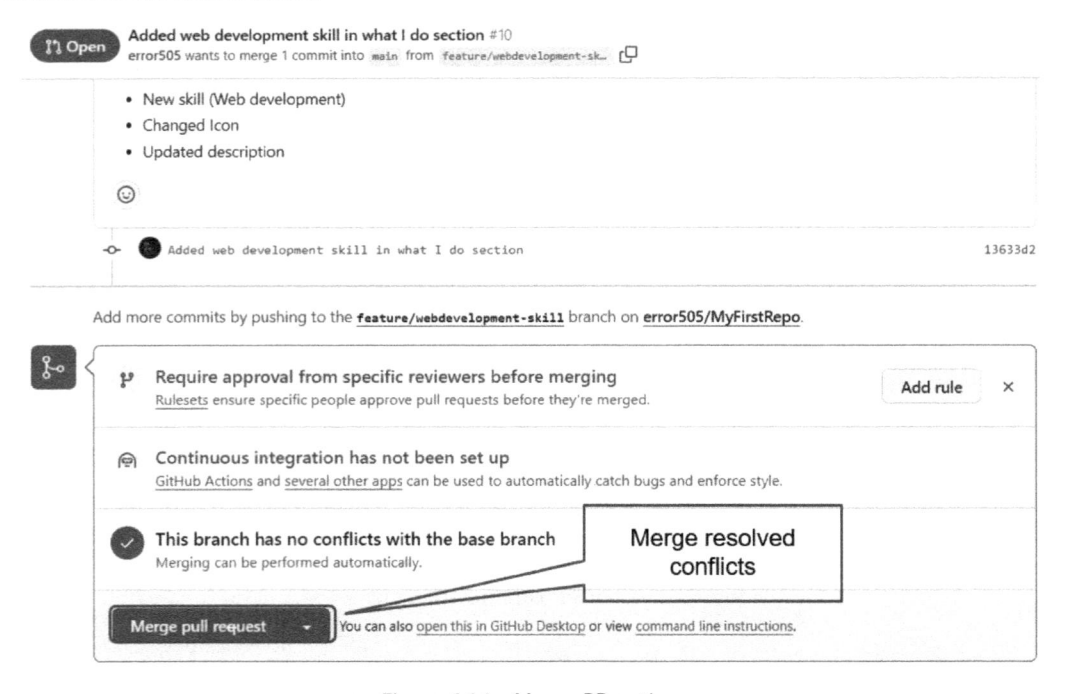

Figure 6.14 – Merge PR option

Having multiple merge conflicts

It's a bit like fixing several misunderstandings in a group chat, one at a time. It can happen when someone merges merge conflicts' markers by mistake directly to the main branch:

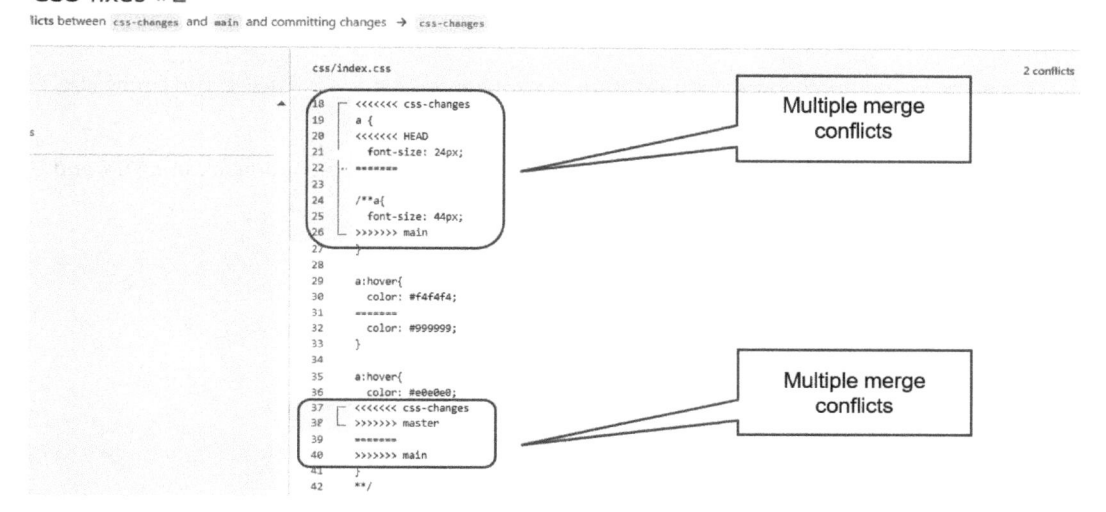

Figure 6.15 – File with multiple merge conflicts

In this example, different collaborators have made different changes to the same parts of a CSS file. We'll go through these steps as if we're cleaning up a messy group project.

To start resolving these issues, you can go to the **Pull Requests** section of the conflicting file. Click on the **Resolve conflicts** button. This takes you into an editor right in your browser, where you can fix things.

In the editor, you'll see markers such as <<<<<<<, =======, and >>>>>>>. These markers separate different changes. It's as if you're seeing both sides of an argument, side by side:

Figure 6.16 – Example of file with multiple merge conflicts

Here, one side (`css-changes`) wants the font size to be 24px, while the other (`main`) wants it to be 44px.

Choosing the right changes

Now, decide which changes to keep. Maybe you think 24px looks better, or maybe 44px, or perhaps something entirely different.

Delete the markers and the changes you don't want. If you choose 24px, you'll delete everything else, including the markers, so it looks like a normal CSS again.

Repeat this for each conflict in the file. In each case, you're choosing the best option, cleaning up, and making sure it looks like a regular piece of code again:

```
18
19    a {
20        font-size: 24px;
21    }
22
23    a:hover{
24        color: #f4f4f4;
25    }
26
```

Figure 6.17 – Resolved multiple conflicts

Resolving the conflicting file and completing the PR

After fixing all conflicts, GitHub will let you mark the file as resolved by clicking on the **Mark as resolved** button we used previously. It's like saying, *"Okay – we've sorted out these issues"*:

Mark as resolved

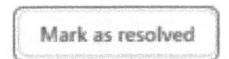

Figure 6.18 – Mark as resolved button

Now, you can complete the PR process, which is like finalizing changes in your group project. Solving multiple merge conflicts in GitHub's UI is like solving one single conflict several times.

Resolving merge conflicts using the command line like a pro

You have already learned what a merge conflict is, how it happens, and how you can solve it using the GitHub UI, but what about resolving conflicts locally using the command line?

Imagine you and a friend are both editing the same part of your project's website – say, the **What I Do** section. You both make different changes to the same lines in this section.

Now, when you both try to combine (or *merge*) your changes into the main project, GitHub gets confused. It's like when two friends try to write different sentences in the same space on a piece of paper at the same time. GitHub needs your help to decide which changes to keep. This is where you can use the command line to resolve conflicts and help GitHub know which changes to merge.

Going to the Pull Request page

On the GitHub **Pull Request** page, you will see a screen that we already have seen in *Figure 6.9* with the statement that this branch has conflicts that must be resolved. From there, you can use the UI by clicking on the **Resolve conflicts** button, which will open the web editor for you, but you can also use the *command line* to resolve conflicts. If you click on the command line, you will get full instructions on how to resolve conflicts with the command line, following the steps in *Figure 6.19*:

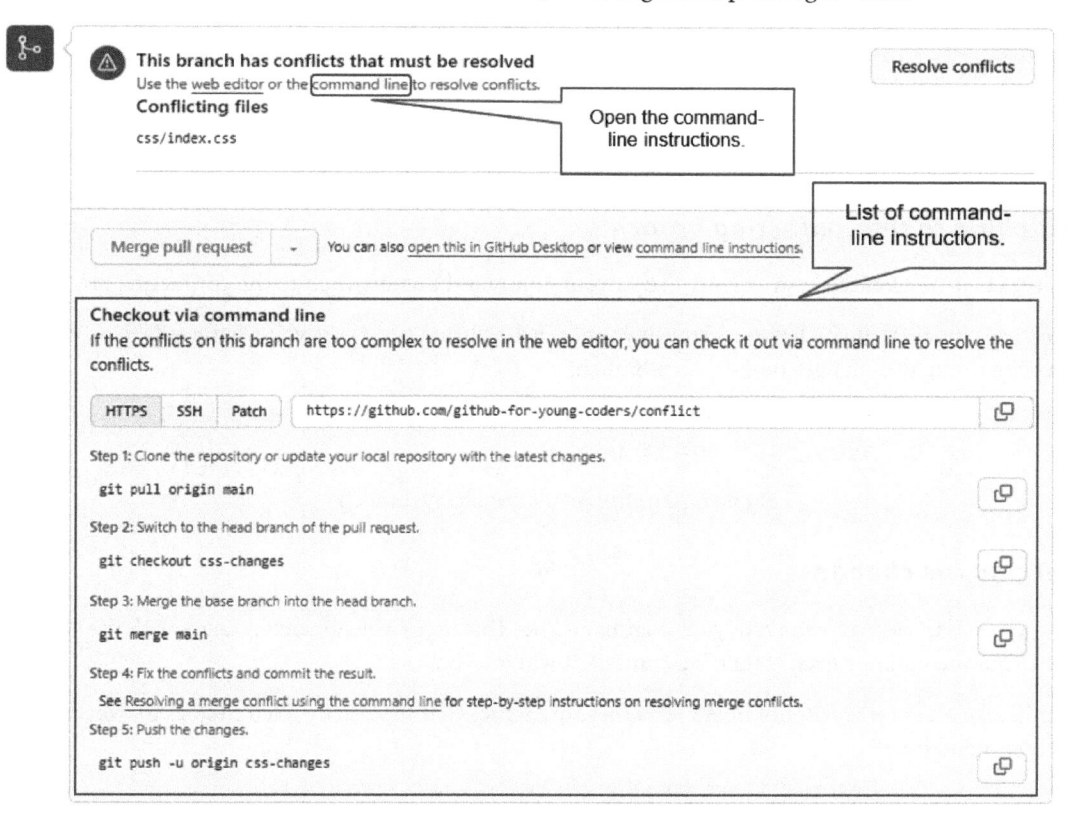

Figure 6.19 – GitHub command-line instructions

Open your favorite command-line tool, such as Git Bash or PowerShell, and navigate to your local project repository that has merge conflicts using the `cd` command and the name of the repository:

```
\GitHub for Young Coders\GitHub Code> cd Chapter 6
```

Figure 6.20 – Entering the conflicting repository

Pulling the latest changes from the main branch

In your project repository folder open the command line and type `git pull origin main`. This command says to GitHub, *"Give me the latest update on the main project"*.

As you can see, GitHub will already tell you that you have conflicts in your branch and that auto merging has failed and is not possible before you fix the merge conflicts manually:

```
From https://github.com/github-for-young-coders/conflict
 * branch            main       -> FETCH_HEAD
Auto-merging css/index.css
CONFLICT (content): Merge conflict in css/index.css
Automatic merge failed; fix conflicts and then commit the result.
```

Figure 6.21 – Conflict message

Switching to the conflicting branch

Type `git checkout css-changes` in your command line to switch to the conflicting branch.

A message saying that you have an error in your branch will appear, telling you what needs to be fixed before merging the changes to the `main` branch:

```
error: you need to resolve your current index first
css/index.css: needs merge
```

Figure 6.22 – Error on the conflicting branch

Merging the changes

Now, type `git merge main` in your command line. This tries to blend your changes with the latest ones from the main project. If there's a conflict, it will tell you.

In this case, the merge will not be done as there are issues that have to be fixed first, as displayed in the command line:

```
error: Merging is not possible because you have unmerged files.
hint: Fix them up in the work tree, and then use 'git add/rm <file>'
hint: as appropriate to mark resolution and make a commit.
fatal: Exiting because of an unresolved conflict.
```

Figure 6.23 – Merge command conflicts

Additionally, you can type `git status` to files that are having conflicts.

Addressing merge conflicts   163

After typing the `git status` command, you will have a list of changed and unmerged files displayed:

```
On branch css-changes
Your branch is up to date with 'origin/css-changes'.

You have unmerged paths.
  (fix conflicts and run "git commit")
  (use "git merge --abort" to abort the merge)

Changes to be committed:
        modified:   README.md
        modified:   manual.html
        modified:   mini.html

Unmerged paths:
  (use "git add <file>..." to mark resolution)
        both modified:   css/index.css
```

Figure 6.24 – Git status list of modified and conflicted files

Open these files in your text editor, such as Visual Studio Code or any other you prefer. You'll see markers such as <<<<<<< and >>>>>>> indicating the conflicting changes. Edit the files to keep the changes you want by removing the conflict markers (<<<<<<<, =======, >>>>>>>):

```
22
23   a:hover{       Do not use empty rulesets
24   <<<<<<< HEAD (Current Change)       } expected
25       color: #f4f4f4;
26   =======        at-rule or selector expected
27       color: #999999;        { expected
28   }   at-rule or selector expected
29
30   a:hover{
31       color: ▢#e0e0e0;
32   >>>>>>> master (Incoming Change)     { expected
33   =======
34       color: #dddddd;
35   }
36   **/
```

Figure 6.25 – Editing merge conflicts in Visual Studio Code

Now, you can save your files and go back in the command line to type `git add .` to stage your changes, and `git commit -m "resolved conflict"` to save them.

Finally, type `git push origin your-branch-name` to send your resolved changes back to GitHub.

Now, as the merge conflict is resolved, you can go back to the GitHub **Pull Request** page and finish the PR by clicking the **Merge pull request** button.

Solving conflicts with removed files

Let's say you decide to remove a file from the project because you think it's not needed anymore. Meanwhile, your friend has made some changes to the same file on their version of the project. When you both try to combine your changes, GitHub gets confused. It's like, *Hey – you're trying to update a file that's not here anymore!*:

```
On branch
Your branch is up to date with 'origin/         '.

You have unmerged paths.
  (fix conflicts and run "git commit")
  (use "git merge --abort" to abort the merge)

Changes to be committed:
        modified:   README.md
        modified:   css/index.css

Unmerged paths:
  (use "git add/rm <file>..." as appropriate to mark resolution)
        deleted by us:   manual.html
```

Figure 6.26 – Status of the deleted file

To resolve this kind of conflict, you can follow the whole procedure we covered with the previous topic, where we explained how to solve issues using the command line.

Removing the file

Now, you will have to decide whether you want to keep the file removed (your change) or bring it back with your friend's changes.

If you want to keep the file removed, just add this decision to your project with `git rm file-name` (replace `file-name` with the actual filename):

```
git remove index.css
```

Figure 6.27 – Removing not needed file

Then, commit this decision with `git commit -m "Decided to keep the file removed"`:

```
git commit -m "Decided to keep the file removed"
```

Figure 6.28 – Committing not needed file

Want to bring the file back?

Use `git checkout --theirs file-name` to bring back the file with your friend's changes and then add this file back to the project with `git add file-name`.

It will update the repository with the file that was previously deleted. Commit this decision with `git commit -m "Brought back the file with changes"`. Now, complete the merge process with `git merge`. This is like telling GitHub, *"Okay – we've resolved the issue; let's move on!"*

Finally, update the main project with your resolved changes using `git push origin branch-name`.

You're informing everyone, *"All sorted! The project is now up to date"*:

```
Enumerating objects: 13, done.
Counting objects: 100% (13/13), done.
Delta compression using up to 8 threads
Compressing objects: 100% (5/5), done.
Writing objects: 100% (5/5), 582 bytes | 582.00 KiB/s, done.
Total 5 (delta 4), reused 0 (delta 0), pack-reused 0
remote: Resolving deltas: 100% (4/4), completed with 4 local objects.
To https://github.com/github-for-young-coders/conflict.git
    cb8731a..afc36dd  manual -> manual
```

Figure 6.29 – Status that everything is up to date

Now, again, you will be able to proceed with the PR and close it. Click on the **Merge pull request** button, and finally, you have finished the PR:

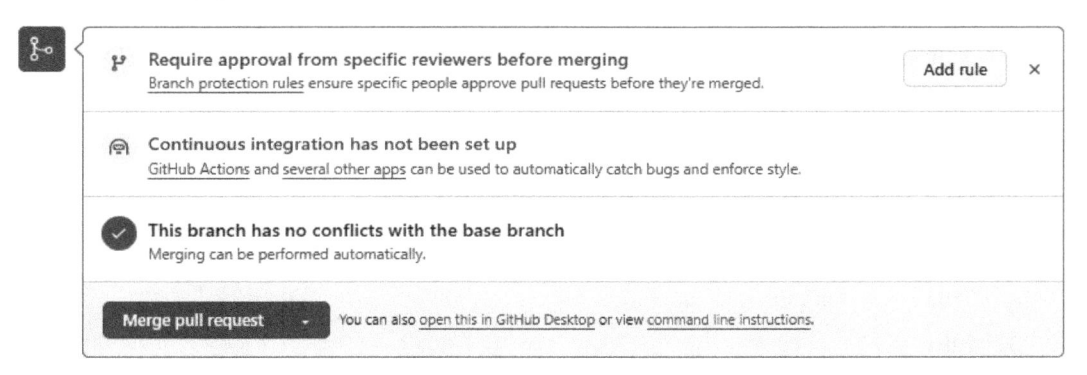

Figure 6.30 – Merging resolved PR

You've now learned how to handle merge conflicts, including what happens during a merge conflict, how to resolve conflicts using GitHub's interface, and how to deal with multiple conflicts and removed files. This knowledge ensures that your code integrates smoothly, even when changes overlap.

Summary

Congratulations on completing *Chapter 6*! You've navigated through the twists and turns of merge conflicts and emerged more skilled and confident.

This chapter was all about mastering merge conflicts. We learned what merge conflicts are and how to resolve them through various methods and in multiple scenarios. With these skills, you can now do the following:

- Identify merge conflicts, understanding them as natural parts of collaborative coding
- Resolve conflicts both using GitHub's UI and the command line, blending different changes into a harmonious final product
- Become a more effective team player, able to handle conflicts in your projects with ease and expertise

As you continue building your one-page portfolio website, remember that merge conflicts are just opportunities to synergize different ideas and improvements. In the next chapter, you'll learn how to navigate, alter, and understand the history of your coding projects in Git. You will look at various key concepts and features, such as `git bisect`, the `diff` command, `git reset`, and cherry-picking.

Quiz

Test your knowledge with the following questions:

1. What is a merge conflict in GitHub?

    A. An error during file download

    B. A disagreement between team members

    C. Overlapping changes made to the same part of a file

    D. A problem with internet connectivity

    Answer: C. Overlapping changes made to the same part of a file

2. Merge conflicts occur only when two people change the same line in a file.

    A. True

    B. False

    Answer: B. False (they can also occur with file deletions, renaming, and so on)

3. In GitHub, you can't use the _____ to manually resolve merge conflicts.

    Answer: command line

4. Why is it important to resolve merge conflicts in a project?

    Answer: To ensure that the project has a coherent and functioning code base, reflecting the intended changes from all contributors.

5. What does the `git merge` command do?

    A. Deletes a branch

    B. Combines changes from different branches

    C. Changes the project name

    D. Creates a new file

    Answer: B. Combines changes from different branches

6. Merge conflicts are always a sign of mistakes in coding.

    A. True

    B. False

    Answer: B. False (they are a natural part of collaboration)

7. What should you do after resolving a merge conflict?

    A. Delete the project

    B. Commit and push the changes to remote

    C. Ignore the changes

    D. Close GitHub

    Answer: B. Commit and push the changes to remote

8. When a merge conflict occurs, GitHub will mark the conflicted area with special symbols such as <<<<<<<, =======, and >>>>>>>. These symbols are known as _____.

    Answer: conflict markers

9. You can only resolve merge conflicts using GitHub's web interface.

    A. True

    B. False

    Answer: B. False (they can also be resolved using the command line)

10. What is one common scenario that can lead to a merge conflict?

    Answer: When two contributors make different changes to the same part of a file or when one deletes a file that the other has modified.

Challenge – Stellar enhancements in Space Explorer!

Fantastic work on mastering merge conflicts in *Chapter 6*! Now that you've got the hang of juggling different changes in GitHub, let's apply those skills to our *Space Explorer* game. It's time to add some cosmic flair with images of stars, a spaceship, and obstacles.

Ready to make your game look out of this world?

Your mission

Enhance the *Space Explorer* game by adding images of stars, a spaceship, and obstacles, while managing and resolving any merge conflicts that arise:

Figure 6.31 – Space Explorer game

Steps to success

1. **Create a new branch**: In your GitHub repository for the *Space Explorer* game, create a new branch named `feature-stellar-graphics`.

    This branch will be your creative workshop for adding cool space images.

2. **Get the images**: I'll provide you with awesome images of stars, a spaceship, and obstacles. Download these images and get ready to integrate them into your game. You can find them in three folders:

   - Ship image: `https://github.com/PacktPublishing/GitHub-for-Next-Generation-Coders/blob/main/Space%20Explorer%20Game/Game%20Obsticles%20-%20Ship%20Picture/rocket.png`

   - Star image: `https://github.com/PacktPublishing/GitHub-for-Next-Generation-Coders/blob/main/Space%20Explorer%20Game/Game%20Obsticles%20-%20Stars%20Picture/star.png`

   - Obstacle image: `https://github.com/PacktPublishing/GitHub-for-Next-Generation-Coders/blob/main/Space%20Explorer%20Game/Game%20Obsticles%20-%20Obstacle%20Picture/asteroid.png`

3. **Integrate the images**: Replace the existing placeholders or simple graphics in your game with these new images. Adjust their sizes and positions to fit perfectly in your cosmic landscape.

4. **Commit your changes**: After integrating each image, commit your changes. Use clear messages such as *Added star image* or *Updated spaceship graphic*.

   Regular commits help track your progress and make it easier to handle merge conflicts.

5. **Simulate a merge conflict**: To get a real feel for resolving conflicts, make a change in the `main` branch that could conflict with your new graphics. For example, change the position or size of an obstacle.

   Then, try to merge your `feature-stellar-graphics` branch into `main`.

6. **Resolve any merge conflicts**: If a merge conflict pops up, use your newly acquired skills to resolve it. Decide which changes to keep, update the files, and finalize the merge.

7. **Final touches and merge**: Once everything looks great and works smoothly, create a PR to merge `feature-stellar-graphics` into the `main` branch.

Describe your enhancements and ask for feedback.

Bonus challenge

Why stop here?

Try adding animated graphics for the spaceship's thrusters or create twinkling effects for the stars.

Part 3: Mastering Git Commands and Tools

In this part, we cover the critical tools and commands in Git that will enhance your ability to revert changes, track history, and use GitHub more effectively. From undoing mistakes with `git revert` to understanding the complete history of your projects, this section empowers you with the knowledge to handle complex scenarios and use advanced Git functionalities.

This part contains the following chapters:

- *Chapter 7, Git History and Reverting Commits*
- *Chapter 8, Helpful Tools and Git Commands*

7
Git History and Reverting Commits

Welcome to *Chapter 7*, where we're about to go on a fascinating journey through the history of your coding projects, guided by Git and GitHub. Understanding how history works in these tools is like having a time machine at your fingertips, giving you the power to revisit, analyze, and modify your project's past.

First, we will navigate history with the `git log` and `git reflog` commands. Next, we will learn how to pinpoint exactly when and where bugs were introduced in your code using the `git bisect` command. We will then master the arts of reverting commits, resetting progress, and cherry-picking changes to keep your project timeline clean and efficient. So, let's start this adventure!

Here's what you will learn about in this chapter:

- Understanding Git and GitHub history – tracking changes to your website
- Explaining `git bisect` – finding the needle in the haystack
- Reverting commits to a previous version
- Unraveling mysteries with `git diff` – the tale of the unseen changes
- Undoing changes with `git reset` and cherry-picking

By the end of this chapter, you'll be adept at navigating the complex timelines of your projects, ensuring your coding history is as smooth and error-free as possible. Let's start this exciting journey through time!

Understanding Git and GitHub history – tracking changes to your website

GitHub acts as a remote repository, a place where your Git history is stored online. This means it's not just on your computer; it's in the cloud, accessible from anywhere and by anyone you allow.

GitHub provides a visual and user-friendly interface where you can view your project's history. You can see who made changes, when they were made, and what those changes were. It's like having a collaborative diary for your project, where everyone writes and shares their part of the story.

Imagine you're putting together a scrapbook for your one-page portfolio website. You're adding pages (such as a new **My PORTFOLIO** section), sticking photos, and writing notes. Now, wouldn't it be great if you could flip through this scrapbook anytime to see all the changes you've made over time? That's what the history in Git and GitHub does!

Figure 7.1 – Git and GitHub history explained as a project's scrapbook

Viewing the history locally in a Git repository

Every time you make a change in Git (such as adding the **My PORTFOLIO** section to your website), it's like adding a new entry to your scrapbook and saving a snapshot at a particular moment in time:

Figure 7.2 – The My PORTFOLIO section

- **git log**: To see this history, you can use the `git log` command. This command is like opening your scrapbook. It shows a list of all the changes (commits), who made them, and when. *Figure 7.3* displays how this log looks in Git:

```
PS C:\Users\igor.iric\Desktop\MyFirstRepo> git log
commit abaac86d1e2c25630a43e7622462681687dc9c85 (HEAD -> feature/webdevelopment-skill, origin/main, origin/HEAD)
Merge: a3266be 296aa78
Author: Igor Iric <        _@hotmail.com>
Date:   Sun Dec 31 15:48:09 2023 +0100

:...skipping...
commit abaac86d1e2c25630a43e7622462681687dc9c85 (HEAD -> feature/webdevelopment-skill, origin/main, origin/HEAD)
Merge: a3266be 296aa78
Author: Igor Iric <        hotmail.com>
Date:   Sun Dec 31 15:48:09 2023 +0100

    Merge pull request #10 from error505/feature/webdevelopment-skill

    Added web development skill in what I do section

commit 296aa784b96a2a83f7df82f9870f850dc31ea55d
Merge: 13633d2 a3266be
Author: Igor Iric <        _@hotmail.com>
Date:   Sun Dec 31 15:36:29 2023 +0100

    Merge branch 'main' into feature/webdevelopment-skill

commit a3266be29826082ed911fd4de7441eeaa5b00fc4
Author: Igor Iric <        _@hotmail.com>
```

Figure 7.3 – Git history

Every snapshot (commit) has a unique ID, like a fingerprint. This ID lets you identify each commit and differentiate it from all others. Along with each snapshot, you write a message. This is like explaining what changes you made and why. Good commit messages are like a diary, helping you (and others) understand the history of your project.

- **git log --oneline**: The `git log --oneline` command is also useful for showing the history of your project.

  This command simplifies the output of `git log`. Instead of showing the full commit message and author information, it displays each commit in a single line with just the commit's ID (abbreviated) and the title of the commit message, as shown in *Figure 7.4*. Think of it as the summary or the *headlines* of your project's history:

```
PS C:\Users\igor.iric\Desktop\MyFirstRepo> git log --oneline
abaac86 (HEAD -> feature/webdevelopment-skill, origin/main, origin/HEAD) Merge pull request #10 from error505/feature/webdevelopment-skill
296aa78 Merge branch 'main' into feature/webdevelopment-skill
a3266be Update index.html
13633d2 Added web development skill in what I do section
a9d0982 Merge pull request #8 from error505/feature/webdevelopment-skill-in-what-i-do-section
87f56dd (feature/webdevelopment-skill-in-what-i-do-section, feature/webdevelopment-skil) Added web development skill in what I do section
1c5d5d5 Added web development skill in what I do section
8580b35 (main) Added a new skill in what I do section
cdadca9 Merge pull request #7 from error505/feature/add-what-i-do-section
1430afa Added a new what I do section
8699309 Merge pull request #6 from error505/feature/leftside-menu
034c65d (feature/leftside-menu) Added left side menu and stylings
015a424 Update styles.css
ad44a19 Update index.html
68789b3 Update README.md
dce3f37 Merge pull request #3 from error505/add-what-i-do
851fa2f Update index.html
e7fecba Create styles.css
19778f0 Create index.html
536c70b Initial commit
```

Figure 7.4 – The result of running the git log --oneline command

- **git log --oneline --graph**: You can use the `git log --oneline --graph` command if you want to have a more graphical depiction of your Git history.

  This command takes the simplicity of `--oneline` and adds a visual representation of the commit history using an ASCII graph. It's useful for seeing the structure of your branches and merges. Imagine it like a family tree but for your commits. *Figure 7.5* shows what this looks like:

```
*   abaac86 (HEAD -> feature/webdevelopment-skill, origin/main, origin/HEAD) Merge pull request #10 from error505/feature/webdevelopment-skill
|\
| * 296aa78 Merge branch 'main' into feature/webdevelopment-skill
| |\
| |/
|/|
* | a3266be Update index.html
* | a9d0982 Merge pull request #8 from error505/feature/webdevelopment-skill-in-what-i-do-section
|\ \
| | * 13633d2 Added web development skill in what I do section
| |/
| * 87f56dd (feature/webdevelopment-skill-in-what-i-do-section, feature/webdevelopment-skil) Added web development skill in what I do section
| * 1c5d5d5 Added web development skill in what I do section
|/
* 8580b35 (main) Added a new skill in what I do section
*   cdadca9 Merge pull request #7 from error505/feature/add-what-i-do-section
|\
| * 1430afa Added a new what I do section
|/
*   8699309 Merge pull request #6 from error505/feature/leftside-menu
|\
| * 034c65d (feature/leftside-menu) Added left side menu and stylings
|/
* 015a424 Update styles.css
* ad44a19 Update index.html
* 68789b3 Update README.md
*   dce3f37 Merge pull request #3 from error505/add-what-i-do
```

Figure 7.5 – The result of running the git log --oneline --graph command

- **git log --oneline --graph --decorate**: The `git log --oneline --graph --decorate` command is useful if you want to see more metadata in your history.

    This command builds upon the previous one by adding *decorations*. In Git, decorations are references such as branch names, tags, or `HEAD` requests pointing to specific commits. Using this command makes it easier to see where the branches are concerning the commits. It's like labeling branches in your commit tree so that you know who's who:

```
PS C:\Users\igor.iric\Desktop\MyFirstRepo> git log --oneline --graph --decorate
*   abaac86 (HEAD -> feature/webdevelopment-skill, origin/main, origin/HEAD) Merge pull request #10 from error505/feature/webdevelopment-skill
|\
| * 296aa78 Merge branch 'main' into feature/webdevelopment-skill
| |\
| |/
|/|
* | a3266be Update index.html
* | a9d0982 Merge pull request #8 from error505/feature/webdevelopment-skill-in-what-i-do-section
|\ \
| | * 13633d2 Added web development skill in what I do section
| |/
| * 87f56dd (feature/webdevelopment-skill-in-what-i-do-section, feature/webdevelopment-skil) Added web development skill in what I do section
| * 1c5d5d5 Added web development skill in what I do section
|/
* 8580b35 (main) Added a new skill in what I do section
* cdadca9 Merge pull request #7 from error505/feature/add-what-i-do-section
|\
| * 1430afa Added a new what I do section
|/
* 8699309 Merge pull request #6 from error505/feature/leftside-menu
|\
| * 034c65d (feature/leftside-menu) Added left side menu and stylings
|/
* 015a424 Update styles.css
* ad44a19 Update index.html
* 68789b3 Update README.md
* dce3f37 Merge pull request #3 from error505/add-what-i-do
```

Figure 7.6 – The result of running the git log --oneline --graph --decorate command

- **git log --oneline --graph --decorate --all**: By incorporating the `git log --oneline --graph --decorate --all` command, you can display even more metadata.

    This is the most comprehensive view. It shows all commits from all branches in your repository, not just the commits that can be reached from the current branch. It provides a complete overview of your entire repository's history, including all its branches and merges. It's like having a bird's-eye view of the entire forest of code. *Figure 7.7* provides a snapshot of this:

Git History and Reverting Commits

```
* abaac86 (HEAD -> feature/webdevelopment-skill, origin/main, origin/HEAD) Merge pull request #10 from error505/feature/webdevelopment-skill
|\
| * 296aa78 Merge branch 'main' into feature/webdevelopment-skill
| |\
| |/
|/|
* | a3266be Update index.html
* | a9d0982 Merge pull request #8 from error505/feature/webdevelopment-skill-in-what-i-do-section
|\ \
| | * 13633d2 Added web development skill in what I do section
| |/
| * 87f56dd (feature/webdevelopment-skill-in-what-i-do-section, feature/webdevelopment-skil) Added web development skill in what I do section
* | 1c5d5d5 Added web development skill in what I do section
|/
* 8580b35 (main) Added a new skill in what I do section
* cdadca9 Merge pull request #7 from error505/feature/add-what-i-do-section
|\
| * 1430afa Added a new what I do section
|/
* 8699309 Merge pull request #6 from error505/feature/leftside-menu
|\
| * 034c65d (feature/leftside-menu) Added left side menu and stylings
|/
* 015a424 Update styles.css
* ad44a19 Update index.html
* 68789b3 Update README.md
* dce3f37 Merge pull request #3 from error505/add-what-i-do
```

Figure 7.7 – The result of running the git log --oneline --graph --decorate --all command

- **git reflog**: Finally, `git reflog` provides a detailed history book of your project. It's a command that shows all the recent changes, each with its own SHA code. An **SHA code** in the commit history is a unique identifier that acts like a fingerprint for each change you have made to a project, ensuring the integrity and history of your work in Git. It's helpful when you need to find a specific point in your project's history to reset or cherry-pick. You'll learn more about cherry-picking in the next *Picking specific changes with cherry-picking* section:

```
PS C:\Users\igor.iric\Desktop\MyFirstRepo> git reflog
32e9f11 (HEAD -> main, origin/main, origin/HEAD) HEAD@{0}: reset: moving to HEAD
32e9f11 (HEAD -> main, origin/main, origin/HEAD) HEAD@{1}: checkout: moving from f5cf36ef1873e5ca88bb81f76218c108726b6bb0 to main
f5cf36e HEAD@{2}: revert: Revert "Update index.html"
560b5b8 HEAD@{3}: checkout: moving from 2b407a88fd13d53bdf4ca2682e79a7a7ce968451 to 560b5b8e63e62f23005e5b5d04e41219ac8f4796
2b407a8 HEAD@{4}: checkout: moving from main to 2b407a88fd13d53bdf4ca2682e79a7a7ce968451
32e9f11 (HEAD -> main, origin/main, origin/HEAD) HEAD@{5}: checkout: moving from de972569397ccd00b5611ded8af57876c1ac64d5 to main
de97256 HEAD@{6}: checkout: moving from 2eed77a02fcaba08b5fc71a46034dcec795bf3e0 to de972569397ccd00b5611ded8af57876c1ac64d5
2eed77a HEAD@{7}: checkout: moving from 2b407a88fd13d53bdf4ca2682e79a7a7ce968451 to 2eed77a02fcaba08b5fc71a46034dcec795bf3e0
2b407a8 HEAD@{8}: checkout: moving from main to 2b407a88fd13d53bdf4ca2682e79a7a7ce968451
32e9f11 (HEAD -> main, origin/main, origin/HEAD) HEAD@{9}: checkout: moving from 2b407a88fd13d53bdf4ca2682e79a7a7ce968451 to main
2b407a8 HEAD@{10}: checkout: moving from main to 2b407a88fd13d53bdf4ca2682e79a7a7ce968451
32e9f11 (HEAD -> main, origin/main, origin/HEAD) HEAD@{11}: checkout: moving from main to main
abaac86 (feature/webdevelopment-skill) HEAD@{12}: checkout: moving from abaac86d1e2c25630a43e7622462681687dc9c85 to main
32e9f11 (HEAD -> main, origin/main, origin/HEAD) HEAD@{13}: checkout: moving from main to abaac86d1e2c25630a43e7622462681687dc9c85
32e9f11 (HEAD -> main, origin/main, origin/HEAD) HEAD@{14}: checkout: moving from main to main
32e9f11 (HEAD -> main, origin/main, origin/HEAD) HEAD@{15}: checkout: moving from 87f56dd4fb4e7d18d18441663d91a94cd6e705b8 to main
87f56dd (feature/webdevelopment-skill-in-what-i-do-section, feature/webdevelopment-skil) HEAD@{16}: checkout: moving from a9d09828ae742a4e30931f087f48fa6c79983887 to 87f56dd4fb4
e7d18d18441663d91a94cd6e705b8
a9d0982 HEAD@{17}: checkout: moving from a3266be29826082ed911fd44e7441eeaa5b00fc4 to a9d09828ae742a4e30931f087f48fa6c79983887
a3266be HEAD@{18}: checkout: moving from abaac86d1e2c25630a43e7622462681687dc9c85 to a3266be29826082ed911fd44e7441eeaa5b00fc4
abaac86 (feature/webdevelopment-skill) HEAD@{19}: checkout: moving from main to abaac86d1e2c25630a43e7622462681687dc9c85
32e9f11 (HEAD -> main, origin/main, origin/HEAD) HEAD@{20}: checkout: moving from 87f56dd4fb4e7d18d18441663d91a94cd6e705b8 to main
87f56dd (feature/webdevelopment-skill-in-what-i-do-section, feature/webdevelopment-skil) HEAD@{21}: checkout: moving from a9d09828ae742a4e30931f087f48fa6c79983887 to 87f56dd4fb4
```

Figure 7.8 – The result of running git reflog

Using these different variations of `git log`, you can get the level of detail and the type of overview that suits your current needs, whether it's a glance at recent changes or a deep dive into the full history of your project.

Branches – parallel universes of Git history

In Git, **branches** allow you to create parallel timelines. It's like having alternate versions of your story running simultaneously. You might have a *main* timeline where everything is stable and other branches for trying out new ideas or features:

```
*   abaac86 (HEAD -> feature/webdevelopment-skill, origin/main, origin/HEAD) Merge pull request #10 from error505/feature/webdevelopment-skill
|\
| * 296aa78 Merge branch 'main' into feature/webdevelopment-skill
| |\
| |/
|/|
* | a3266be Update index.html
* | a9d0982 Merge pull request #8 from error505/feature/webdevelopment-skill-in-what-i-do-section
|\ \
| | * 13633d2 Added web development skill in what I do section
```

Figure 7.9 – git log branching and merging

Once you've written a part of your story in a branch that you're happy with, you can bring it into your main timeline. This process of merging can sometimes lead to conflicts, as we saw in the previous chapter, but it's a powerful way to weave different strands of your story together.

The Activity page – a chronicle of your repository's journey

The **Activity** page on GitHub is a place where all the actions and movements are visible and where you can observe the hustle and bustle of your repository's life – from code changes and discussions to the contributions of each team member.

So, how do you get there? On GitHub, go to your website's repository and click on the **Activity** link. This can be found in the right-hand side menu:

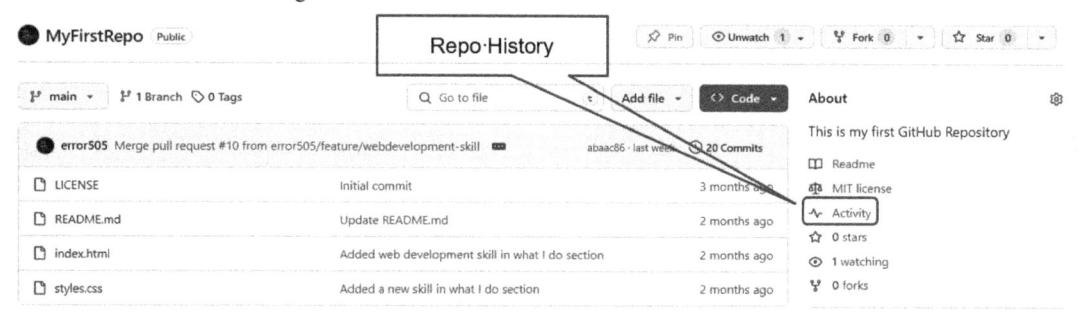

Figure 7.10 – History activity on GitHub

You can also click on **Branches** and then select a specific branch to view its unique activities:

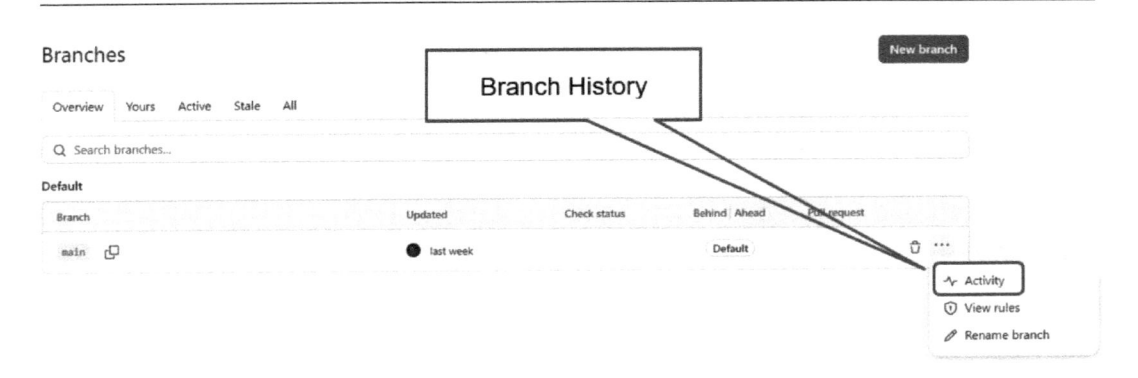

Figure 7.11 – History activity on GitHub – Branches

Here, you'll see a list of all the changes that have been made, just like in Git, but in a more user-friendly way.

But what can you find on the Activity page? The **Activity** page shows a detailed history of changes, including pushes (updates), merges (combining code), and more. Each change is linked to specific commits and the users who made them. This can be seen in *Figure 7.12*:

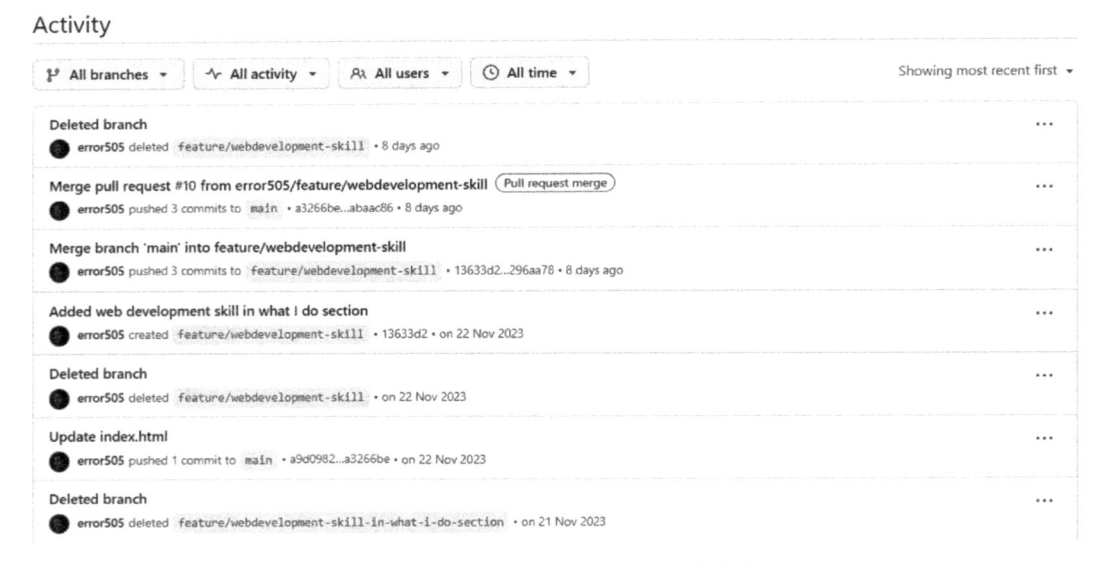

Figure 7.12 – The Activity page on GitHub

You can filter the activities by branch, user, time period, or activity type. This helps you focus on the changes most relevant to what you're looking for. For instance, you may only want to see the changes that have been made to the *develop* branch in the last week:

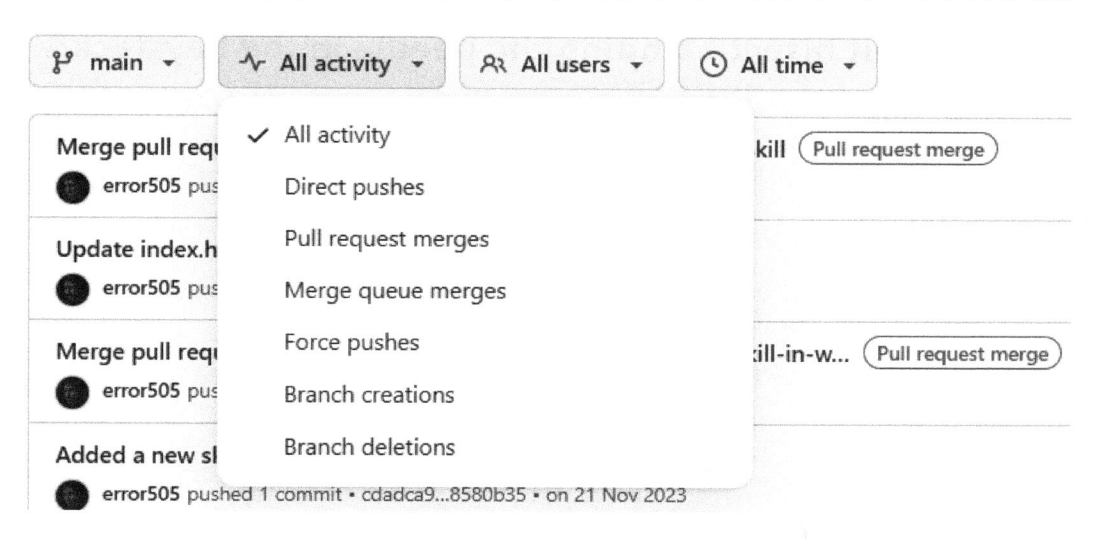

Figure 7.13 – The Activity page's filter options

For each activity, you have the option to view the exact changes made. Clicking **Compare changes** shows you what was added, removed, or altered, providing a clear picture of each update:

Figure 7.14 – Compare changes

It's fascinating to see how your website evolved, from the first line of code to the latest addition – that is, the **My PORTFOLIO** section. If you're working with friends, you can see who did what.

Maybe your friend added a cool design to the **My PORTFOLIO** section. And if something goes wrong, or if you're just curious, you can review the whole activity and see exactly what was changed and when.

Having the **Activity** page on GitHub is like having a detailed diary of your entire project's journey, allowing you and your team to stay informed, aligned, and productive. Empowered with this knowledge, let's look at another important topic: `git bisect`.

Explaining git bisect – finding the needle in the haystack

You're working on a one-page portfolio website and suddenly, something goes wrong. A section of your website, let's say **My PORTFOLIO**, stops working or disappears. You're not sure when or how the problem started. You know it was there before, but now it's gone. What happened? This is where **git bisect** comes to the rescue. It's like a detective game to find out when and where things went wrong in your project.

The `git bisect` command is like a time machine. It helps you find the exact change (commit) in your project that caused the issue.

Let's add some code for the **My PORTFOLIO** section for the one-page portfolio website:

```html
<!-- Portfolio Section -->
<section class="portfolio-section" id="portfolio-section">
    <h1>My PORTFOLIO</h1>
    <div class="portfolio-menu">
        <button class="active">ALL</button>
        <button>DESIGN</button>
        <button>DEVELOPMENT</button>
        <button>GRAPHICS</button>
        <button>Templates</button>
    </div>
    <div class="portfolio-grid">
        <!-- Repeat this for each item in the grid -->
        <div class="portfolio-item" id="development">
            <img src="https://via.placeholder.com/400x250" alt="Description">
            <div class="overlay">
                <h3>3D Graphics</h3>
                <h4>Templates</h4>
            </div>
        </div>
        <div class="portfolio-item" id="development">
          <img src="https://via.placeholder.com/400x250" alt="Description">
          <div class="overlay">
              <h3>3D Graphics</h3>
              <h4>Templates</h4>
          </div>
        </div>
          <div class="portfolio-item" id="development">
            <img src="https://via.placeholder.com/400x250" alt="Description">
            <div class="overlay">
```

Figure 7.15 – Code addition to the website

Now, let's add some CSS styling for this new section:

```css
.portfolio-section {
  padding: 50px;
  background-color: ☐#f8f8f8;
  text-align: center;
}

.portfolio-menu {
  margin: 20px 0;
  text-align: center;
}

.portfolio-menu button {
  padding: 10px 20px;
  border: none;
  background-color: transparent;
  transition: background-color 0.3s;
  cursor: pointer;
}

.portfolio-menu button.active {
  background-color: var(--main-theme-color);
  color: ☐white;
}
```

Figure 7.16 – Code added to the My PORTFOLIO section

The rest of the HTML code and stylings for this section can be found at https://github.com/PacktPublishing/GitHub-for-Next-Generation-Coders/tree/main/Chapter%207.

A git bisect mystery – the case of the missing CSS styling

Your mission: to uncover the enigma behind the missing CSS styling for the **My PORTFOLIO** section of your website. It's a crucial part of your one-page portfolio website, and something's gone bad. The once beautifully styled section now looks bare and unstyled. It's time to unravel this mystery with `git bisect`:

184 Git History and Reverting Commits

My PORTFOLIO

| ALL | DESIGN | DEVELOPMENT | GRAPHICS | Templates |

3D Graphics
Templates

400 x 250

Figure 7.17 – My PORTFOLIO problem

This is where you will start your investigation. Begin by telling Git that you have a mystery to solve by typing `git bisect start`. This is like opening your detective toolkit.

You know the current state is bad since the CSS is not applying correctly. You mark this as a bad commit with the `git bisect bad <SHA-of-bad-commit >` command. Using `git log --oneline`, you can start looking at the history of the commits to try to point out the bad commit. You know that the last time you pulled the code, there was a problem, so you take the last commit as the bad one:

```
PS C:\Users\igor.iric\Desktop\MyFirstRepo> git log --oneline
32e9f11 (HEAD -> main, origin/main, origin/HEAD) Merge pull request #14 from error505/feature/chnage_portfolio_item
560b5b8 Update index.html
2b407a8 Merge pull request #13 from error505/feature/change_portfolio
2eed77a Update index.html
de97256 Merge pull request #12 from error505/feature/portfolio_section
adb5920 Adding Portfolio Section
abaac86 (feature/webdevelopment-skill) Merge pull request #10 from error505/feature/webdevelopment-skill
296aa78 Merge branch 'main' into feature/webdevelopment-skill
a3266be Update index.html
13633d2 Added web development skill in what I do section
a9d0982 Merge pull request #8 from error505/feature/webdevelopment-skill-in-what-i-do-section
87f56dd (feature/webdevelopment-skill-in-what-i-do-section, feature/webdevelopment-skil) Added web development skill in what I do section
1c5d5d5 Added web development skill in what I do section
8580b35 Added a new skill in what I do section
cdadca9 Merge pull request #7 from error505/feature/add-what-i-do-section
1430afa Added a new what I do section
8699309 Merge pull request #6 from error505/feature/leftside-menu
034c65d (feature/leftside-menu) Added left side menu and stylings
015a424 Update styles.css
ad44a19 Update index.html
68789b3 Update README.md
```

Figure 7.18 – git bisect log

Then, you recall a commit where everything looked perfect and the CSS was on point. You mark this as your good commit by running `git bisect good <SHA-of-good-commit>`:

```
PS C:\Users\igor.iric\Desktop\MyFirstRepo> git bisect start
status: waiting for both good and bad commits
PS C:\Users\igor.iric\Desktop\MyFirstRepo> git bisect bad 32e9f11
status: waiting for good commit(s), bad commit known
PS C:\Users\igor.iric\Desktop\MyFirstRepo> git bisect good adb5920
Bisecting: 1 revision left to test after this (roughly 1 step)
[2b407a88fd13d53bdf4ca2682e79a7a7ce968451] Merge pull request #13 from error505/feature/change_portfolio
```

Figure 7.19 – git bisect – good and bad

Like a trusty sidekick, Git transports you to a commit halfway between good and bad. This is your first checkpoint and acts as a potential scene of the crime.

It's time to inspect. You open the HTML file. Is the **My PORTFOLIO** section good here? If it's still not displaying properly, it's a *bad* commit. If it's styled properly, then it's *good*.

You can type `git bisect bad` or `git bisect good` based on what you find. In our case, we found that the **My PORTFOLIO** section still hasn't been implemented.

So, continue by typing `git bisect bad`:

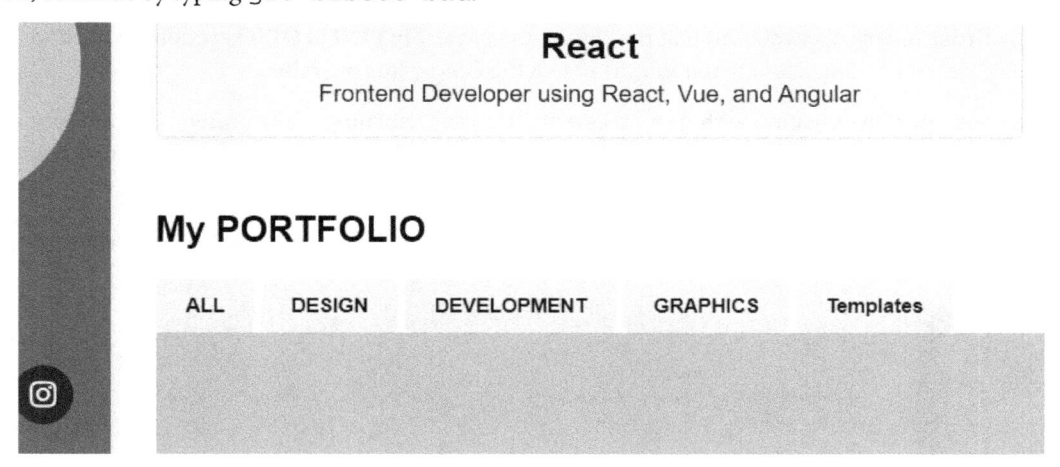

Figure 7.20 – git bisect bad

Each response helps Git eliminate half of the remaining suspects. It's like a high-stakes game of *Guess Who?* Here, Git whisks you to another commit. After a few rounds, you hit a commit where you see the first time the class name changed and caused the **My PORTFOLIO** section to look bad. *"Aha!"*, you exclaim. This is where it all went wrong:

Git History and Reverting Commits

```
● PS C:\Users\igor.iric\Desktop\MyFirstRepo> git bisect bad 32e9f11
  status: waiting for good commit(s), bad commit known
● PS C:\Users\igor.iric\Desktop\MyFirstRepo> git bisect good adb5920
  Bisecting: 1 revision left to test after this (roughly 1 step)
  [2b407a88fd13d53bdf4ca2682e79a7a7ce968451] Merge pull request #13 from error505/feature/change_portfolio
● PS C:\Users\igor.iric\Desktop\MyFirstRepo> git bisect bad
  Bisecting: 0 revisions left to test after this (roughly 1 step)
  [2eed77a02fcaba08b5fc71a46034dcec795bf3e0] Update index.html
● PS C:\Users\igor.iric\Desktop\MyFirstRepo> git bisect bad
  Bisecting: 0 revisions left to test after this (roughly 0 steps)
  [de972569397ccd00b5611ded8af57876c1ac64d5] Merge pull request #12 from error505/feature/portfolio_section
● PS C:\Users\igor.iric\Desktop\MyFirstRepo> git bisect good
  2eed77a02fcaba08b5fc71a46034dcec795bf3e0 is the first bad commit
  commit 2eed77a02fcaba08b5fc71a46034dcec795bf3e0       ← Bad Commit
  Author: Igor Iric <iric_i@hotmail.com>
  Date:   Wed Jan 10 11:10:46 2024 +0100

      Update index.html

      Small change

  index.html | 6 +++---
  1 file changed, 3 insertions(+), 3 deletions(-)
```

Figure 7.21 – git bisect found a bad commit

Git triumphantly announces the first bad commit. You've found the culprit. The commit message gives away the developer's name and the timestamp, the exact moment of the misdeed. By looking at the HTML of your index page, you find that the class name of the **My PORTFOLIO** section changed to `portfolio-sections` and that it doesn't match the CSS styling anymore.

You wrap up your investigation with `git bisect reset`, returning to the present. The mystery has been solved, but there's work to be done. It's time to correct that class name and restore order to your **My PORTFOLIO** section.

Once you've fixed these problems, you find that your **My PORTFOLIO** section looks good again, as shown in *Figure 7.22*:

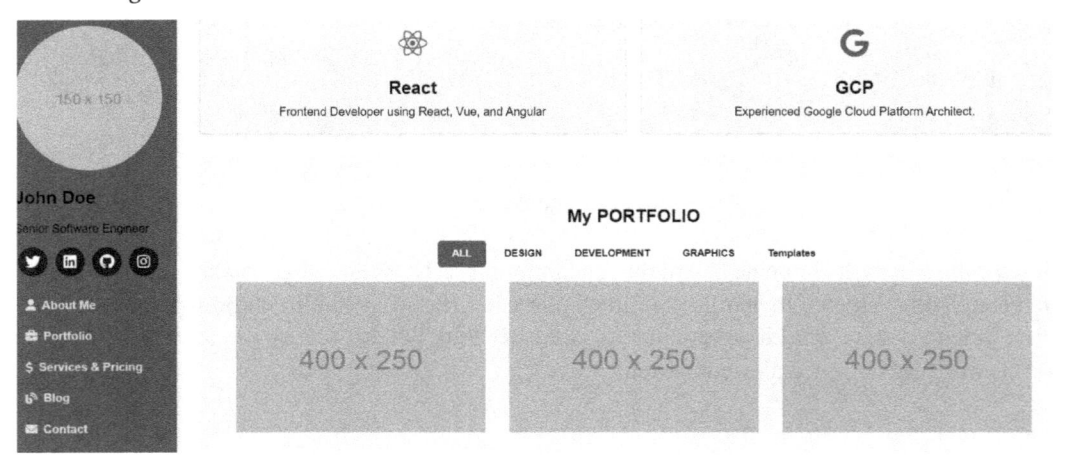

Figure 7.22 – Fixed bad code

Not only did you solve the mystery, but you also mastered a powerful tool in your developer toolkit. The **My PORTFOLIO** section shines again, all thanks to your detective skills and `git bisect`.

The short way of using git bisect – let Git do the detective work

You can tell Git both the bad and good commit right at the start with the `git bisect start <bad-SHA> <good-SHA>` command.

You can use the `git bisect run ls index.html` command to automate the test. Git will use this to check each commit.

If the `index.html` file has changed, the script exits with a code of 0 (all good). If not, it exits with a code between 1 and 127 (problem found):

```
PS C:\Users\igor.iric\Desktop\MyFirstRepo> git bisect start 32e9f11 adb5920
Bisecting: 1 revision left to test after this (roughly 1 step)
[2b407a88fd13d53bdf4ca2682e79a7a7ce968451] Merge pull request #13 from error505/feature/change_portfolio
PS C:\Users\igor.iric\Desktop\MyFirstRepo> git bisect run ls index.html
running 'ls' 'index.html'
index.html
Bisecting: 0 revisions left to test after this (roughly 0 steps)
[560b5b8e63e62f23005e5b5d04e41219ac8f4796] Update index.html
running 'ls' 'index.html'
index.html
32e9f1175311f332861b9908ec77e03610488114 is the first bad commit
commit 32e9f1175311f332861b9908ec77e03610488114
Merge: 2b407a8 560b5b8
Author: Igor Iric <iric_i@hotmail.com>
Date:   Wed Jan 10 11:12:41 2024 +0100

    Merge pull request #14 from error505/feature/chnage_portfolio_item

    Update index.html

 index.html | 4 ++--
 1 file changed, 2 insertions(+), 2 deletions(-)
bisect found first bad commit
```

Figure 7.23 – git bisect found a bad commit

Once Git finds the first bad commit, reset everything with `git bisect reset`.

Instead of manually checking every change, `git bisect` quickly identifies the problem. It finds the exact change that caused the issue, which is super helpful when you're working with lots of code and updates.

Using `git bisect` is like being a detective with a time machine. For your portfolio website, it's a fast way to find out when a problem started and fix it, ensuring your website always shows off your best work! Next, let's learn about another cool feature.

Reverting commits to a previous version

Imagine yourself as a coder with a secret power: the ability to travel back in time within your projects. Your mission: to undo the mishap of the misnamed `portfolio-section` by using the magical `git revert`. It's like having *Ctrl + Z* for your code on a grand scale, allowing you to reverse a mistake and restore to the point where everything was working great:

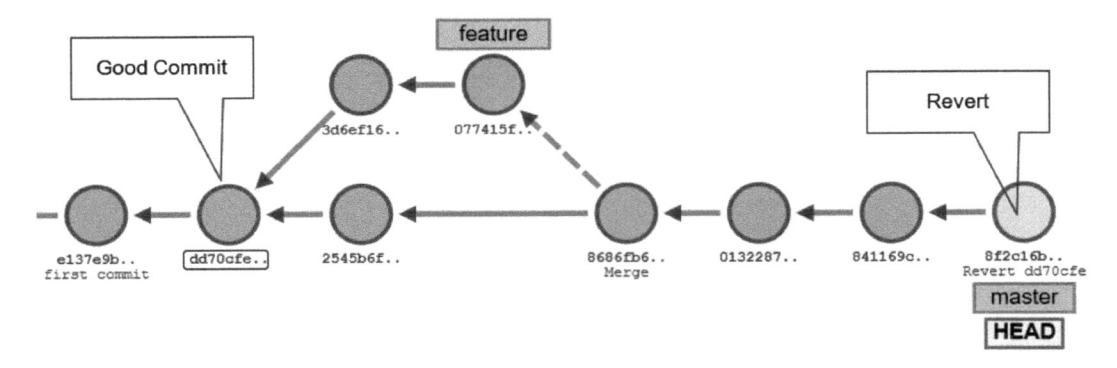

Figure 7.24 – git revert

Your recent adventure with `git bisect` led you to the exact moment when the class name was changed incorrectly. Now, you decide to use `git revert` to rewind history and correct this mistake.

Your detective work with `git bisect` has paid off. You have the commit ID where everything was last perfect, right before the error was introduced.

In your terminal, the gateway to your time machine, type `git revert <SHA-of-bad-commit>`.

This command is like setting the coordinates in your time machine, pinpointing the exact moment you want to undo:

```
PS C:\Users\igor.iric\Desktop\MyFirstRepo> git revert 2eed77a
Auto-merging index.html
hint: Waiting for your editor to close the file...
```

Figure 7.25 – git revert SHA commit

As you hit *Enter*, Git starts its magic. It doesn't delete the bad commit; instead, it creates a new commit. This new commit is the mirror opposite of the bad one, effectively undoing the change:

```
PS C:\Users\igor.iric\Desktop\MyFirstRepo> git revert 2eed77a
Auto-merging index.html
[detached HEAD f5cf36e] Revert "Update index.html"
 1 file changed, 3 insertions(+), 3 deletions(-)
```

Figure 7.26 – git revert finished

It's like adding an anti-paragraph to a novel that cancels out a paragraph you didn't like. If there are conflicts between the present and the past you're trying to restore, Git will ask you to resolve them. If not, it will open a text editor, where you can change the commit message:

```
Revert "Update index.html"

This reverts commit 2eed77a02fcaba08b5fc71a46034dcec795bf3e0.

# Please enter the commit message for your changes. Lines starting
# with '#' will be ignored, and an empty message aborts the commit.
#
# HEAD detached at 560b5b8
# You are currently reverting commit adb5920.
#
# You are currently bisecting, started from branch 'main'.
#
# Changes to be committed:
#    modified:   index.html
#
```

Figure 7.27 – Git commit message

Once resolved, you complete the reversion with a commit. It's your note in the time-travel logbook. With the reversion complete, your code is now as it was in the good old days – `portfolio-section` is back to its original glory.

You check your project, and everything looks just as you remember it, with the CSS styling perfectly in place:

```html
<!-- Portfolio Section -->
<section class="portfolio-section" id="portfolio-section">
    <h1>My PORTFOLIO</h1>
    <div class="portfolio-menu">
        <button class="active">ALL</button>
        <button>DESIGN</button>
        <button>DEVELOPMENT</button>
        <button>GRAPHICS</button>
        <button>Templates</button>
    </div>
```

Figure 7.28 – Git reverted the code to its previous state

You've just done something incredible! You changed the past without altering your journey to the present.

The power of `git revert` lies in its ability to undo while keeping history intact. It's like having a safety net, ensuring that you can always go back and fix mistakes, no matter how far along you are in your project.

You've not only fixed a bug but also mastered a vital skill in your coding toolkit. Each command in Git is like a spell, giving you control over the fabric of your project's history. With `git revert`, you've experienced firsthand the power of responsible time travel in coding.

Now that you've harnessed the power of `git revert` to fix a mistake without rewriting history, let's move on and explore the power of `git diff`, which helps you find differences between code.

Unraveling mysteries with git diff – the tale of the unseen changes

You're working on your one-page portfolio website and you've just added a new section called **My PORTFOLIO**. Now, you're curious to see what changes you've made compared to the previous version. In the world of Git, this is where the `diff` command comes into play. It's like having a magnifying glass to spot the differences between your older website and the new one within the **My PORTFOLIO** section.

What is the diff command?

The `diff` command in Git shows you the differences between your current code and the last committed version. It's like comparing two photos of your room, one before and one after a redecoration, to see exactly what's changed.

Your command line is the place where you will use the `diff` command to investigate the changes on your one-page portfolio website. Make sure you're in the directory (folder) of your website project on your computer.

Type `git diff` and press *Enter*. This command asks Git to show you the differences between your last saved (committed) version and what you have now.

What you'll see is a list of changes: lines added will be marked with a plus (+) and lines removed will have a minus (-). The output might look a bit confusing at first, but it shows you exactly what's different.

For example, if you added a new paragraph about your skills in the **My PORTFOLIO** section, `git diff` will show this new paragraph with a plus sign next to it, as shown in *Figure 7.29*:

```
 PS C:\Users\igor.iric\Desktop\MyFirstRepo> git diff
diff --git a/index.html b/index.html
index c649e73..a365090 100644
--- a/index.html
+++ b/index.html
@@ -123,16 +123,16 @@
             <!-- ... Add other services similarly ... -->
         </section>
         <!-- Portfolio Section -->
-       <section class="portfolio-sections" id="portfolio-section">
+       <section class="portfolio-section" id="portfolio-section">^M
            <h1>My PORTFOLIO</h1>
-           <div class="portfolio-menus">
+           <div class="portfolio-menu">^M
                <button class="active">ALL</button>
                <button>DESIGN</button>
                <button>DEVELOPMENT</button>
                <button>GRAPHICS</button>
                <button>Templates</button>
:...skipping...
diff --git a/index.html b/index.html
index c649e73..a365090 100644
```

Figure 7.29 – The result of running the git diff command

Other diff commands

Now, let's learn about some other `diff` commands that can help you discern the differences between two versions of code:

- `git diff --staged`: This command shows the changes that you've staged (added to your next commit) but haven't committed yet. It's like looking at a list of items you've packed for a trip but haven't put in your suitcase yet:

```
        </div>
        <div class="portfolio-item" id="design">
           <img src="https://via.placeholder.com/400x250" alt="Description">
           <div class="overlay">
-              <h3>3D Graphics</h3>
+              <h3>IaC</h3>^M
               <h4>Templates</h4>
           </div>
        </div>
        <div class="portfolio-item" id="design">
           <img src="https://via.placeholder.com/400x250" alt="Description">
           <div class="overlay">
-              <h3>3D Graphics</h3>
+              <h3>DevOps</h3>^M
               <h4>Templates</h4>
           </div>
        </div>
```

Figure 7.30 – The result of running the git diff --staged command

- `git diff HEAD`: This shows all the changes in your working directory since the last commit, whether they're staged or not. *HEAD* refers to the latest commit on your current branch. It's like taking a step back to see all the changes you've made in your room since the last time you cleaned it.

- `git diff <REF-1> <REF-2>`: You can use this command if you want to compare your current website to an older version with two references (such as commit IDs or branch names). For example, `git diff main my-portfolio` compares the main branch with a branch called `my-portfolio`:

```
PS C:\Users\igor.iric\Desktop\MyFirstRepo> git diff main feature/webdevelopment-skill-in-what-i-do-section
diff --git a/index.html b/index.html
index c649e73..5be96e9 100644
--- a/index.html
+++ b/index.html
@@ -1,186 +1,111 @@
-<!DOCTYPE html>
-<html lang="en">
-  <head>
-    <meta charset="UTF-8" />
-    <meta name="viewport" content="width=device-width, initial-scale=1.0" />
-    <title>John Doe - Senior Software Engineer</title>
-    <link
-      rel="stylesheet"
-      href="https://cdnjs.cloudflare.com/ajax/libs/font-awesome/6.0.0-beta3/css/all.min.css"
-    />
-    <link rel="stylesheet" href="styles.css" />
-  </head>
```

Figure 7.31 – The result of running the git diff <REF-1> <REF-2> command

- `git diff origin/main main`: This command compares local and remote branches, which lets you see the difference between your local *main* branch and the *main* branch on GitHub (remote). It's like comparing the blueprint of your house that's with you (local) and the one you sent to your architect (remote).

Alternatively, let's say you want to compare specific commits. Sometimes, you want to compare two specific points in your project's history. You can use their commit IDs (those long alphanumeric codes) to see what's changed between those two points.

Each variation of the `git diff` command offers a unique way to inspect the changes in your one-page portfolio website. Whether you're looking at recent edits, preparing for your next big update, or comparing different versions, these commands help you stay on top of your project's evolution. You've not only uncovered the mysteries hidden in your code but also gained a deeper understanding of your project's evolution.

Now that we've explored how to spot the differences in your project with `git diff`, let's shift gears. Coming up, we'll navigate how to step back and tweak those changes, or even pick out the best ones, using `git reset` and cherry-picking. It's about fine-tuning your project, ensuring every addition fits perfectly into your website's big picture.

Undoing changes with git reset and cherry-picking

Imagine that you're working on adding a **My PORTFOLIO** section to your one-page website, but you realize you need to undo some changes or pick specific changes from your project's history. This is where Git's **reset** and **cherry-pick** features come in handy. It's like a time machine that takes your project back to a specific point in history before certain changes were made.

Types of git reset

There are three different kinds of `git reset`. We will explain all of them:

- **Soft reset**: The `git reset --soft` command moves your project back but keeps your recent changes in the staging area (like keeping your unsaved work). You can then re-commit if you want.
- **Mixed reset**: The `git reset --mixed <SHA>` command is the default reset. It moves your project back to a previous state (the state is identified by a unique code called SHA) and puts your changes in the working directory, unsaved. You can then decide what to keep and what to discard.
- **Hard reset**: The `git reset --hard <SHA>` command is a complete rewind. It takes your project back to a previous state and permanently removes all changes made after that point. It's like hitting a reset button and losing unsaved changes.

Picking specific changes with cherry-picking

Imagine going through a photo album and picking out specific photos. Cherry-picking in Git lets you select specific changes from the project's history and apply them to your current version. You can use the unique SHA code of the change you want to apply by running `git cherry-pick <SHA>`.

Let's quickly explain the concept of **cherry-picking** in Git. It is a version control system that's used for tracking changes in files and coordinating work among multiple people. *Figure 7.32* demonstrates this further:

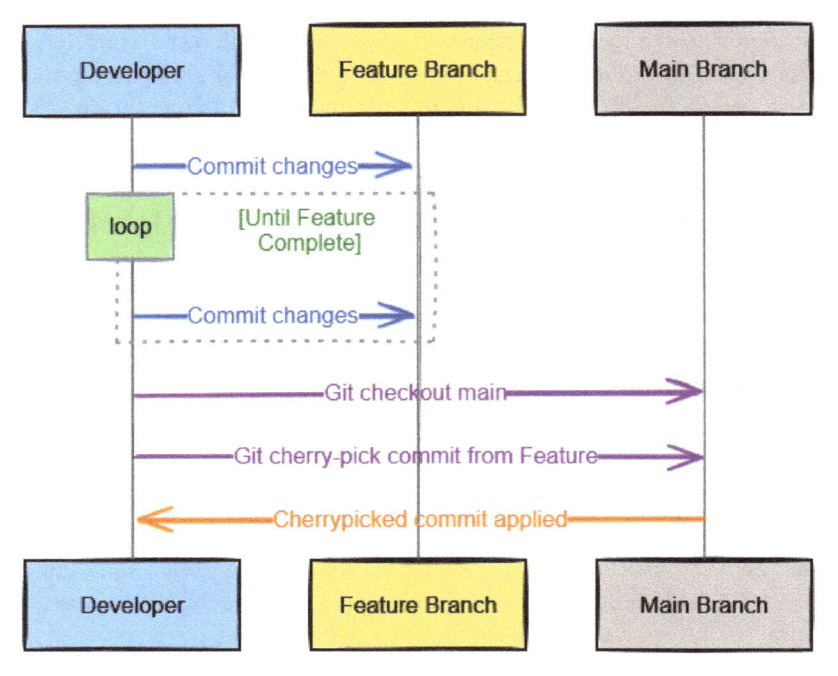

Figure 7.32 – Git cherry-picking

Let's decode the terms shown in the preceding figure in detail:

- The **feature branch** is like having a separate workspace or a sketchpad where a developer is drawing or writing something new – in this case, a new feature for your one-page portfolio website.

- The **main branch** is the final draft of your website that everyone sees. It's the polished version where all the fully finished features are displayed.

The sequence of events is as follows:

1. The developer works in their feature branch, making changes (called **commits**). Each commit is like saving a new version of your website design. As the developer makes more changes, they save more commits. They're in a **loop** of making changes and saving commits until the feature is complete.

2. At some point, the developer or their team decides that one specific change (a single commit) from the feature branch is ready to be added to the main branch.

3. The developer switches to the main branch using the `git switch main` command. This is like saying, "*Okay, let's go to the final draft.*"

4. Then, they use the `git cherry-pick` command with the specific commit from the feature branch. This is like saying, "*Let's take this one good idea from my sketchpad and put it into the final draft.*"

5. The cherry-picked commit is applied to the main branch. It's like pasting a sticker from the sketchpad onto the final draft.
6. The main branch now has the new change, and the feature branch still exists with all the other changes that are perhaps still being worked on.

The *loop* signifies that the development process is iterative. Developers often make many changes and may cherry-pick any of those changes at any time to add to the main branch.

Cherry-picking can be used to quickly include a specific fix or feature from a series of commits without merging all the changes from the feature branch. This can be useful for managing what is ready to be included in the main branch of your website.

Summary

Congratulations on completing *Chapter 7*! You've just journeyed through some of the most powerful aspects of Git, learning to control and navigate the history of your coding projects. Let's take a moment to reflect on the key skills and concepts you've mastered.

Here's a recap of what you've accomplished.

- **Understanding Git's history**: You dove into the depths of `git log`, exploring various ways to view your project's timeline. You've seen how `git log --oneline`, `git log --graph`, and `git log --decorate` can transform your understanding of project history from a simple list to a detailed and informative map.
- **git bisect – a detective tool**: The `git bisect` command became your detective tool, helping you to efficiently track down the exact commit that introduced a bug.
- **Reverting and resetting**: You've learned the art of undoing changes with `git revert` and `git reset`.
- **Exploring changes with diff**: The `diff` command opened up new ways to compare changes between commits, branches, and your working directory. You've gained the ability to spot differences at a granular level, enhancing your insight into every modification.
- **Cherry-picking – selective integration**: Cherry-picking taught you to selectively integrate changes from different branches.

As you close this chapter, remember that the skills you've acquired are not just about managing code – they're about managing time and change.

The next chapter will open the door to a treasure trove of tools and commands that will elevate your Git and GitHub experience. We will cover a varied range of important topics, such as Sourcetree, GitHub Codespaces, GitHub Desktop, Visual Studio Code, and DevHub notifications.

Quiz

Answer the following questions:

1. What does `git log` do in Git?

    A. Deletes old commits

    B. Shows the history of commits

    C. Changes the latest commit

    D. Creates a new branch

    Answer: B. Shows the history of commits.

2. True or false: The `git reflog` command can be used to see a list of actions (such as commits and merges) made in a repository.

    A. True

    B. False

    Answer: A. True.

3. `git _____` is a tool that's used to find the commit that introduced a bug by automatically testing commits.

    Answer: `bisect`.

4. What is the purpose of reverting a commit in Git?

    Answer: Reverting a commit is used to undo changes made by a specific commit, bringing the project back to its previous state without altering the project's history.

5. What does the `git diff` command show?

    A. Differences between branches

    B. Differences between commits

    C. Current changes in your working directory

    D. All of the above

    Answer: D. All of the above.

6. True or false: Once a commit is reverted, it is completely removed from the project's history.

    A. True

    B. False

    Answer: B. False (the commit is still in the history, but its changes are undone).

7. What is cherry-picking in Git?

    A. Deleting specific commits

    B. Choosing and applying specific commits from one branch to another

    C. Creating a new repository

    D. Renaming a branch

    Answer: B. Choosing and applying specific commits from one branch to another.

8. To undo changes made by a commit and keep those changes in the working directory, you would use `git reset --_____`.

    Answer: `soft`.

9. True or false: The `git bisect` command automatically fixes the bugs it finds in code.

    A. True

    B. False

    Answer: B. False (it helps identify the problematic commit, but it doesn't fix the bug).

10. What's the main difference between `git reset` and `git revert`?

    Answer: The `git reset` command is used to undo changes by moving the current branch to a specific commit, potentially altering the project's history, while `git revert` creates a new commit that undoes the changes of a previous commit without altering the project's history.

Enhanced challenge – Space Explorer splash screen, score tracking, and mastering Git commands

Implement a splash screen and score tracking in *Space Explorer* while practicing advanced Git commands for reverting, resetting, and cherry-picking:

Figure 7.33 – Space Explorer start screen

Here are your updated steps to success:

1. **Initial setup**: Just as before, create a new branch (`feature-splash-and-score`) and start implementing the splash screen and score tracking.

2. **Intentional "mistakes"**: After your initial implementation, make a couple of intentional "mistakes" or suboptimal changes in your code. Commit these changes.

3. **Use git revert**: Identify one of the intentional mistakes you made and use `git revert` to undo that specific commit. This will help you understand how to backpedal changes that have been recorded in Git's history.

4. **Experiment with git reset**: For a different mistake, use `git reset` to undo the changes. You can try using `git reset --soft` or `git reset --hard` (be cautious with `--hard` as it will permanently delete your changes) to understand how they differ in handling the project history.

5. **Cherry-pick a feature**: Create a new branch (for example, `additional-feature`) and add a small but noticeable new feature or enhancement to the splash screen or score system.

   Once committed, switch back to your `feature-splash-and-score` branch and use `git cherry-pick` to selectively apply the new feature from `additional-feature` to your main working branch.

6. **Final test and merge**: Thoroughly test all features, including the splash screen functionality and score tracking, to ensure everything works seamlessly.

   Resolve any merge conflicts if they arise when merging into the main branch using your newly acquired skills.

7. **Reflection and documentation**:

   I. Document your process and challenges in a README file or in commit messages.

   II. Reflect on how using these Git commands aided in your development process.

You can download the code from `https://github.com/PacktPublishing/GitHub-for-Next-Generation-Coders/tree/main/Space%20Explorer%20Game/Game%20Obsticles%20-%20Splash%20screen%20and%20player%20name`.

8
Helpful Tools and Git Commands

Welcome to *Chapter 8*, where we'll venture deeper into the world of Git and GitHub, unlocking new levels of efficiency and collaboration for your one-page portfolio website. You've learned the basics, and now it's time to harness the full potential of some powerful tools and commands.

In this chapter, we will learn about helpful Git commands that can create shortcuts, trace a file's evolution, rearrange your commits, and perform more such tasks. We will then look at Sourcetree and how it can transform the complexity of Git into a visual map of your project's journey. We will learn how to code from anywhere using Codespaces, manage repositories with GitHub Desktop, and set up DevHub, a personal desktop assistant that keeps you informed about all the activity in your GitHub projects.

Here's a sneak peek at what we'll explore:

- Crafting shortcuts and keeping it clean – advanced Git commands
- Streamlining your website work with Sourcetree
- Navigating your website project with GitHub Desktop
- Crafting your website with GitHub Codespaces
- Managing your project's buzz with DevHub

Crafting shortcuts and keeping it clean – advanced Git commands

While building your website, you might find yourself repeating some long Git commands. Wouldn't it be easier to have shortcuts? Or to clean up unnecessary files with just a command? This section will cover some advanced commands that can do just that and more.

Working on your one-page portfolio website can sometimes feel like juggling. You've got new ideas you want to try, you're fixing bugs, and at the same time, you're keeping track of all the changes. Here are some Git commands that act like clever tools to help you juggle more efficiently.

- **Creating your own shortcuts with git alias**: Use `git config --global alias.[alias_name] [git_command]` to create shortcuts for commands you use often. For example, if you frequently check the status, you can set up an alias such as `git config --global alias.st status`.

   Now, instead of typing `git status` each time, you just type `git st`:

   ```
   PS C:\Users\iric_\Desktop\GitHub for Young Coders\MyFirstRepo> git st
   On branch main
   Your branch is up to date with 'origin/main'.

   You are currently bisecting, started from branch 'main'.
     (use "git bisect reset" to get back to the original branch)

   Changes to be committed:
     (use "git restore --staged <file>..." to unstage)
           modified:   index.html
   ```

   Figure 8.1 – Git alias status

- **Tracking file history across changes with git log follow**: When you want to see the changes made to a specific file over time, use `git log --follow [file]`. It's like having a detailed timeline for a single piece of your website's code.

   For example, let's see the log of the `index.html` file and type `git log --follow index.html`. What you will get is the detailed history of the `index.html` file, as shown in *Figure 8.2*:

```
PS C:\Users\iric_\Desktop\GitHub for Young Coders\MyFirstRepo> git log --follow index.html
commit 560b5b8e63e62f23005e5b5d04e41219ac8f4796 (refs/bisect/good-560b5b8e63e62f23005e5b5d04e41219ac8f4796)
Author: Igor Iric <iric_i@hotmail.com>
Date:   Wed Jan 10 11:12:22 2024 +0100

    Update index.html

    Changing Portfolio Items

commit 2eed77a02fcaba08b5fc71a46034dcec795bf3e0
Author: Igor Iric <iric_i@hotmail.com>
Date:   Wed Jan 10 11:10:46 2024 +0100

    Update index.html
```

Figure 8.2 – git log follow

- **Rearranging your work with git rebase**: The `git rebase [branch]` command allows you to move or combine a sequence of commits to a new base commit. It's a bit like rearranging the pages in your website's story to make the narrative flow better.

- **Cleaning up untracked files with git clean**: To remove untracked files from your working directory, `git clean -fd` is your go-to command. Think of it as keeping your workspace clean by removing all the scraps of paper and unused sketches that didn't make it to your website.

- **Saving work in progress with git stash**: Imagine you're making new designs on your website and someone suddenly asks you to add some new feature. You need a quick way to hide your design and clean your local repo without losing what you have already done. `git stash` is like that *save* option; it takes all your changes and temporarily hides them, leaving you with a clean working directory.

- **Bringing back stashed changes with git stash pop**: After your feature is implemented and you want to get back to designing, `git stash pop` removes that cover and brings back your work. It's like uncovering your CSS design and images and getting right back to where you left off.

- **Marking milestones with git tag**: When you reach a significant point in your website development, such as completing the **About** section, you might want to mark it. `git tag [tagname]` is like putting a sticky note on that point in your timeline so you can easily find it later.

- **Finding out who made changes with git blame**: If you notice something has changed on your website and you're not sure who did it, `git blame [file]` will show you who last edited each line of a file and when. It's like doing detective work to figure out who moved or removed your design.

Let us, for example, try the `index.html` file to see what result we will get. After typing `git blame index.html`, you will see the whole file and each line with the name of the person who made additions or changes:

```
PS C:\Users\iric_\Desktop\GitHub for Young Coders\MyFirstRepo> git blame index.html
adb5920d (Igor Iric    2024-01-10 11:02:21 +0100   1) <!DOCTYPE html>
adb5920d (Igor Iric    2024-01-10 11:02:21 +0100   2) <html lang="en">
adb5920d (Igor Iric    2024-01-10 11:02:21 +0100   3)   <head>
adb5920d (Igor Iric    2024-01-10 11:02:21 +0100   4)     <meta charset="UTF-8" />
adb5920d (Igor Iric    2024-01-10 11:02:21 +0100   5)     <meta name="viewport" content="width=device-width, initial-scale=1.0" />
adb5920d (Igor Iric    2024-01-10 11:02:21 +0100   6)     <title>John Doe - Senior Software Engineer</title>
adb5920d (Igor Iric    2024-01-10 11:02:21 +0100   7)     <link
adb5920d (Igor Iric    2024-01-10 11:02:21 +0100   8)       rel="stylesheet"
adb5920d (Igor Iric    2024-01-10 11:02:21 +0100   9)       href="https://cdnjs.cloudflare.com/ajax/libs/font-awesome/6.0.0-beta3/css/all.min.css"
adb5920d (Igor Iric    2024-01-10 11:02:21 +0100  10)     />
adb5920d (Igor Iric    2024-01-10 11:02:21 +0100  11)     <link rel="stylesheet" href="styles.css" />
adb5920d (Igor Iric    2024-01-10 11:02:21 +0100  12)   </head>
adb5920d (Igor Iric    2024-01-10 11:02:21 +0100  13)
adb5920d (Igor Iric    2024-01-10 11:02:21 +0100  14)   <body>
adb5920d (Igor Iric    2024-01-10 11:02:21 +0100  15)     <div class="container">
adb5920d (Igor Iric    2024-01-10 11:02:21 +0100  16)       <aside class="sidebar">
adb5920d (Igor Iric    2024-01-10 11:02:21 +0100  17)         <img src="https://via.placeholder.com/150" alt="John Doe" class="profile-image" />
adb5920d (Igor Iric    2024-01-10 11:02:21 +0100  18)         <h2>John Doe</h2>
adb5920d (Igor Iric    2024-01-10 11:02:21 +0100  19)         <p>Senior Software Engineer</p>
adb5920d (Igor Iric    2024-01-10 11:02:21 +0100  20)         <!-- Social Media Icons -->
```

Figure 8.3 – The git blame result

- **Getting updates from all branches with git fetch --all**: The `git fetch --all` command is like checking all your mailboxes for messages. It gets the latest updates from all the branches in your GitHub repository without merging them into your work. You can see what everyone else has been doing before deciding what to merge into your own branch.

- **Checking your remote repositories with git remote**: To see a list of the remote connections you have to other versions of your repository, `git remote -v` is like having an address book. It shows you where you can push your changes (send your finished feature) or fetch from (get a new feature).

These advanced commands give you more flexibility and efficiency in managing your website project. They're like having a set of power tools in your workshop, making your work cleaner and your efforts more streamlined.

They make sure you can save your work, keep track of contributions, mark important milestones, and stay up to date with everyone's work.

It's all about making your website development process as smooth and stress-free as possible! In the next section, we will learn more tools that can help us with managing Git and repositories.

Streamlining your website work with Sourcetree

While you're creating your one-page portfolio website, instead of manually handling all the updates and changes using command-line Git commands, you have a tool that visually lays everything out for you. That tool is Sourcetree by Atlassian.

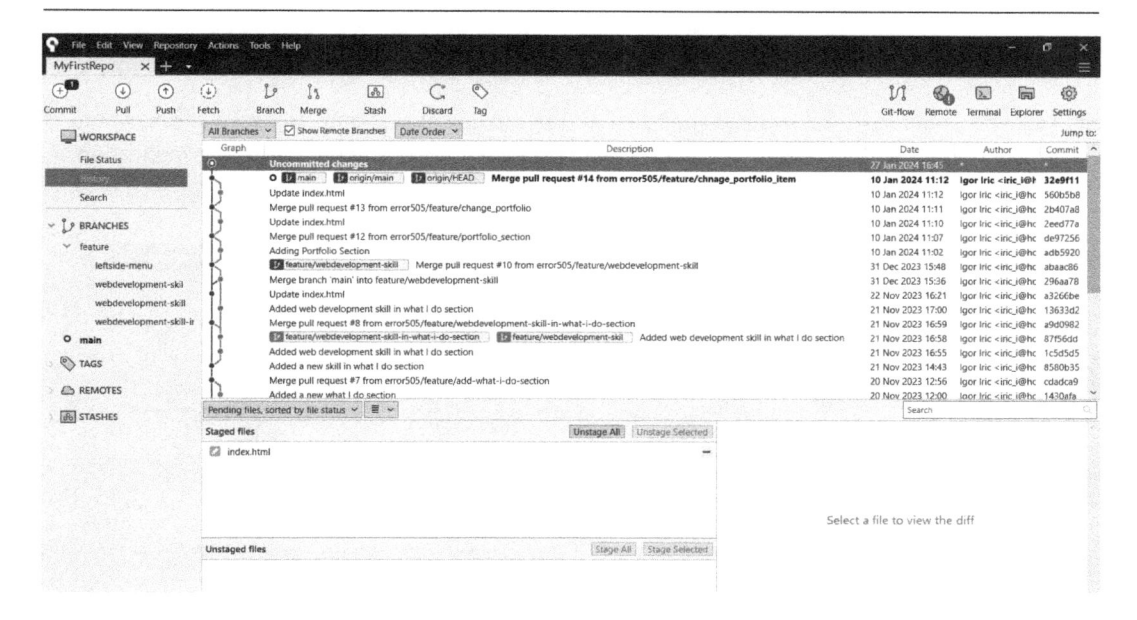

Figure 8.4 – Sourcetree

Why should you use Sourcetree?

Sourcetree presents your project in a way that's easy to understand. It's like having a map of all your work, from the big picture down to the details.

If typing out Git commands feels like trying to order coffee in a foreign language, Sourcetree is like having a friendly barista who knows your regular order. Just point and click.

Before you commit, you can review your changes in Sourcetree, ensuring you know exactly what's being updated. Sourcetree also makes it easy to see who's working on what, making team collaboration a breeze.

With all actions being visual and clear, there's less room for error. You're less likely to accidentally merge the wrong branch or commit the wrong file.

Sourcetree can turn the complex web of Git commands into a visually manageable process, perfect for focusing on building a stunning one-page portfolio website without getting tangled up in command-line syntax. It's a tool that can make managing your project's development as easy and intuitive as editing your website's design in a visual editor.

Getting started with Sourcetree

First of all, you would have to download and install it. You can download Sourcetree if you head over to the Sourcetree website (https://www.sourcetreeapp.com/) and download the application. Installing it is as straightforward as setting up any other software on your computer.

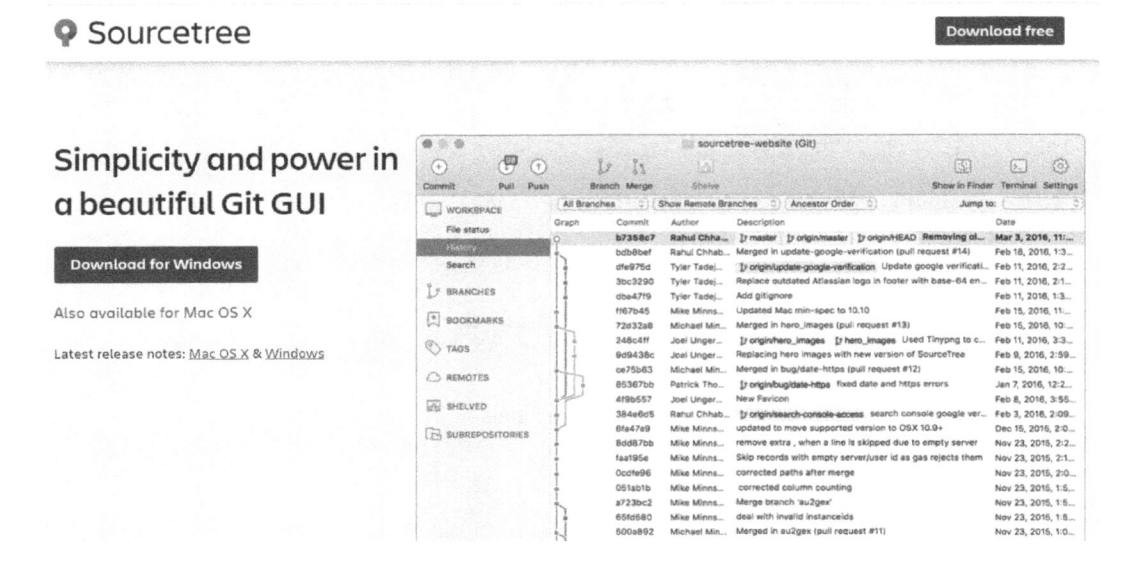

Figure 8.5 – Downloading Sourcetree

Connecting your repository

Once Sourcetree is up and running, you can connect it to your GitHub account, which holds your website's repository. It's like telling your navigation app where you want to go.

Streamlining your website work with Sourcetree   205

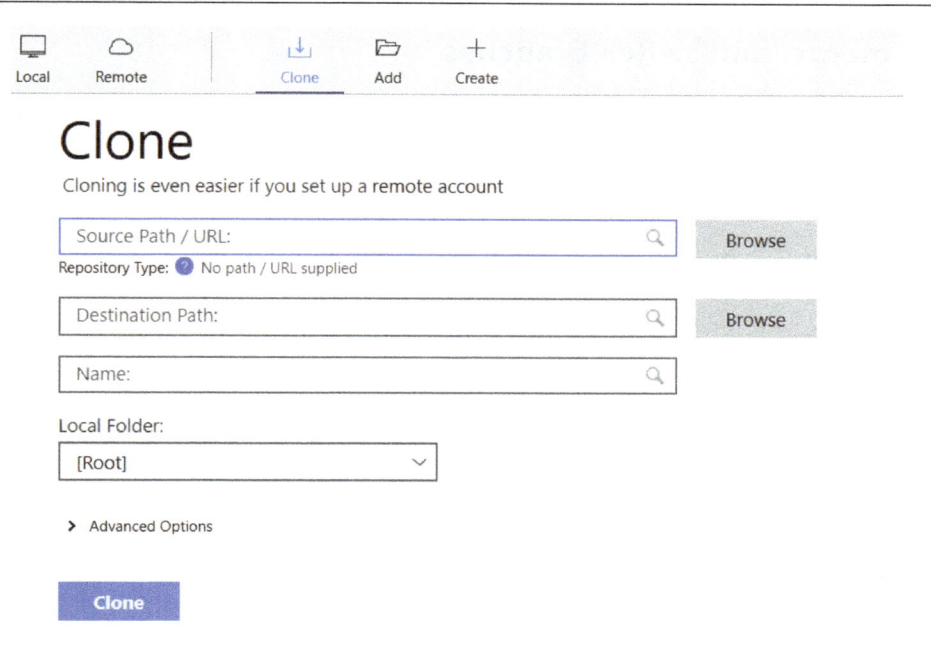

[Figure 8.6 – Setting up your repo with Sourcetree]

Viewing your project timeline

Once you've set up your repo, Sourcetree will show you the history of your project. Each commit, branch, and tag is displayed in an interactive timeline. It's like having a storyboard of your website's development.

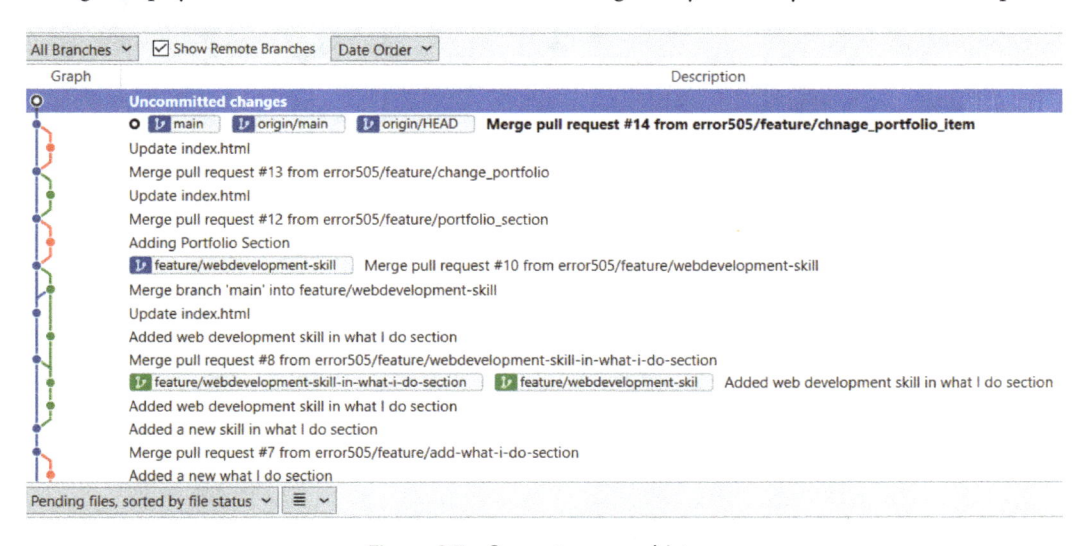

Figure 8.7 – Sourcetree repo history

Create, merge, and switch branches

Want to work on a new feature or fix a bug? The options to undertake the processes of **create**, **merge**, and **switch** branches are easy to find, as shown in *Figure 8.8*.

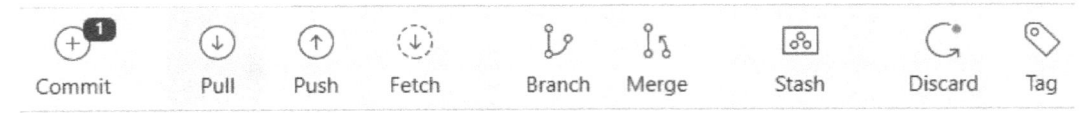

Figure 8.8 – Sourcetree quick commands

In Sourcetree, you can create new branches by clicking on the **Branch** button, as shown in *Figure 8.9*.

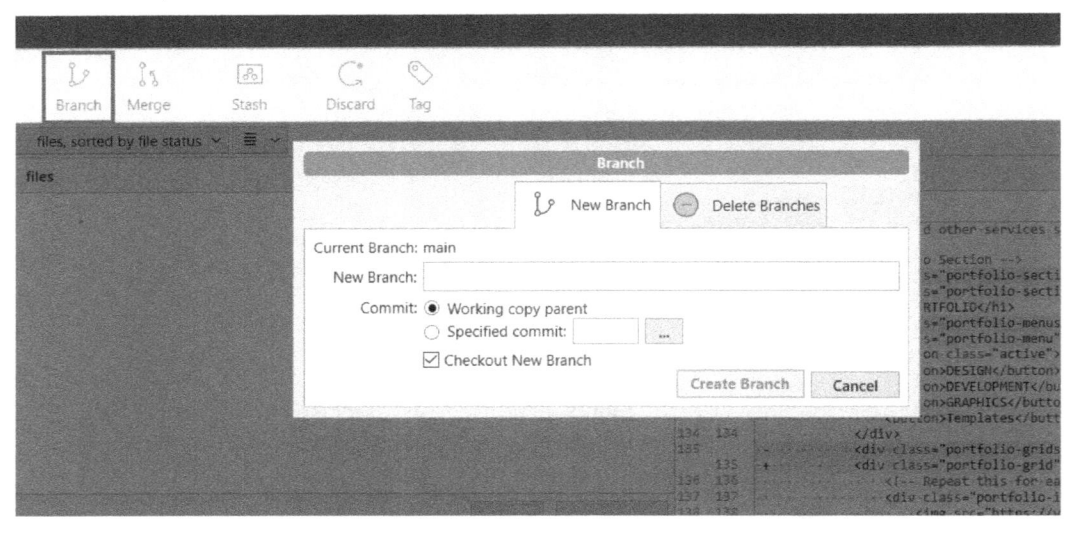

Figure 8.9 – Creating a new branch in Sourcetree

Merging them is just as easy, by right-clicking on the branch you want to merge into your current branch, as shown in *Figure 8.10*.

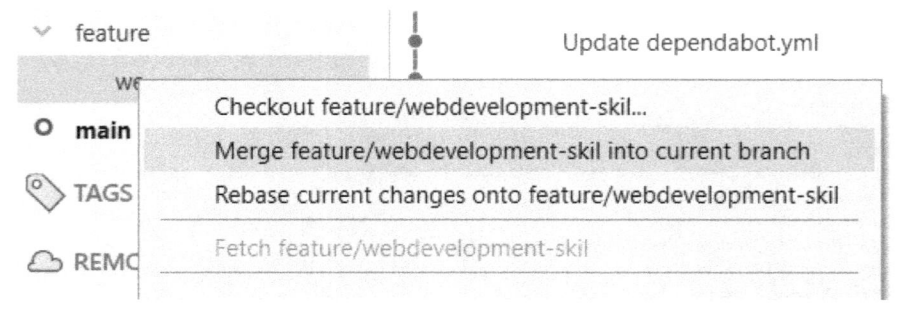

Figure 8.10 – Merging the target branch into the current branch

Switching between branches is as simple as switching TV channels; you can do it simply by double-clicking on the branch you want to switch. Simple as that.

Staging and committing changes

When you make changes to your website, such as adding a new section or updating a picture, Sourcetree helps you organize these changes in a simple way. Imagine you're sorting items into a box before mailing them. Sourcetree shows you all the changes you've made, allowing you to choose (or *stage*) the ones you're ready to keep. Once you've picked all the items you want to send, hitting the **Commit** button in Sourcetree is like sealing the box and mailing it. This action updates your project with your latest changes, easily and efficiently.

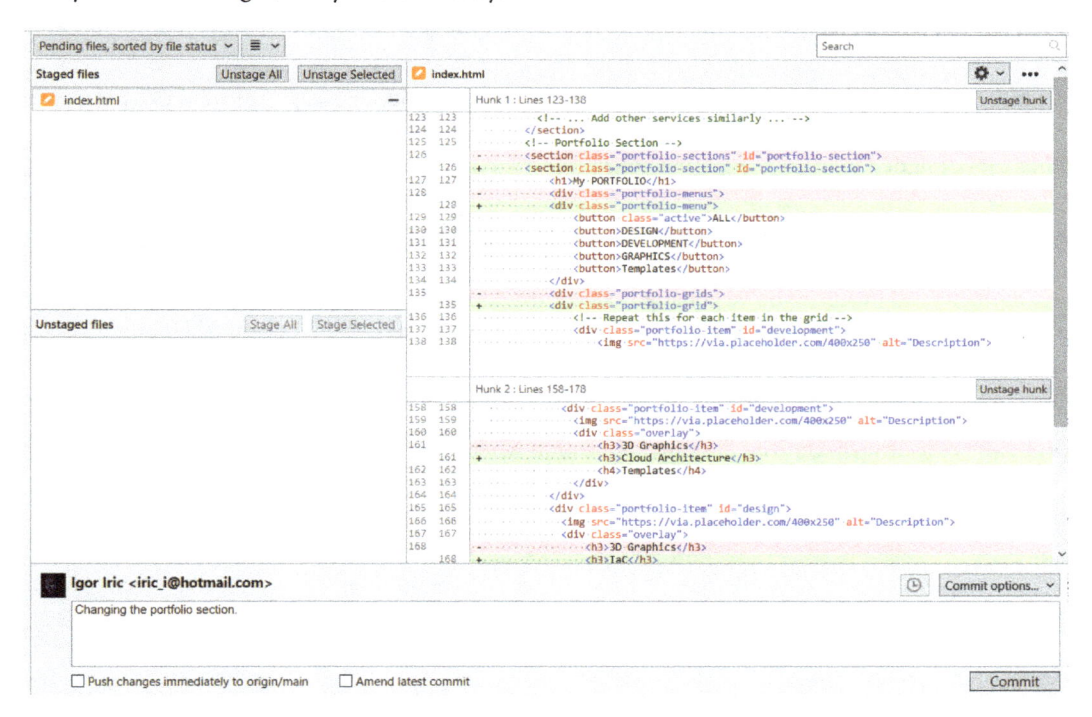

Figure 8.11 – Sourcetree stage and commit changes

Pull, Fetch, and Push

Need to update your local repository with changes from others? Or send your updates to the team? Sourcetree turns these actions into simple button clicks.

The **Pull** button in Sourcetree performs the same action as executing the `git pull` command in Git Bash, updating your local repository with changes from the remote. Similarly, the **Push** button automates the `git push` command, sending your local updates to the remote repository. For retrieving updates without merging them into your local branch, the **Fetch** button in Sourcetree mirrors the `git fetch` command, allowing you to see what's new before deciding to integrate those changes.

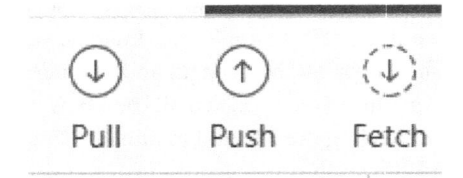

Figure 8.12 – Sourcetree Pull, Push, and Fetch buttons

Now that you've explored how Sourcetree can simplify managing your one-page portfolio website with its visual interface and straightforward controls, let's continue. Next, we'll look at GitHub Desktop, another tool designed to make navigating your project's development easier.

Navigating your website project with GitHub Desktop

Take a moment and imagine you've got a remote control for your one-page portfolio website project. This isn't just any remote; it's one that lets you manage all the updates, changes, and collaborations as if you were flipping through TV channels. That's what GitHub Desktop is like.

GitHub Desktop is like a friendly guide in the world of Git. It takes away the complexity of command lines and offers a clean, organized desktop application where you can manage your project easily.

Setting up GitHub Desktop

Visit the GitHub Desktop website (`https://desktop.github.com/`) and download the app. You can install it as you do any new app on your computer. Then, follow these steps:

1. **Connect to your GitHub repository**: Once installed, open GitHub Desktop and sign in with your GitHub account. It's like logging in to your email on a new device.

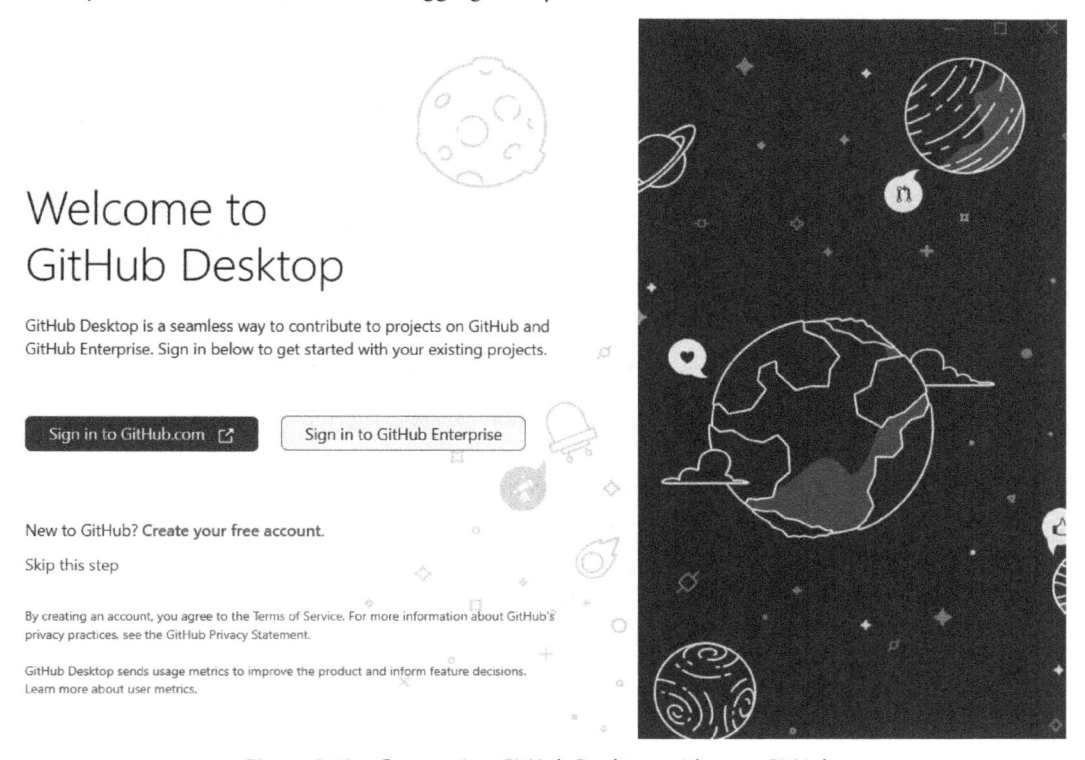

Figure 8.13 – Connecting GitHub Desktop with your GitHub

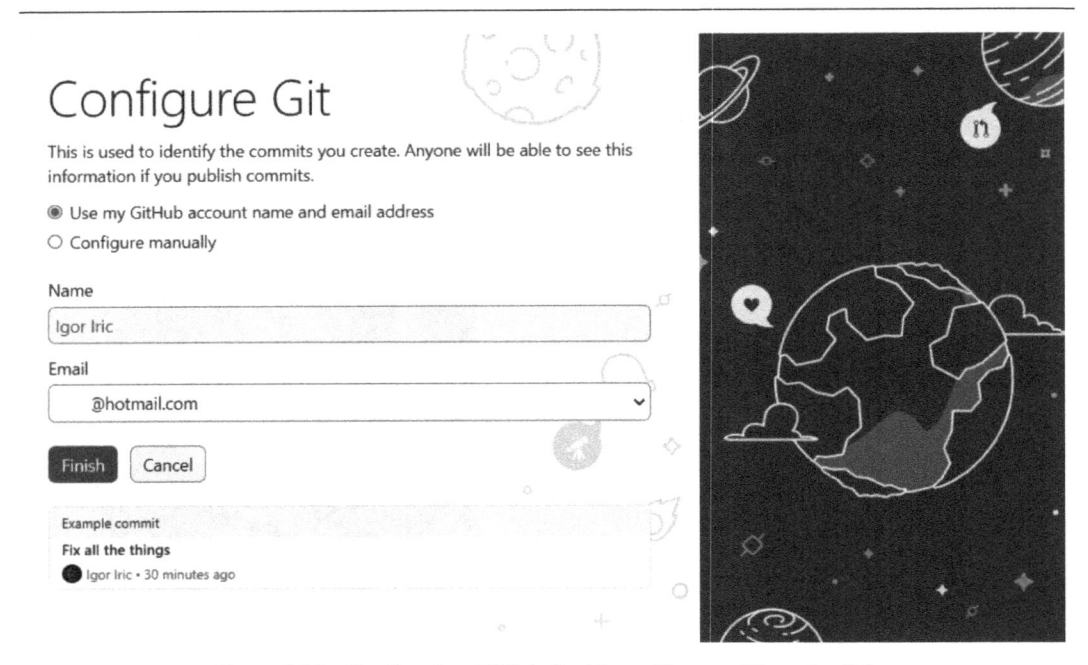

Figure 8.14 – Configuring GitHub Desktop with your Git credentials

2. **Clone your repository**: Find the **Clone a repository** option and select your one-page portfolio website. This will create a copy of your project on your local machine, much like saving a document to work on it offline.

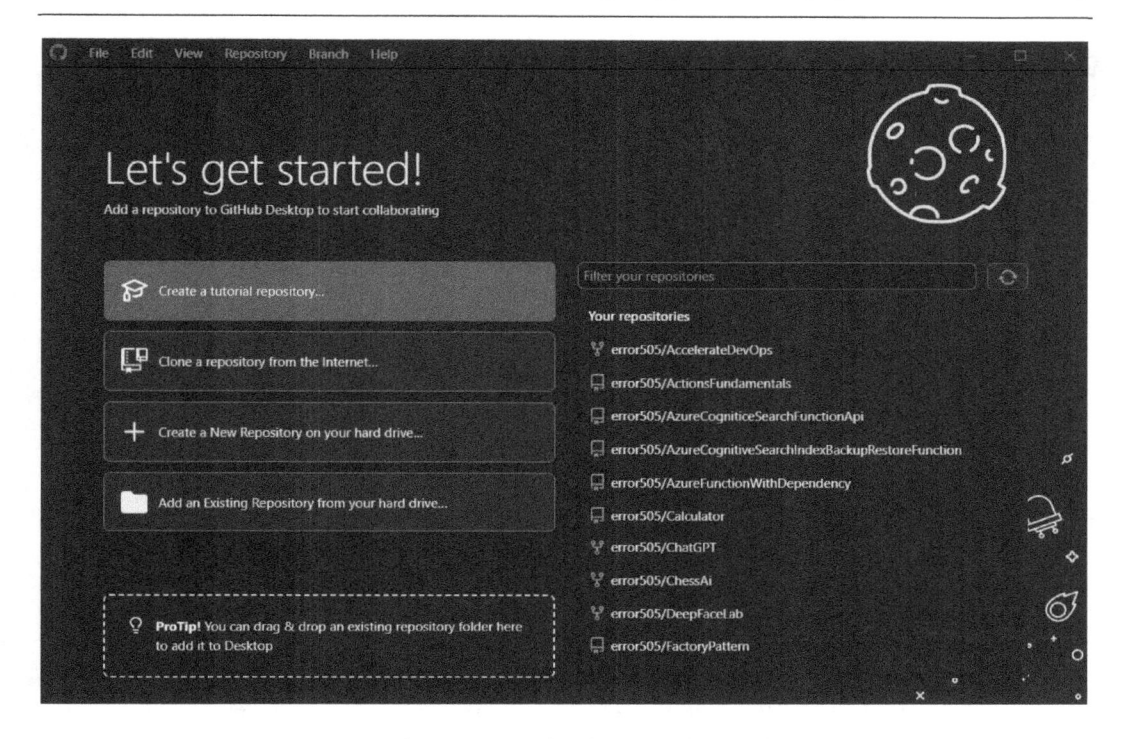

Figure 8.15 – Choosing repo for cloning with GitHub Desktop

What can you do with GitHub Desktop?

There are numerous features offered by GitHub Desktop. Let's look at some of these:

- **Check your changes**: Did you make some updates to your website? GitHub Desktop shows you a list of files you've changed and exactly what's different in each one. It's like tracking changes in a Word document.

212　Helpful Tools and Git Commands

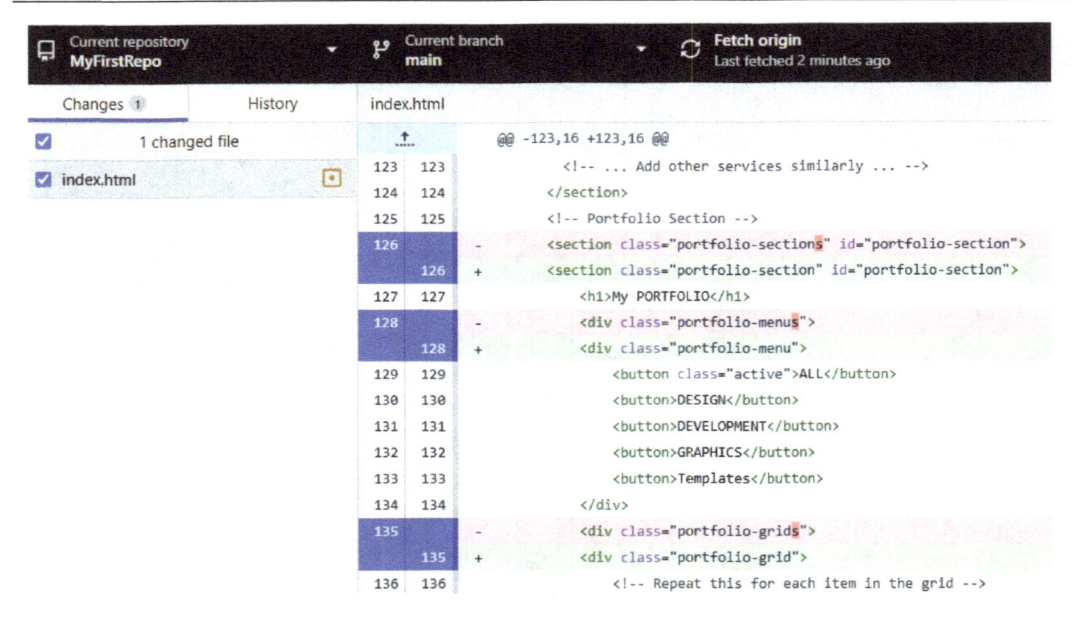

Figure 8.16 – Cloning repo with GitHub Desktop

- **Commit to your history**: When you're ready to save your changes, add a descriptive message and commit. This is like signing off on your work, saying, "*This is ready to be added to the project.*"

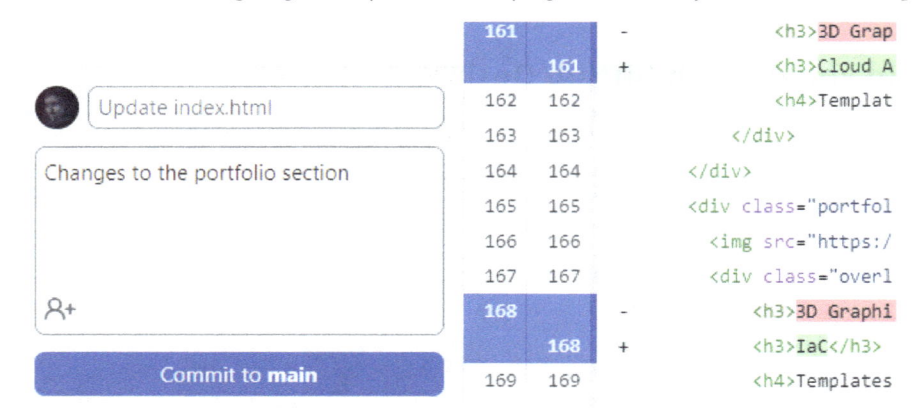

Figure 8.17 – Committing changes

- **Sync with GitHub**: Use the **Fetch**, **Pull**, and **Push** buttons to sync your local work with the online repository. It's as easy as syncing your music library across your devices.

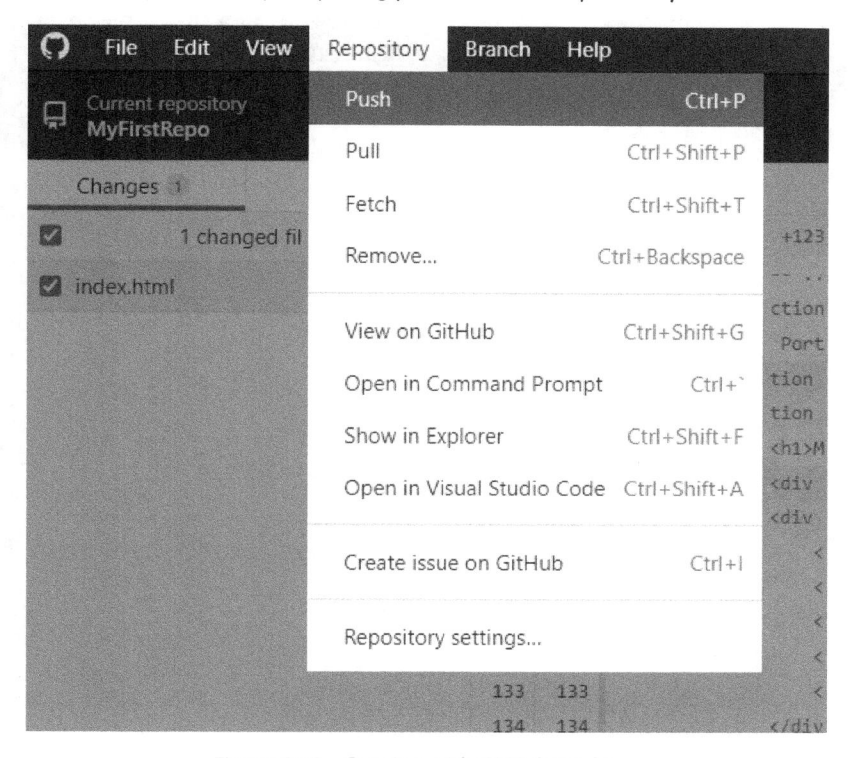

Figure 8.18 – Syncing with GitHub Desktop

- **Branch out, merge, and conclude; stay collaborative**: Want to try a new idea? Create a new branch in GitHub Desktop. It's like opening a new, clean feature branch to code on while keeping your original code safe.

When your new feature is ready, you can merge it into the main project. GitHub Desktop guides you through the process to ensure it's a smooth transition.

Pull requests are a key part of teamwork. GitHub Desktop integrates this feature so you can review, discuss, and merge work from team members within the app.

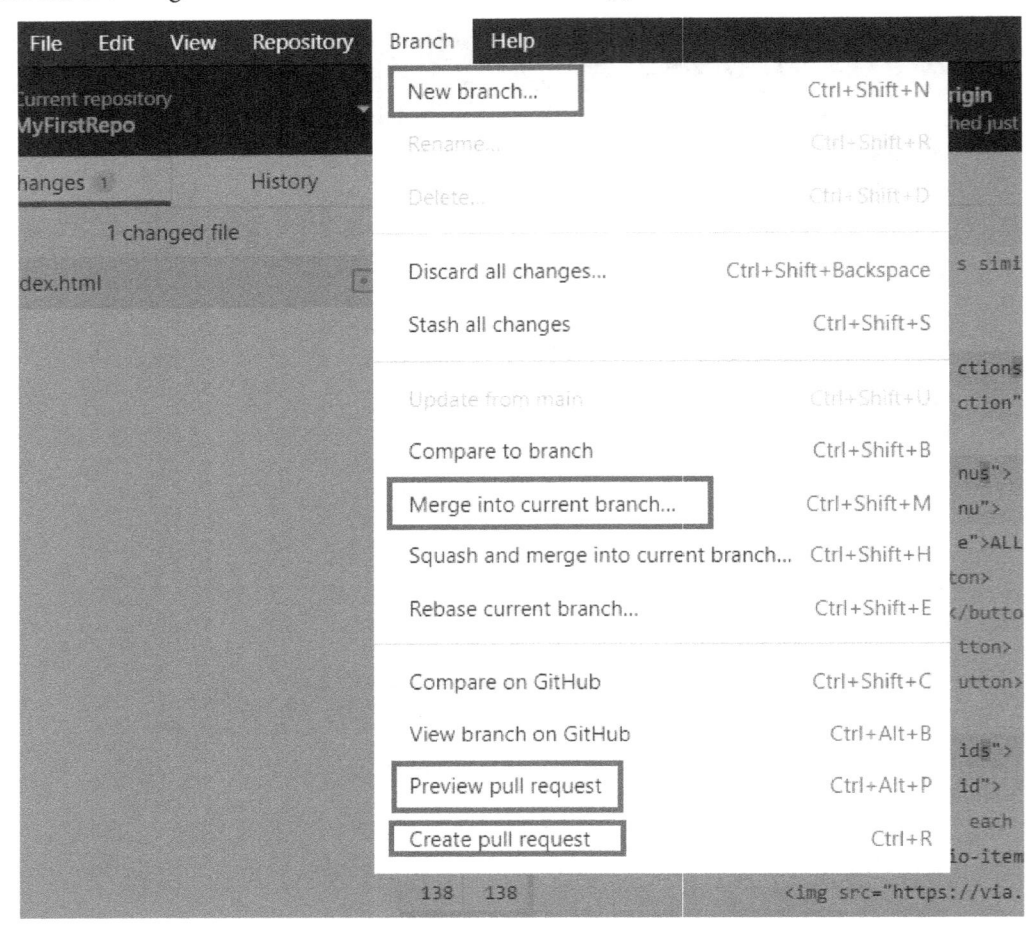

Figure 8.19 – Branch in GitHub Desktop

Next, let's look at some benefits of GitHub Desktop:

- It enables you to see a visual representation of your project's progress
- You can share your work and incorporate others' contributions without touching a command line
- Its clean interface helps prevent mistakes that are easy to make in the command line

Using GitHub Desktop for your one-page portfolio website is about simplifying your development process. It's a tool that makes version control feel less like rocket science and more like a regular part of your creative workflow. It's perfect for those who want the power of Git with the simplicity of a point-and-click interface. Next, let's move on and see how you can develop your website directly in the browser using GitHub Codespaces. This is like having your entire development studio within your web browser, giving you the tools to write, run, and debug code without needing to switch applications.

Crafting your website with GitHub Codespaces

Take a moment and picture yourself walking into a fully equipped workshop where you can start creating right away, without having to set up any tools or machines. That's what GitHub Codespaces offers for your one-page portfolio website project. It's a complete, ready-to-go development environment right in your browser.

Why would you use Codespaces?

No need to spend time setting up your development environment on your local machine, you can work on your website from any device with internet access. If you've got a browser, you're ready to go. You don't have to worry about whether your device is powerful enough. Codespaces provides you with a coding environment that's hosted online (in the cloud).

Codespaces comes with the features of **Visual Studio Code**, a top-notch editor. It's like having a high-end computer with professional software at your disposal.

How to get started with Codespaces

It's actually easier than you think. With one click, you can create a new codespace directly from your repository. Just head to your website's GitHub repository and click on the <> **Code** button. From there, you will be able to create new codespaces for your repo. GitHub will set up everything in the background.

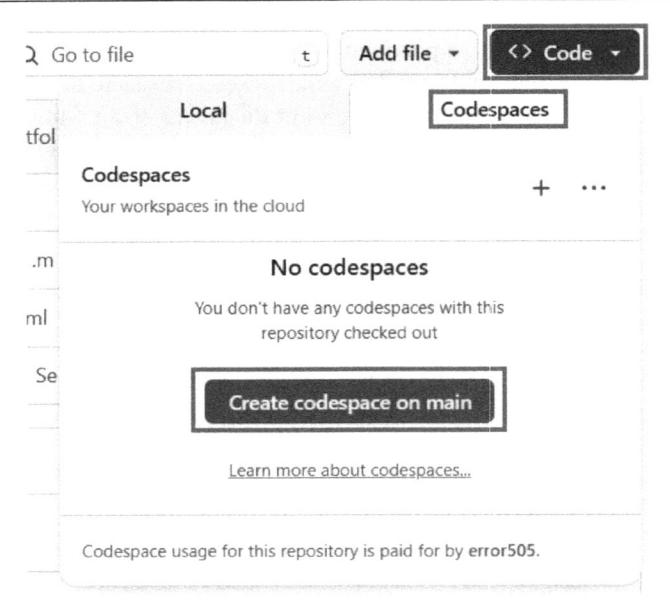

Figure 8.20 – Creating GitHub Codespaces

Once your codespace is ready, you can access a full coding environment in your web browser, complete with a terminal, a code editor, and even debugging tools.

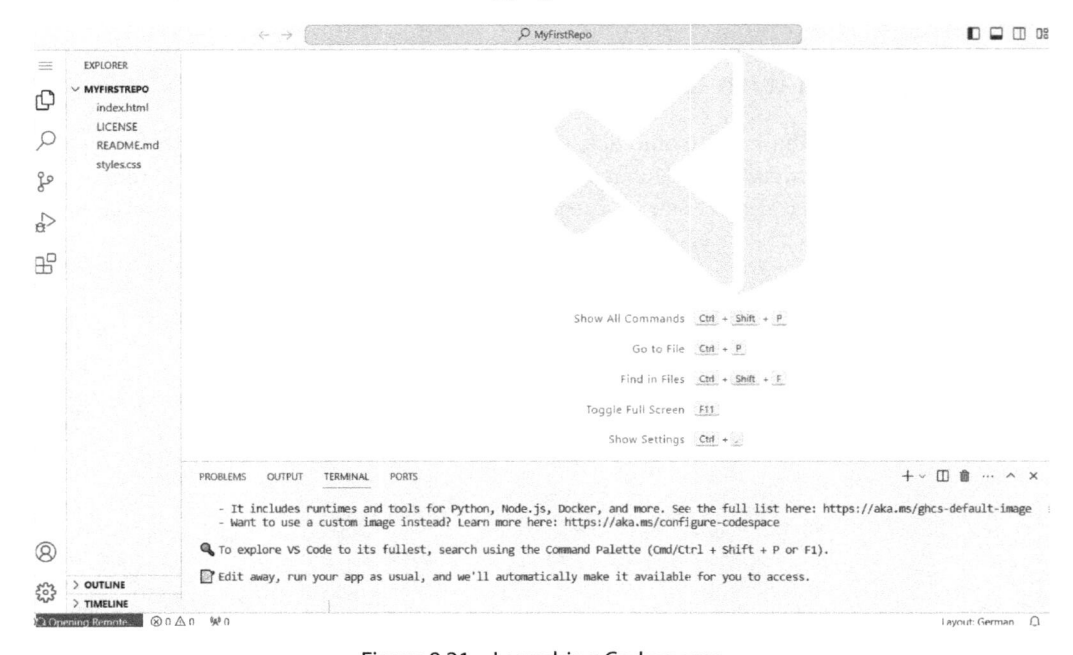

Figure 8.21 – Launching Codespaces

Now, you can work on your website just as you would on your local computer. Edit files, run your site to test changes, and commit your updates.

Everything you need is right there in your browser. Share your codespace with collaborators so they can jump in and code with you.

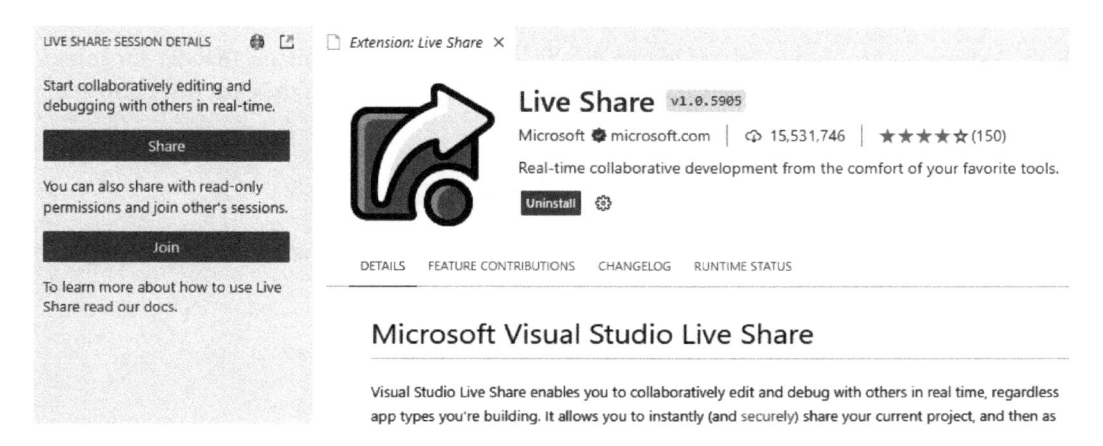

Figure 8.22 – Sharing your codespace with collaborators

With Codespaces, you can jump right back into your project, picking up where you left off, anytime and anywhere.

Figure 8.23 – Codespaces source control

Helpful Tools and Git Commands

GitHub Codespaces is like having a portable, virtual workshop that you can enter from anywhere to work on your one-page portfolio website. It's designed to make the development process smoother, letting you focus on creating and collaborating without any setup hassles.

Deleting Codespaces

GitHub Codespaces are always deleted automatically if you stop it or it remains inactive for some period of time but, you can also delete it manually whenever you want with the click of a button, as shown in *Figure 8.24*.

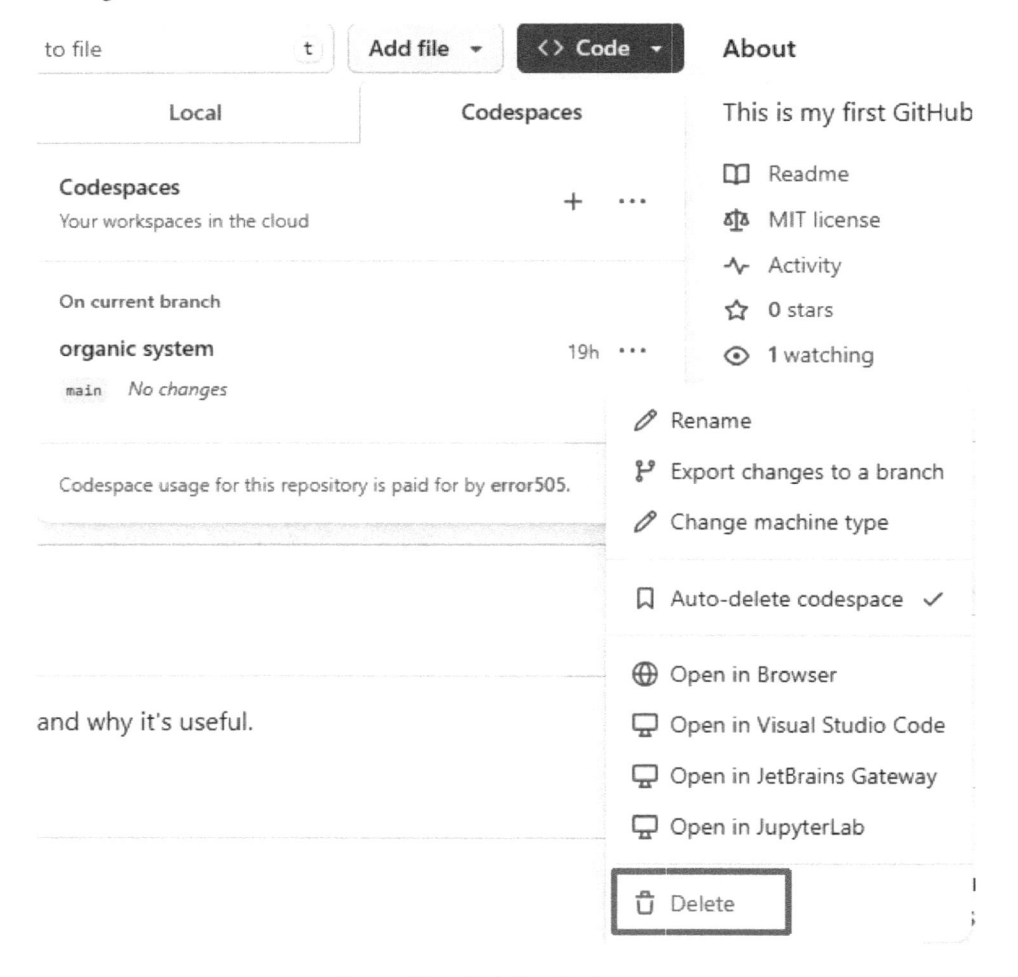

Figure 8.24 – Deleting Codespaces

Now that we've explored how GitHub Codespaces turns any device into a powerful development studio for your website, let's move on. Next up, we'll look at how to keep track of all your project updates and conversations. With DevHub, you'll find managing your project's notifications a piece of cake, right from your desktop or web browser.

Managing your project's buzz with DevHub

So you've got your one-page portfolio website up, and there's a lot happening. Friends are suggesting changes, there are discussions about new features, and updates are being made. With all this activity, your GitHub notifications are buzzing like crazy. This is where DevHub steps in. It's like having a personal assistant to help you manage the buzz.

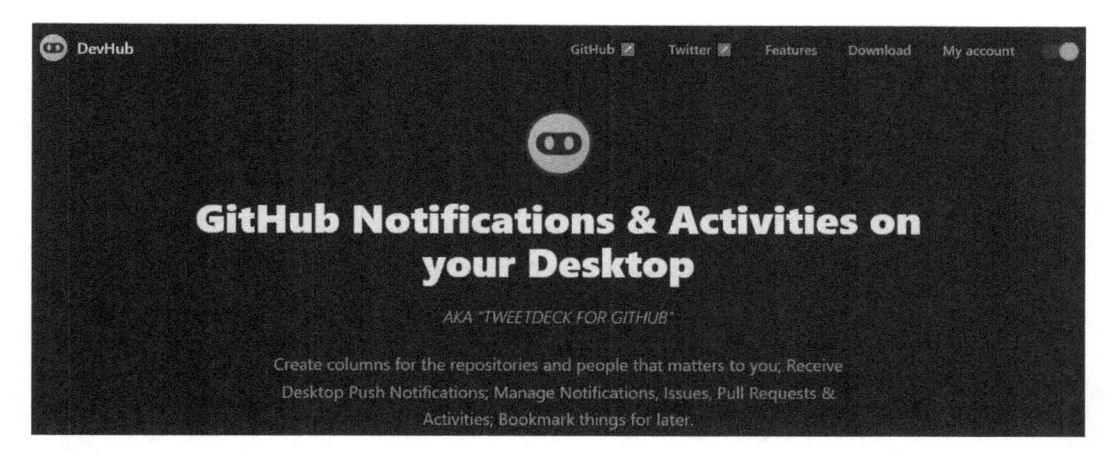

Figure 8.25 – DevHub

DevHub is a tool that brings all your GitHub notifications and activities straight to your desktop. You can also use the web version's control panel where you can see everything that's happening in your project without having to log in to GitHub.

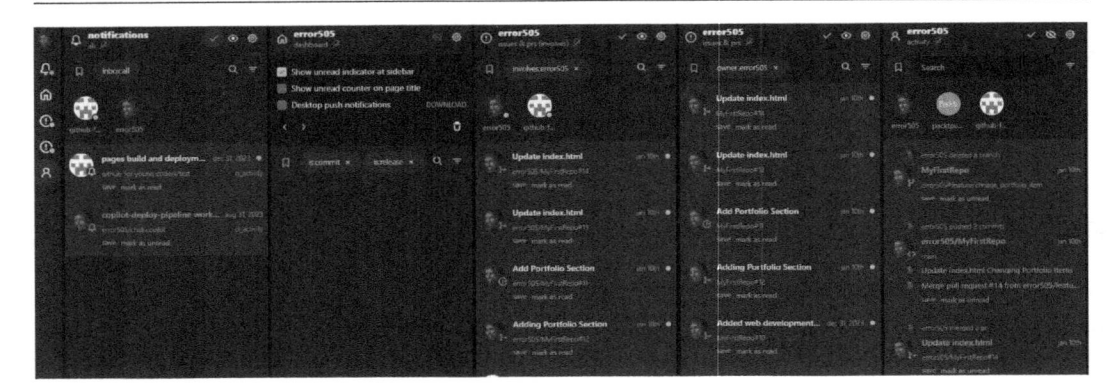

Figure 8.26 – DevHub dashboard

How do you set up DevHub?

You can use the web version by navigating to https://devhubapp.com/ and then simply clicking on the **Use web version** button. You will see the **Sign in with GitHub** button to connect DevHub to your GitHub account. After you click on that button, a new pop-up window will open where you will have to sign in to your GitHub account. Next, you will have to agree that you want to connect DevHub with your GitHub account. Now, you are ready to use DevHub in your browser.

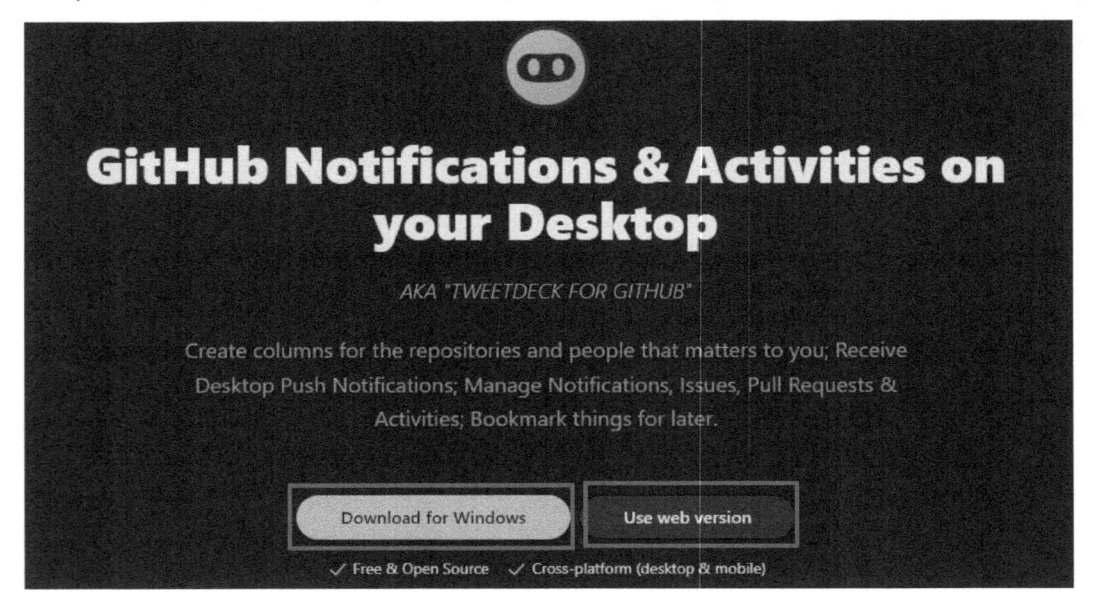

Figure 8.27 – DevHub versions

If you want your notification manager directly on your desktop, then click on the **Download for Windows** button and download DevHub for your desktop. It's like getting a new gadget that's going to make life a whole lot easier. Similar to setting up the web version, you just have to connect your GitHub account and you are good to go.

Why should you use DevHub in your daily workflow?

DevHub organizes your notifications by repository and type, which is like having a mailbox with different sections for bills, letters, and ads, so you can easily find what you're looking for. With all your notifications and activities in one spot, you can handle things faster and more efficiently.

You can configure DevHub to show you what's most important, to only ping you for the notifications you care about. Instead of getting lost in a sea of notifications on GitHub, DevHub helps you focus on the tasks at hand.

Whether you're responding to pull requests, checking out issues, or merging changes, DevHub lets you do it from one place. DevHub keeps you in sync with your team's activities, making collaboration smooth and continuous, almost like a group chat dedicated to your project's development.

Having learned how DevHub can simplify keeping track of your project's activity and notifications, you are now ready to move to the next chapter where you will learn a lot of cool stuff about GitHub Actions and build automation.

Summary

Congratulations on completing *Chapter 8*! By now, you've expanded your toolkit and are equipped to manage your one-page portfolio website project with greater ease and sophistication. Let's recap the powerful skills and tools you've now mastered.

With the help of some advanced Git commands, such as `alias` and `rebase`, you learned how to streamline your Git workflow, trace the detailed history of a single file's changes, and reorganize your commits for neater project history. Then, you discovered how Sourcetree can make complex histories understandable at a glance. Furthermore, you enhanced your skills with Codespaces, Desktop, and DevHub to channel GitHub into a friendly desktop interface and manage your notifications.

Throughout this chapter, you've not only acquired new skills but also learned how these tools can make your collaboration more effective and your project management more efficient. You're now ready to tackle your development tasks with confidence, knowing you have the right tools to help you along the way.

In the next chapter, you will discover how GitHub Actions can automate your work, how to set up automated sequences called workflows, set up tests that run automatically, and troubleshoot your workflows and fine-tune them for peak performance.

Chapter 9 is all about making your project work smarter, not harder. Get ready to automate your way to a more productive and stress-free development experience!

Quiz

Answer the following questions:

1. What is the primary use of `git config`?

    A. To change the project's settings

    B. To configure user information in Git

    C. To create a new branch

    D. To merge two branches

    Answer: B. To configure user information in Git

2. GitHub Desktop is an application that allows users to manage Git repositories through a graphical interface.

    A. True

    B. False

    Answer: A. True

3. What allows you to create, view, and edit files directly in GitHub's interface?

    Answer: GitHub Codespaces

4. What is the purpose of using Visual Studio Code with Git?

    Answer: Visual Studio Code integrates with Git, allowing users to manage Git operations, view changes, and make commits directly from the editor.

5. What is Sourcetree used for?

    A. Tracking project issues

    B. Managing Git repositories visually

    C. Automated code testing

    D. Cloud-based development

    Answer: B. Managing Git repositories visually

6. DevHub Notifications is a tool for managing notifications and activities on GitHub.

    A. True

    B. False

    Answer: A. True

7. What is `git blame` is used for?

    A. Finding out who last modified a line of code

    B. Deleting a branch

    C. Renaming files in a repository

    D. Cloning a repository

    Answer: A. Finding out who last modified a line of code

8. The `git _____ -fd` command is used to clean the working directory by removing untracked files.

    Answer: `clean`

9. The `git stash` command is used to permanently delete changes from a branch.

    A. True

    B. False

    Answer: B. False (it temporarily stores changes)

10. What is the purpose of `git fetch --all`?

    Answer: It fetches all branches from the remote repository, ensuring you have all the latest updates.

Challenge – Crafting a Game Over screen for Space Explorer

In this chapter, you've unlocked new skills with Git and GitHub tools. Let's put these to the test by creating a captivating **Game Over** screen for your *Space Explorer* game, complete with a cool background.

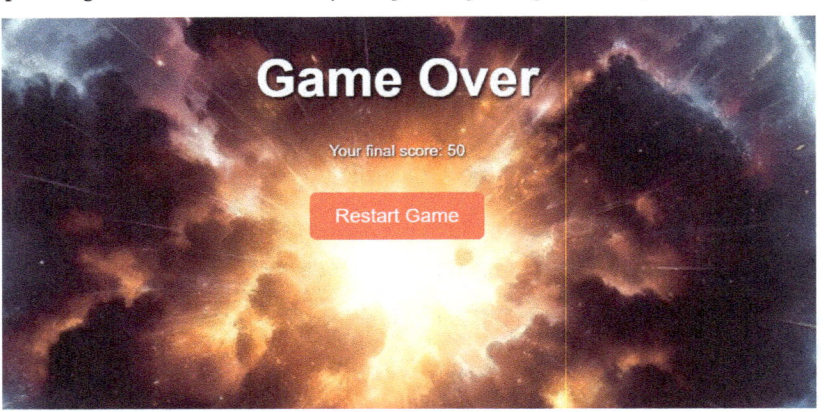

Figure 8.28 – Space Explorer Game Over screen

Your mission is to develop a **Game Over** screen using new Git commands and tools and integrate a unique background into this screen.

You can download the code from here: `https://github.com/PacktPublishing/GitHub-for-Next-Generation-Coders/tree/main/Space%20Explorer%20Game/Game%20Obsticles%20-%20Game%20Over%20screen`.

Here is how you can accomplish your mission:

1. **Branch creation**: Using GitHub Desktop or Sourcetree, create a new branch named `feature-game-over-screen`. This visual tool simplifies the branching process.
2. **Designing the screen**: Design a **Game Over** screen with a cool space-themed background. Get creative with your design – this is the player's last view after their adventure!
3. **Implementing the screen**:
    I. Code the new screen in your project. Ensure it displays when the game ends.
    II. Experiment with Codespaces and Git features to commit your changes as you work.

4. **Using advanced Git commands**:
    I. If you make a mistake or want to try a different design, use `git revert` or `git reset` to undo changes.
    II. Share your progress by pushing your branch to GitHub.
5. **Pull requests and reviews**:
    I. Use GitHub Codespaces or GitHub Desktop to open a pull request for your new branch.
    II. Ask peers for feedback or review the changes yourself, using the pull request features of GitHub.
6. **Merge your changes**: Once you're satisfied with your **Game Over** screen, merge your branch into the main project. Handle any merge conflicts that might arise using the skills you've learned.
7. **Reflect and document**: Write about your experience in a README file or document. How did using these new tools and commands enhance your development process?

Part 4: Advanced GitHub Functionalities

Here, you'll learn how to automate workflows using GitHub Actions, secure your projects with advanced security measures, and engage effectively with the open source community. Each chapter provides insights into making your GitHub projects more dynamic, secure, and community-focused.

This part contains the following chapters:

- *Chapter 9, Leveraging GitHub Actions for Automation*
- *Chapter 10, Enhancing GitHub Security Measures*
- *Chapter 11, Engaging with the Open Source Community*

9
Leveraging GitHub Actions for Automation

In this chapter, we'll be diving into the world of GitHub Actions, an automation powerhouse that can significantly streamline your project workflows. Whether you're a solo developer working on a personal project or part of a team developing complex applications, understanding and utilizing GitHub Actions can transform the way you build, test, and deploy your software.

By the end of this chapter, you'll not only grasp the fundamentals of GitHub Actions but also learn how to create your own workflows to automate tasks, conduct automated testing, and manage deployments directly within GitHub.

We will create your own GitHub Actions workflows, complete with practical demos, and discover how to securely manage sensitive information within our workflows to protect our project and streamline configuration. Lastly, we will learn about tips and strategies to troubleshoot workflows and optimize our GitHub Actions for efficiency and effectiveness. We'll explore practical examples to illustrate these concepts, providing you with the knowledge to apply these techniques to your projects.

In this chapter, we're going to cover the following main topics:

- Understanding a GitHub Actions workflow
- Creating action workflows
- Automated testing and deployment with GitHub Actions
- Managing secrets and environment variables
- Troubleshooting and optimizing GitHub Actions

Technical requirements

For some activities, we'll be using marketplace GitHub Actions. Ensure that you're familiar with YAML syntax, as it's essential for customizing GitHub Actions.

Additionally, download a GitHub folder named `Chapter 9` from `https://github.com/PacktPublishing/GitHub-for-Next-Generation-Coders/tree/main/Chapter%209` where you can find a `static.yml` file inside the `.github/workflow` folder, which you can use to follow along.

Understanding a GitHub Actions workflow

You and your friends have built a cool one-page portfolio website. Now, you want to add a new section called **MY ARTICLES** to showcase your blog posts. You may have heard that GitHub Actions can help automate some tasks for you, but what does that mean? Let's break it down simply.

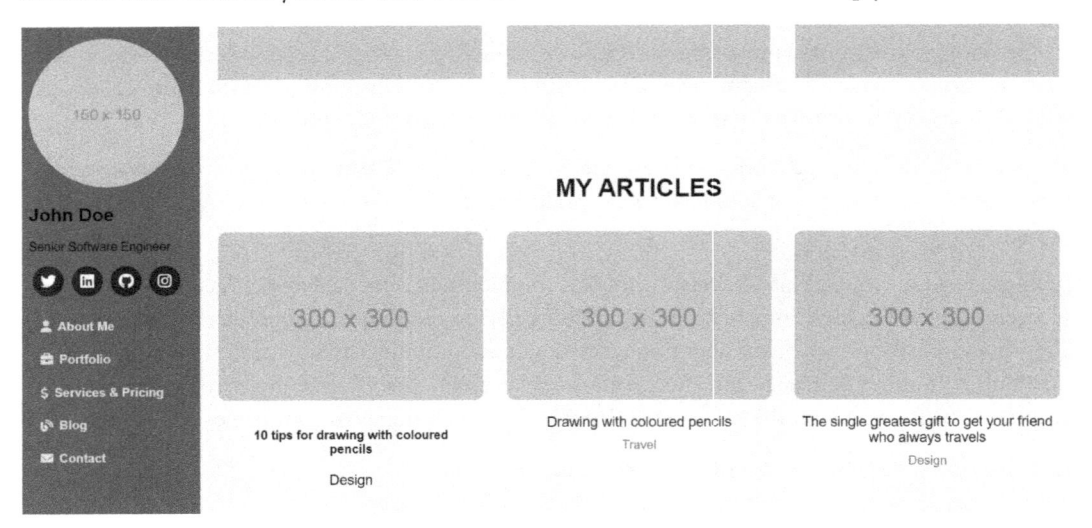

Figure 9.1 – The MY ARTICLES section

What are GitHub Actions? Think of **GitHub Actions** as a helpful robot that can do tasks for you. For example, every time you want to add a new article to your **MY ARTICLES** section, this robot can automatically update your website with a new post. It's like having a little helper that takes care of repetitive tasks without you needing to do anything manually. But what if you're not sure how to write the instructions for that helpful robot we talked about? Good news – you don't have to start from scratch. GitHub has a place called the **Marketplace** where you can find actions that other smart folks have already made. It's like going to a store to pick out a robot instead of building one yourself.

GitHub Actions workflow files are written in **YAML**, which stands for **YAML Ain't Markup Language**. It's a human-readable data serialization standard, which means it's designed to be easy for us humans to read and write. A **workflow** is an automated process that you define in your GitHub repository and store in the .github/workflows directory of your repository. A workflow can be triggered by different events, such as pushing code to a repository, creating a pull request, or on a schedule.

Let's quickly look at the syntax of the workflow to deploy your website to GitHub Pages in *Figure 9.2*.

```
1   name: Deploy static content to Pages
2   on:
3     push:
4       branches: ["main"]
5
6     workflow_dispatch:
7
8   permissions:
9     contents: read
10    pages: write
11    id-token: write
12
13  concurrency:
14    group: "pages"
15    cancel-in-progress: false
16
17  jobs:
18    deploy:
19      environment:
20        name: github-pages
21        url: ${{ steps.deployment.outputs.page_url }}
22      runs-on: ubuntu-latest
23      steps:
24        - name: Checkout
25          uses: actions/checkout@v4
26        - name: Setup Pages
27          uses: actions/configure-pages@v4
28        - name: Upload artifact
29          uses: actions/upload-pages-artifact@v3
30          with:
31            path: '.'
32        - name: Deploy to GitHub Pages
33          id: deployment
34          uses: actions/deploy-pages@v4
```

Figure 9.2 – The GitHub Actions Marketplace

Let's understand each of the terms in the preceding figure in greater detail:

- **Events**: An event in GitHub Actions is a specific activity that triggers an automated workflow. For example, when someone pushes code to a repository or creates a pull request, it can automatically start actions such as testing code. It's like setting off a domino effect, with one action leading to another. Here are some of the events that you can use in your GitHub Actions:

- `on`: This part tells us when to run this workflow. It's like setting up specific times to start cooking.
- `push`: This says to the GitHub action, "*Start the workflow when someone pushes code to the main branch*". In the context of cooking, it's like saying, "*When the main dish is ready, it's time to serve.*"
- `workflow_dispatch`: This allows you to manually start the workflow. It's like having a button in the kitchen to start the oven whenever you're ready.

- **Jobs**: This outlines what needs to be done.
- **Runners**: This tells us on which system the actions will be executed.
- **Steps**: Steps in GitHub Actions are individual tasks that make up a job in a workflow. It's like steps in a recipe, where each step does one specific thing, such as mixing ingredients or baking them, to achieve the overall goal of making a cake. Here are some of the steps you can use:

  - `Checkout`:
    - **What it does**: This step gets the latest code from your repository.
    - **How it works**: The `uses: actions/checkout@v4` line tells GitHub to use the `checkout` action, which is a pre-built action from the GitHub Marketplace. This action checks out your repository so that the workflow can access the code.

  - `Setup Pages`:
    - **What it does**: Prepares GitHub Pages with the necessary settings.
    - **How it works**: The `uses: actions/configure-pages@v4` line indicates the use of the `configure-pages` action from the Marketplace. This action configures your repository to be ready for GitHub Pages deployment.

  - `Upload artifact`:
    - **What it does**: Gathers all the files from your repository for deployment.
    - **How it works**: The `uses: actions/upload-pages-artifact@v3` line specifies the use of the `upload-pages-artifact` action from the marketplace. This action uploads the files that need to be deployed. The `with` keyword allows you to pass parameters to this action. Here, `path: '.'` means that all files in the current directory are included.

  - `Deploy to GitHub Pages`:
    - **What it does**: Deploys your website to GitHub Pages.
    - **How it works**: The `uses: actions/deploy-pages@v4` specifies line the use of the `deploy-pages` action from the Marketplace, which handles the deployment. `id: deployment` gives this step an identifier that can be referred to later if needed.

- **Actions**: Actions are like automated helpers that can do tasks for you in your GitHub project, which we will learn more about in the next section.

This workflow is a set of instructions that tells GitHub how to update your website automatically whenever you make changes to your project, specifically when you push new changes to the main branch or manually trigger a deployment. It's all about making the process of updating your website as easy as pressing a button.

Now that you've gotten a taste of what GitHub Actions can do to streamline updating your one-page portfolio website, it's time to go deeper. Next, we're going to explore how to create your own action workflows. This is where you'll learn to write the script for your helpful robot, giving it the detailed instructions it needs to automatically build up your site whenever you add a new feature.

Creating action workflows

On your GitHub repository page, there's a tab labeled **Actions**. When you click on it, you'll see an option to search for workflow on the GitHub Starter page. You also can search for GitHub Actions on the GitHub Marketplace and configure the one you need. Imagine walking into a robot store and seeing a catalog of all the different robots you can use. Actions are like those different little robots that you can use to automatically run different parts of your project, such as checking your code for errors or deploying your website online whenever you make changes to it.

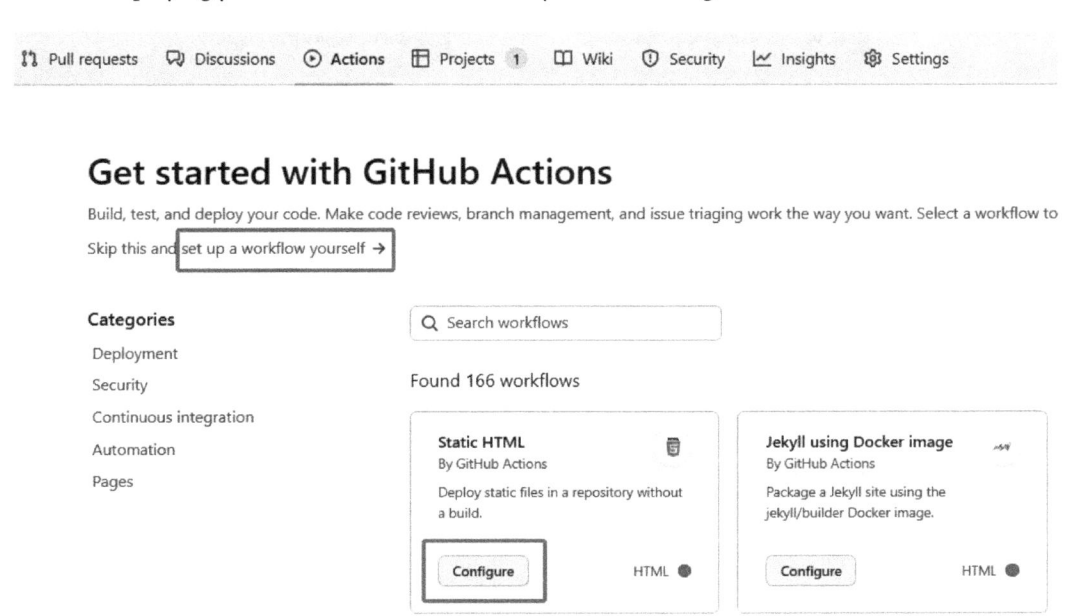

Figure 9.3 – The GitHub Actions starter page

Exploring prebuilt actions

In the Marketplace UI shown in *Figure 9.4*, you're presented with a variety of prebuilt actions. It's like each robot has a special skill. Some are great at organizing your blog posts, others excel at checking your website's layout, and so on. You're looking for one that can automatically rebuild your website whenever you add a new blog post.

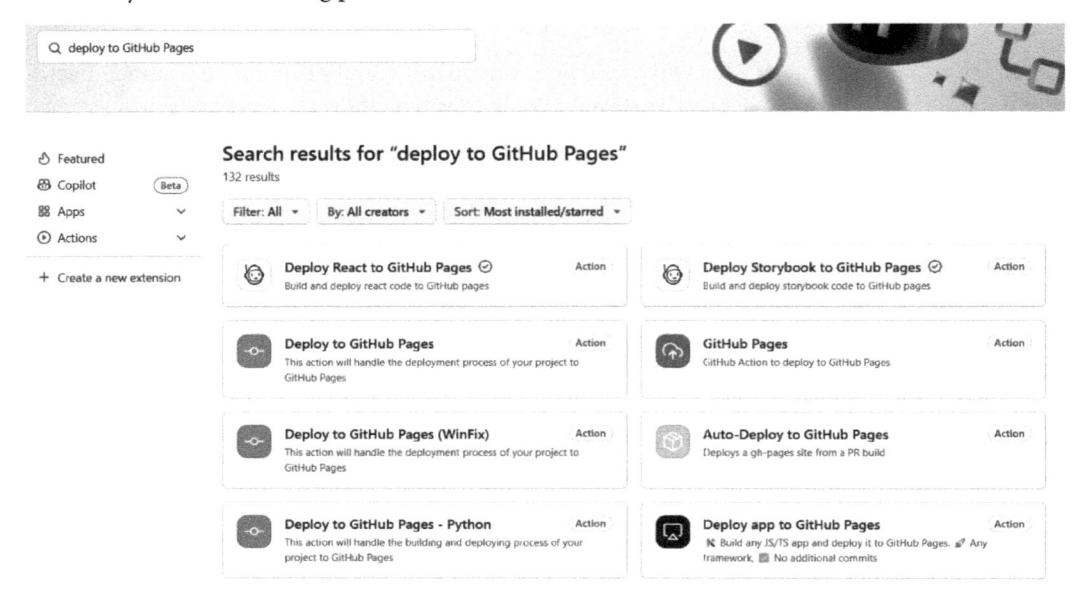

Figure 9.4 – The GitHub Actions Marketplace page

Here's how to find and use actions from the Marketplace:

1. Search for `Actions`.
2. Go to the GitHub Marketplace (`https://github.com/marketplace?type=actions`).
3. Use the search bar to find actions that suit your needs (e.g., `deploy to GitHub Pages`), as shown in *Figure 9.4*.
4. Copy the usage snippet. (Each action in the marketplace provides a usage snippet.)
5. Paste this snippet into your workflow file.

Each action in the Marketplace comes with its own set of instructions and reviews. It's like reading the back of the robot's box to see what it can do and checking online reviews to see how well it works for others.

Once you've found the perfect prebuilt action, it's time to add it to your workflow by clicking on the **Use latest version** button and copying the code snippet to your workflow.

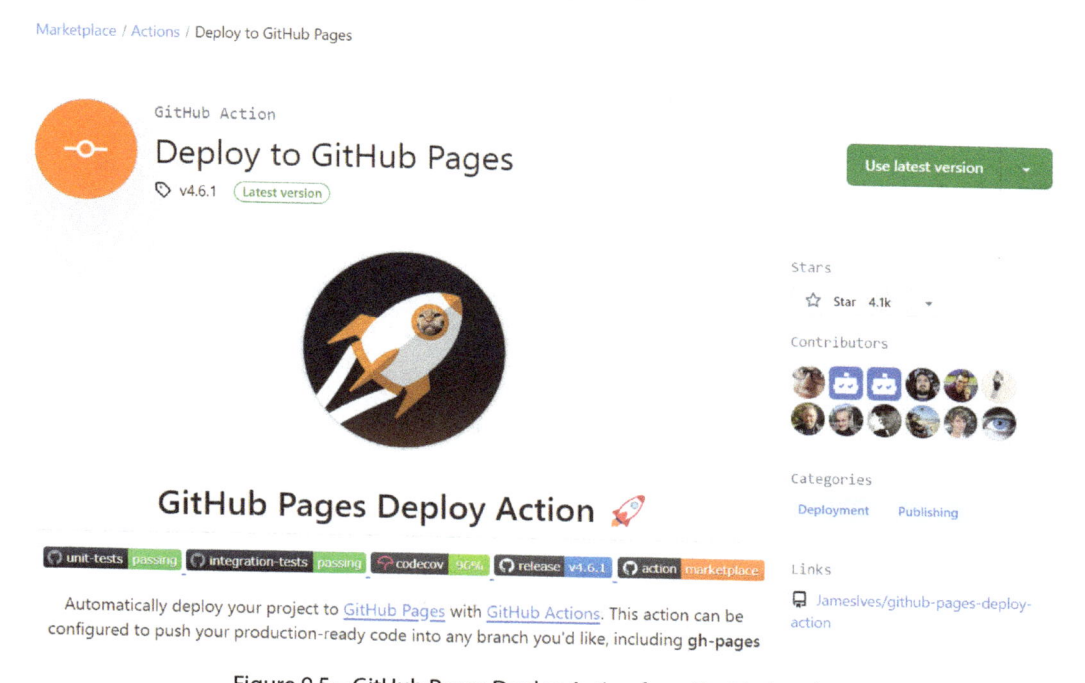

Figure 9.5 – GitHub Pages Deploy Action from the Marketplace

Some actions might require a bit of setup, such as telling it exactly where your website is stored or how you want it to be built. It's like giving your robot specific instructions to make sure it does the job right.

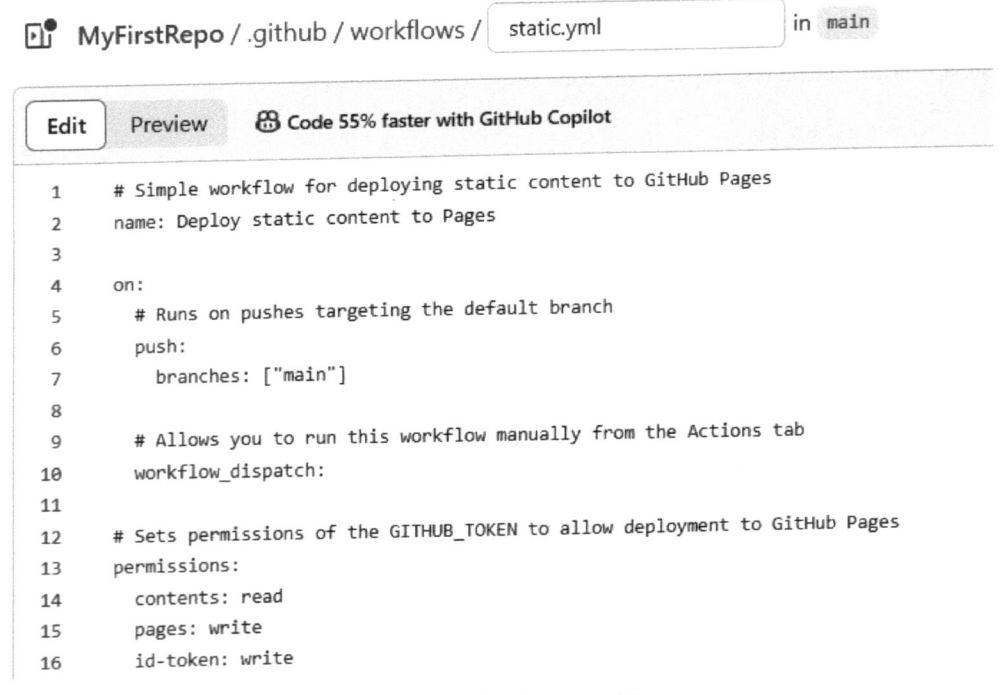

```yaml
# Simple workflow for deploying static content to GitHub Pages
name: Deploy static content to Pages

on:
  # Runs on pushes targeting the default branch
  push:
    branches: ["main"]

  # Allows you to run this workflow manually from the Actions tab
  workflow_dispatch:

# Sets permissions of the GITHUB_TOKEN to allow deployment to GitHub Pages
permissions:
  contents: read
  pages: write
  id-token: write
```

Figure 9.6 – Configuring a workflow

Many actions allow you to pass parameters using the `with` keyword. For example, if your project is built with Node.js, you'll need an action to set up the Node.js environment before you can run tests or build your project. This typically looks like this:

```
steps:
- uses: actions/checkout@v2
- name: Set up Node.js
  uses: actions/setup-node@v2
  with:
    node-version: '14' # Specify the Node.js version you want to use
```

After adding and customizing your chosen action, you'll save the changes to your repository. It's the same as turning on your robot and letting it get to work.

Figure 9.7 – Committing your Workflow

Your new workflow will be placed in the `.github/workflows` folder of your repository. You can always go there and edit the workflow if you need to.

Understanding the uses keyword

The `uses` keyword in a GitHub Actions workflow file specifies that a step should use an action. Actions are reusable units of code that can be shared across workflows, making it easy to automate tasks. The `uses` keyword indicates which action to use where and to fetch it from.

Here's how the `uses` keyword works:

- **Specifying an action**: When you use the `uses` keyword, you specify the location of the action. This can be an action from the GitHub Marketplace, a public repository, or even a local action within your repository.

- **Format**: The format for the `uses` keyword typically includes the repository where the action is located and the version of the action. For example, `actions/checkout@v2` refers to the checkout action from version 2 of the `actions` repository.

Seeing it in action

Now, every time you push a new blog post to your repository, the action you chose from the Marketplace gets to work, updating the **MY ARTICLES** section on your website automatically. You've essentially partially automated the task of maintaining your website with a little help from the GitHub community.

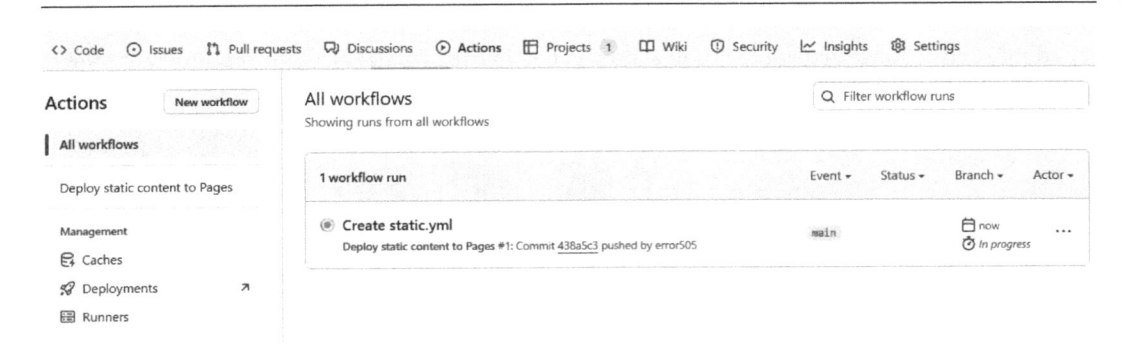

Figure 9.8 – Running the workflow

What are reusable workflows in GitHub Actions?

Reusable workflows in GitHub Actions allow you to define a set of actions and steps once and then use them across multiple workflows and repositories. This helps maintain consistency, reduce duplication, and simplify the management of workflows across your projects.

There are several reasons why you would use reusable workflows:

- **Consistency**: Ensures that the same steps and actions are used across different workflows, reducing the risk of errors
- **Efficiency**: Saves time by avoiding the need to rewrite the same steps for different workflows
- **Maintainability**: Easier to update and manage common workflows in a single place

How to create and use a reusable workflow

A reusable workflow is defined in a separate YAML file in the `.github/workflows` directory of your repository, just like regular workflows. For example, you might create a file named `reusable-workflow.yml`.

To use this reusable workflow in another workflow, you can call it using the `uses` keyword, just as we used it with the actions from the marketplace.

Now that you've seen how to pick and set up a helpful action from the GitHub Marketplace to keep your website updated, let's move on to the next exciting step. You're going to learn how to make sure your website not only updates automatically but also works perfectly every time you make a change. Next, we'll look into how to use GitHub Actions for automated testing and deployment. This means you'll set up more *robot helpers* to check your work and put it live on the web, making sure your website always looks great and functions flawlessly.

Automated testing and deployment with GitHub Actions

Let's imagine you're working on your one-page portfolio website again. This time, you want to make sure that every time you add something new, like a **MY ARTICLES** section for your blog posts, everything works perfectly before anyone else sees it. Plus, once you're sure it's all good, you want it to appear on your website without doing extra work. This is where GitHub Actions can be your superhero, helping with automated testing and deployment.

What's automated testing?

Think of **automated testing** as having a robot that checks your new **MY ARTICLES** section to ensure that everything works as it should. For example, it makes sure that when someone clicks on an article, it actually appears. It's like the robot is clicking around your website for you, looking for mistakes. Automated testing saves you a lot of time and drastically reduces the chance of errors slipping through, making sure your website always delivers the best experience to your visitors without you having to manually check every detail.

Deployment

Deployment is when your robot not only checks your work but also puts it up on your website for you. So, after the robot makes sure all the links in **MY ARTICLES** work, it then updates your live website with this new section. It's exactly what we did previously when we picked up the workflow to build and deploy static HTML.

How do we set this up with GitHub Actions?

To ensure your one-page portfolio website not only deploys smoothly but also adheres to best coding practices, let's integrate testing into your GitHub Actions workflow. We'll use ESLint for JavaScript, Stylelint for CSS, and HTMLHint for HTML. This step ensures that your code is clean, follows set conventions, and is free from common errors before it goes live.

Before doing the setup, let's understand the testing tools:

- **ESLint**: Checks your JavaScript files (`scripts.js`) for coding errors and style issues. It's like having a friend who's really good at JavaScript point out improvements.
- **Stylelint**: Reviews your CSS file (`styles.css`) to make sure it looks good and follows best practices. Imagine a fashion expert advising on how to match your outfit.
- **HTMLHint**: Examines your HTML file (`index.html`) for mistakes and best practices. It's like a grammar check for your HTML code.

Now that you're familiar with the tools, let's move to the setup part:

1. **Editing the workflow and adding an action**: Start by opening a list of instructions (a workflow file) for your robot in the `.github/workflows` directory of your GitHub repository, called `static.yml`.

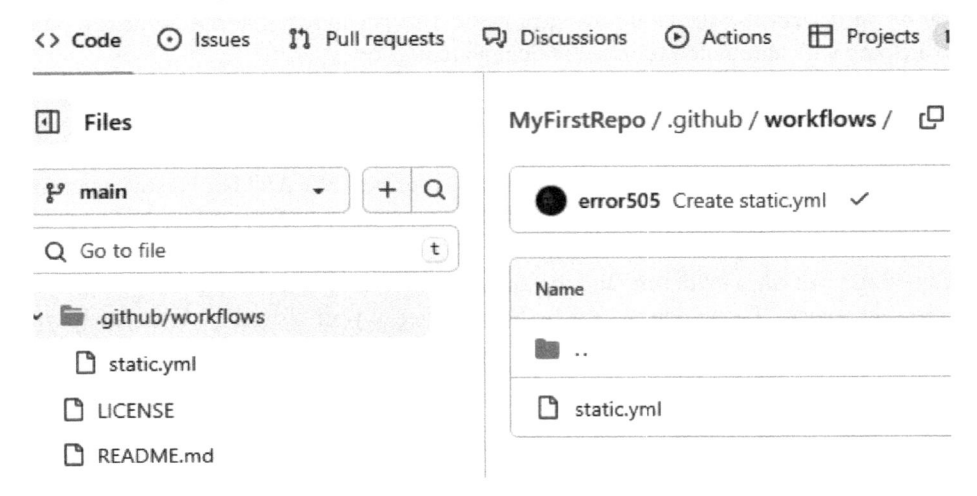

Figure 9.9 – The workflow .yml file location

2. **Writing the instructions**: In your `static.yml` file, you'll write down what you want your robot to do.

3. **Tweaking the workflow to include testing**: In the `jobs` section, before deploying, we add a test job that sets up Node.js (since our linters require Node.js), installs the necessary packages for ESLint, Stylelint, and HTMLHint, and then runs these tools in your respective files.

```yaml
jobs:
  test:
    runs-on: ubuntu-latest
    steps:
      - name: Checkout
        uses: actions/checkout@v4

      - name: Set up Node.js
        uses: actions/setup-node@v2
        with:
          node-version: '14'

      - name: Install Dependencies
        run: |
          npm install eslint stylelint htmlhint --save-dev
          npm install

      - name: Run ESLint
        run: npx eslint scripts.js

      - name: Run stylelint
        run: npx stylelint styles.css

      - name: Run HTMLHint
        run: npx htmlhint index.html
```

Figure 9.10 – Adding new steps

4. The **deploy job** now needs the **test job** to succeed before it can run. This means that if there's an issue caught by the linters, deployment won't proceed until those issues are resolved.

```yaml
deploy:
  needs: test
  environment:
    name: github-pages
    url: ${{ steps.deployment.outputs.page_url }}
  runs-on: ubuntu-latest
  steps:
    - name: Checkout
      uses: actions/checkout@v4

    - name: Setup Pages
      uses: actions/configure-pages@v4

    - name: Upload artifact
      uses: actions/upload-pages-artifact@v3
      with:
        path: '.'

    - name: Deploy to GitHub Pages
      id: deployment
      uses: actions/deploy-pages@v4
```

Figure 9.11 – Adding new actions

By adding these tests, you're ensuring that your **MY ARTICLES** section and the rest of your website not only look good to visitors but also have clean, error-free code behind them. This step is crucial to maintain a professional and polished website.

Now, when you look at your updated GitHub Action, you will see that it already found some errors in the CSS files, which should be fixed (*Figure 9.12*).

Figure 9.12 – The test results

You can go over the `styles.css` file and fix the errors. Additionally, you could use the linter directly in your Visual Studio to fix some of the errors for you automatically.

You've now seen how automated testing and deployment with GitHub Actions can make your development process smoother and more reliable. These tools help ensure that your code is always in good shape and can be deployed automatically. Next, we'll move on to learning how to manage secrets and environment variables. This is important for keeping sensitive information, like passwords and API keys, secure while still making them accessible to your workflows. We'll explore how to handle these elements safely in your projects.

Managing secrets and environment variables

Imagine you're building a magic box for your one-page portfolio website, specifically for a section called **MY ARTICLES**. This magic box can automatically update your website with new articles you write. However, to work, it needs a special key that accesses your blog posts without letting anyone else see this key. This is where secrets and environment variables come into play in GitHub Actions.

What are secrets and environment variables?

Secrets are like special or keys that you don't want others to find. For your website, a secret could be a password or a special code that lets GitHub Actions access your blog posts safely.

Environment variables are settings that can change based on different conditions when GitHub Actions run. They help customize how your workflows operate without changing the actual code.

Why use them?

Using secrets and environment variables has the following benefits:

- **Keeping sensitive info safe**: You don't want to leave your special keys (passwords and API keys) out in the open for anyone to grab. Secrets keep them safe.
- **Flexibility**: Environment variables allow your magic box (GitHub Actions) to work a little differently depending on the situation, such as showing more articles on your home page or changing how they're displayed without changing the main spell (your code).

Using secrets and environment variables for MY ARTICLES

Before we go into the details, let's talk about a smart way to use secrets and environment variables to update your **MY ARTICLES** section on GitHub. This process helps keep sensitive information safe while allowing your website to automatically update with new content. Here's how you can set this up:

1. **Setting up secrets**:

    I. Go to your GitHub repository for the website.

    II. Find **Settings**, and then look for **Secrets and variables** in the left-side menu under the **Security** section.

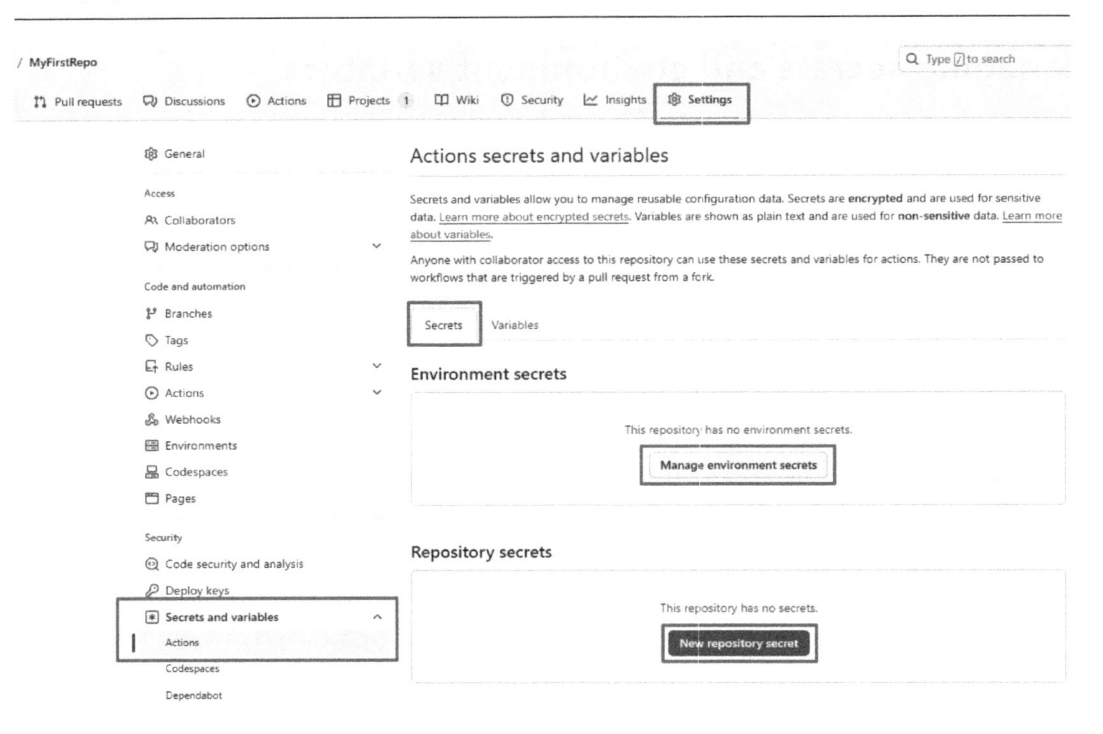

Figure 9.13 – Managing GitHub Action secrets

III. Click **New repository secret**. Name it something like BLOG_API_KEY and put your special key (API key) as the value. It's like hiding your key in a secret compartment that only GitHub Actions can open.

Figure 9.14 – Adding GitHub Action secrets

Managing secrets and environment variables 245

Repository secrets		New repository secret
Name ↕	Last updated	
🔒 BLOG_API_KEY	now	✏️ 🗑️
🔒 MYSUPERSECRET	5 minutes ago	✏️ 🗑️

Figure 9.15 – Viewing GitHub Action secrets

2. **Using environment variables**: In the `.github/workflows` folder, in your `static.yml` workflow file, you can set environment variables. For example, you might set an environment variable for the number of articles to display on your home page.

It's like telling your magic box, *"Use this spell to show five articles on the home page."*

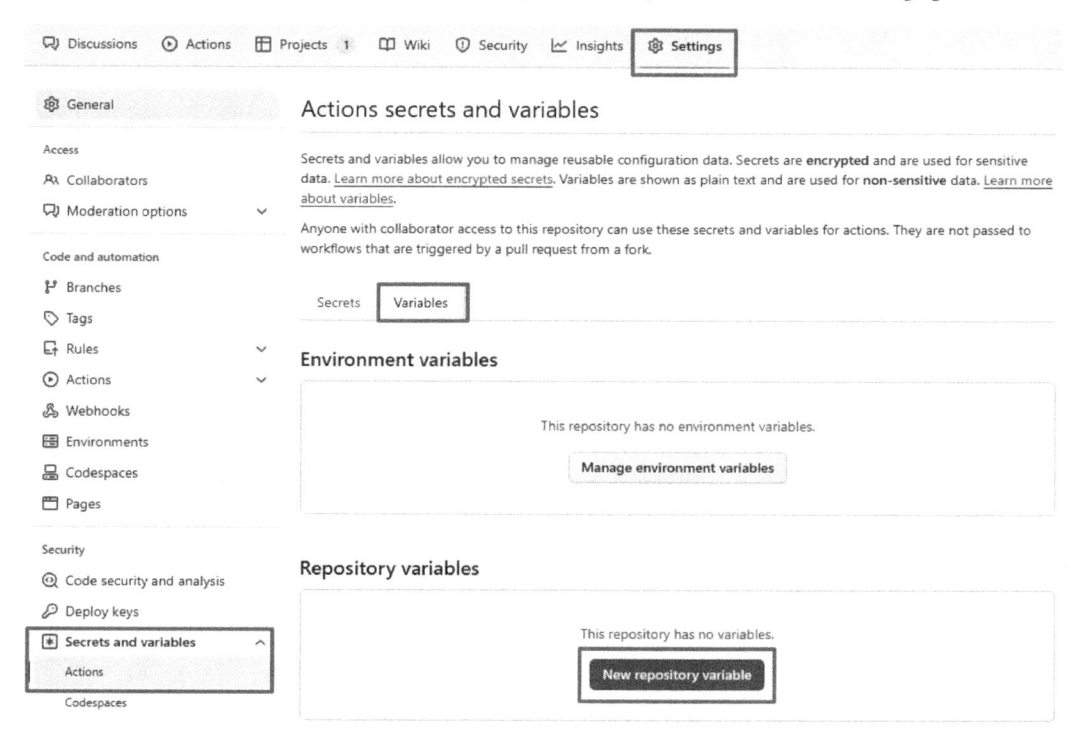

Figure 9.16 – Managing a GitHub Action variable

3. **Accessing secrets in your workflow**: Within your `static.yml` file, you can tell GitHub Actions to use your secret when it needs the key to fetch your blog posts. You'll refer to it by its name, such as `${{ secrets.BLOG_API_KEY }}`.

It's as if you're whispering the secret location of your key to the magic box so that it can fetch your articles safely.

```
steps:
  - name: Checkout
    uses: actions/checkout@v4

  - name: Add Blog API
    uses: blogapiaction/add-blog-api@v1
    with: # Set the secret as an input
      super_secret: ${{ secrets.BLOG_API_KEY }}
    env: # Or as an environment variable
      super_secret: ${{ secrets.BLOG_API_KEY }}

  - name: Set up Node.js
    uses: actions/setup-node@v2
    with:
      node-version: '^18.17.0 || >=20.5.0'
      cache: 'npm'
```

Figure 9.17 – Updating a GitHub Action with secrets and variables

4. **Adjusting environment variables as needed**: You can change these variables directly in your workflow file whenever you need. For example, if you want to showcase more articles, you just update the number in the environment variable.

With secrets and environment variables, you're making sure that your website stays updated with your latest writings without exposing any sensitive information, and you can adjust the settings as needed without rewriting your entire spell. It's a smart, secure way to keep your portfolio fresh and engaging.

Now that you've learned how to use secrets and environment variables to safely update the **MY ARTICLES** section of your website, let's move forward. Next, you'll find out how to troubleshoot and optimize your GitHub Actions, ensuring everything runs smoothly and efficiently. Think of it as learning how to quickly fix any problems with your robot.

Troubleshooting and optimizing GitHub Actions

Let's say you've set up GitHub Actions to automatically update, build, test, and deploy your one-page portfolio website every time you write a new line of code. But what if it doesn't work as expected, or you want to make it work better? This section will explain how you can troubleshoot and optimize it, explained as if you're learning to fix a bike with no prior experience.

Troubleshooting GitHub Actions

Imagine you've tried to ride your bike, but it won't go. The first thing you'd do is check whether there's something obviously wrong. In GitHub Actions, you have logs for this. Go to the **Actions** tab in your GitHub repository and find the run that didn't work. Here, you can see a detailed log of what it tried to do and where things might have gone wrong, like noticing your bike chain is off.

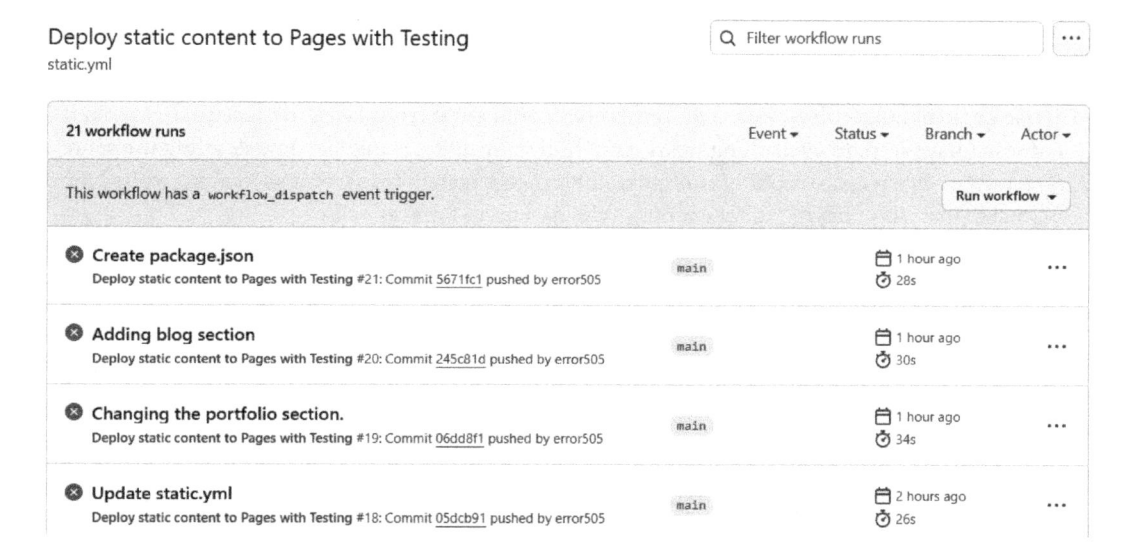

Figure 9.18 – A GitHub Action log

Common issues to look for

It's a good idea to keep an eye for common issues, such as the following:

- **Syntax errors**: Checking for syntax errors in your .yml file means ensuring that the instructions you've given are clear and in a language the GitHub Action understands.
- **Access issues**: Sometimes, a GitHub Action doesn't have the right keys to access certain parts of your project. In this case, you should check for permission issues.
- **External services**: If your script relies on external services (such as fetching data from somewhere else), ensure those services are up and running. It's like checking the weather before you plan a bike ride.

Optimizing GitHub Actions

It's important to know how to optimize GitHub Actions because it helps improve your development process, saving time and resources and making sure your projects run efficiently. The following are some methods that will aid you in doing so:

- **Reduce runtime**: If your workflow takes a long time to update your website, look for ways to make its tasks simpler or more direct. Maybe you're asking it to check too many things. It's like planning the most efficient route for your bike ride to save time.
- **Use caching**: GitHub Actions can remember some steps from last time (cache them), so it doesn't have to redo everything from scratch. If your updates involve downloading the same resources every time, caching can make the process faster. Think of it as keeping your bike outside instead of having to take it out of the basement for every ride.
- **Optimize for costs**: If you're using paid services or resources in your GitHub Actions, look for ways to minimize their use. It's like making sure you're not taking longer bike routes that require more energy bars than necessary.
- **Regularly update and review**: Just like how bike designs improve over time, GitHub Actions and the tools you use with it get updated. Regularly check whether there's a newer, more efficient way to do things.

Every now and then, take a step back and review your GitHub Actions setup to see whether it still meets your needs as your project grows.

In this section, you've learned how to spot and fix issues with GitHub Actions, kind of like troubleshooting a bike that's not ready to ride. We looked at checking logs to find out what went wrong. We also covered common problems such as syntax errors, access issues, and reliance on external services, which can all stop your GitHub Actions from working smoothly. Just like getting your bike ready for a smooth ride, optimizing your GitHub Actions makes your development process more efficient, saving time and effort.

Summary

This chapter delved into GitHub Actions, showcasing how to automate project workflows. It covered creating workflows, automated testing, deployment, managing secrets, and optimization tips. Learning GitHub Actions simplifies building, testing, and deploying code, essential for efficient project management. The skills you acquired in this chapter will help you streamline collaboration and enhance code quality, readying projects for smooth deployment.

Chapter 10, *Enhancing GitHub Security Measures*, advances into securing your coding environment. It will cover repository permissions, two-factor authentication, vulnerability scanning with Dependabot, secure coding practices, and protecting branches, marking a critical step toward safeguarding your projects on GitHub.

Quiz

Answer the following questions:

1. What is the primary purpose of GitHub Actions?

    A. Automating workflows

    B. Storing code

    C. Manual testing

    D. Social networking

    Answer: A. Automating workflows

2. GitHub Actions can be used to automate tests every time someone pushes code to a repository.

    A. True

    B. False

    Answer: A. True

3. Fill in the blank – in GitHub Actions, a _____ defines a set of jobs that run in response to GitHub events.

    Answer: workflow

4. Which file format is used to define GitHub Actions workflows?

    A. JSON

    B. YAML

    C. XML

    D. HTML

    Answer: B. YAML

5. True or false – secrets stored in GitHub Actions are encrypted and can be used in workflows for sensitive operations.

    A. True

    B. False

    Answer: A. True

6. What GitHub feature is used to automatically deploy code to a hosting service whenever the main branch is updated?

   A. GitHub Codespaces
   B. GitHub Pages
   C. GitHub Actions
   D. GitHub Desktop

   Answer: C. GitHub Actions

7. Fill in the blank – GitHub Actions workflows are triggered by specific _____, such as pushing to a branch or opening a pull request.

   Answer: events

8. Which GitHub Actions feature allows for the reuse of workflows across different repositories?

   A. Actions Composer
   B. Workflow Templates
   C. Reusable Workflows
   D. Shared Actions

   Answer: C. Reusable Workflows

9. True or false – you can use GitHub Actions to automatically update dependencies in your project.

   A. True
   B. False

   Answer: A. True

10. What is required to trigger a workflow in GitHub Actions?

    A. A commit to any branch
    B. A specific event defined in a workflow file
    C. A manual start by a user
    D. A pull request approval

    Answer: B. A specific event defined in a workflow file

10
Enhancing GitHub Security Measures

In this exciting chapter, we'll be diving into the heart of securing our one-page portfolio website on GitHub. As we continue to collaborate with friends on this project, understanding and implementing secure development strategies becomes paramount. We're not just building a website; we're ensuring it's a fortress against potential threats. You'll learn how to set permissions, safeguard your site with **Two-Factor Authentication** (**2FA**), utilize Dependabot for vulnerability scanning, and employ *CODEOWNERS* for added protection.

By the end of this chapter, you'll be equipped to apply best practices for security on GitHub. You'll know how to keep your website's data safe, ensure only authorized users can make changes, and maintain a history of those changes. This knowledge is crucial, not just for the safety of your project but also for the security of your visitors' information.

In this chapter, we're going to cover the following main topics:

- Setting up collaboration in your website repo
- Setting up 2FA for your portfolio website
- Keeping your website safe – vulnerability scanning with Dependabot
- Protecting your one-page website with CODEOWNERS
- Applying secure coding practices

Technical requirements

To follow along with the chapter and implement the security features discussed, you'll need the following:

- A GitHub account
- Access to your one-page portfolio website repository on GitHub
- Basic familiarity with navigating GitHub's interface

For the code examples and more detailed instructions, visit the GitHub repository created specifically for this chapter: `https://github.com/PacktPublishing/GitHub-for-Next-Generation-Coders/tree/main/Chapter%2010`.

This repository will contain all the necessary files and further explanations to help you secure your one-page portfolio website effectively.

Setting up collaboration in your website repo

You're still working on a one-page portfolio website with some friends but you have some security concerns. Now, not everyone should be able to change everything, right? Maybe you want some friends to give ideas and feedback but not mess with the website's design directly. That's where setting up permissions and access controls in your GitHub repository comes in handy. It's similar to deciding who will get the keys to your house and who will have to ring the doorbell.

Let's first understand repository permissions.

- **Owner**: That's you. You can do anything on your repo, from inviting friends to help, setting rules for who can do what, and changing any part of the website.
- **Collaborators**: These are friends you've invited to help with your website. You can decide whether they get a key to the house (make big changes) or just a pass to the yard (suggest ideas).

How to set up permissions

1. Go to your website's GitHub page repository.
2. Locate the **Settings** button of your repository and click on it. Usually, you'll find it in the top-right area.

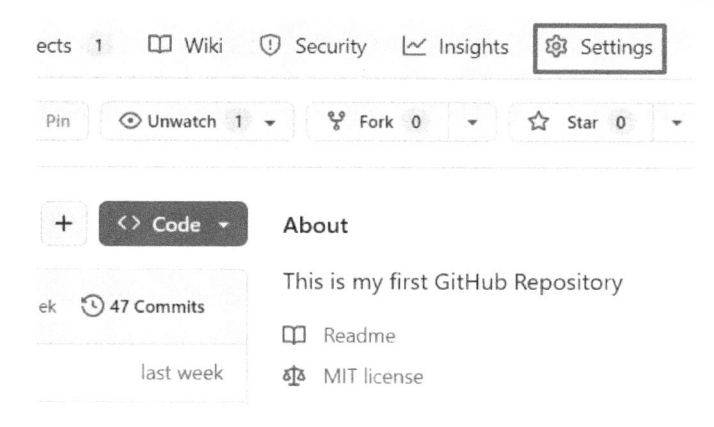

Figure 10.1 – The GitHub Settings button

3. **Access collaborators and teams (organization account repo)**: In the **Access** section, click on **Collaborators and teams.** This is the overview of all your friends who are collaborating with you on the one-page website and their permissions.

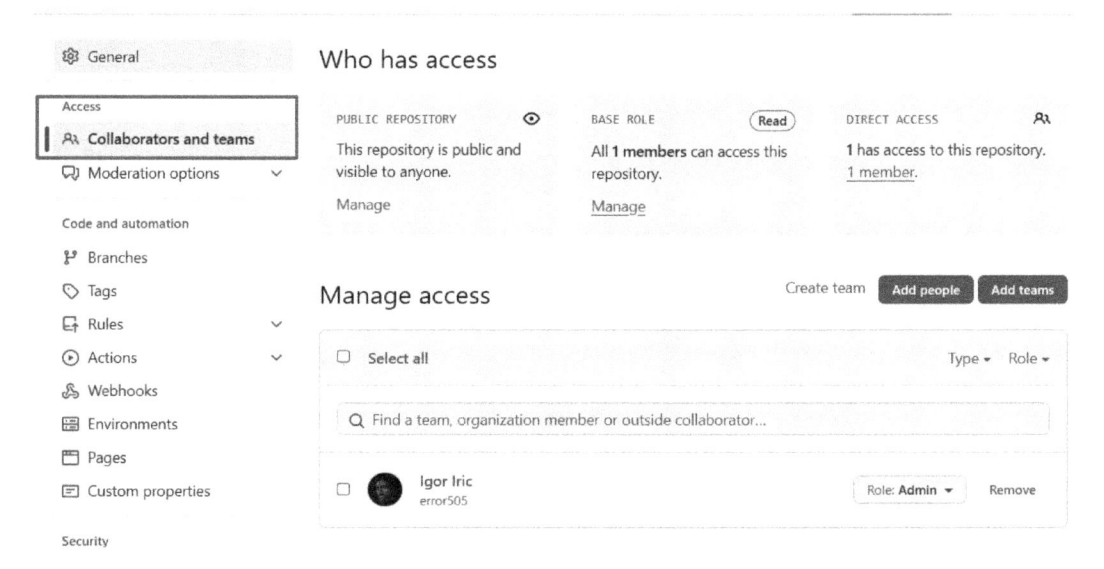

Figure 10.2 – GitHub collaborators and teams

4. **View and adjust permissions**: In the screenshot shown in *Figure 10.2*, you can see that you can also invite friends by their GitHub usernames. For each friend, you can set their repo access level or remove access. A personal GitHub repository has only two permissions:

   - **Ownership**, which gives full control of the repository.
   - **Collaborators**, which applies to users who can read the contents of the repository and write to it by making pull requests.

   If you need more granular access to your repository, you should convert the repository to an organization that has many more roles.

5. **Teams and more (organization account repo)**: If you're working with a lot of people, you can create teams, such as a *design team* or a *content team*, and give each one a different key. Click on **Create team** to start creating new teams.

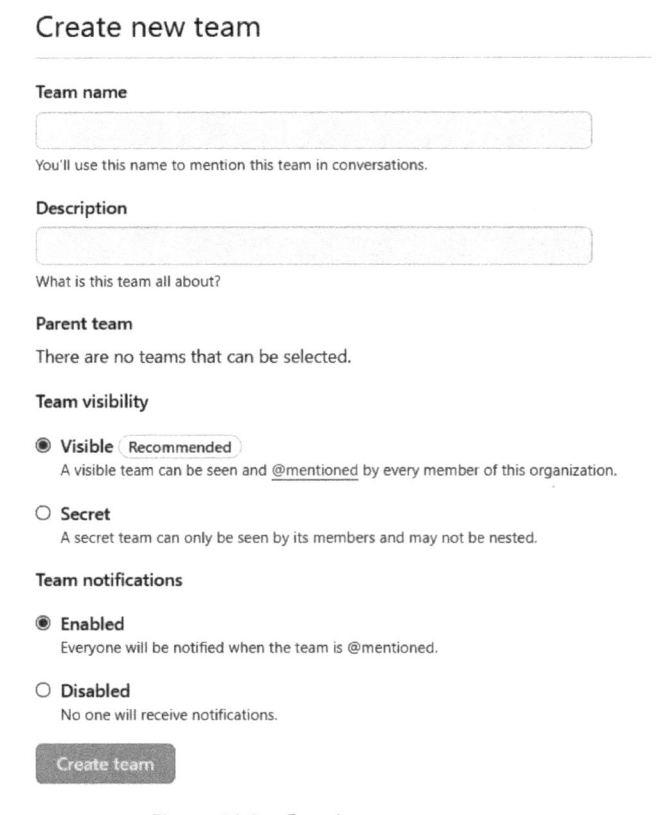

Figure 10.3 – Creating new teams

What are access controls good for?

By setting the access control on your repo, you ensure that only the right people can make changes.

Access control also helps you keep a project organized and know who's responsible for what task in it. It will give you peace of mind to know that your friends helping you on the project won't accidentally remove the main section of your website.

Now that you've learned how to manage who can do what in your website project, let's take the next step to learn how to keep your site even more secure. We'll look into how to set up 2FA for your portfolio website GitHub repository. This is like adding a double lock on your door, making sure that only the right people can get in, even if they have a key.

Setting up 2FA for your portfolio website

You've built a cool one-page portfolio website and stored it on GitHub, a place where you can keep all your website files safe and sound. However, just like how you'd want a good lock on your house to keep your treasures secure, you also want to make sure your GitHub account is extra-protected. That's where 2FA comes in.

What is 2FA?

Think of 2FA as a double-check process to make sure it's really you trying to access your GitHub account. It's like having a door with two locks. The first lock opens with your key (your password), and the second lock opens with a special code (the second factor), which changes every time.

Why use 2FA for your website repo?

Being extra safe has no disadvantages, and it's the same case when it comes to using 2FA. Some of its benefits are:

- 2FA makes it much harder for someone else to sneak into your GitHub account and mess with your website project
- You can be more relaxed knowing your website's code and content are well-protected

How to set up 2FA

To set up 2FA on GitHub, follow these steps:

1. **Go to the settings of your GitHub account**: After logging into GitHub, click on your profile picture in the top-right corner and select **Settings**. It's like going into the control panel of your account.

Enhancing GitHub Security Measures

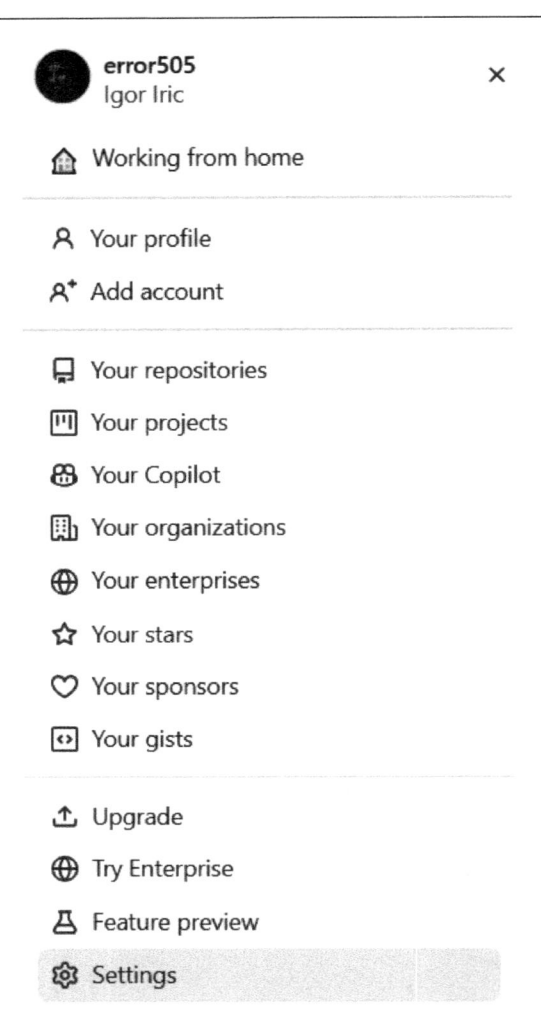

- error505
  Igor Iric
- Working from home
- Your profile
- Add account
- Your repositories
- Your projects
- Your Copilot
- Your organizations
- Your enterprises
- Your stars
- Your sponsors
- Your gists
- Upgrade
- Try Enterprise
- Feature preview
- **Settings**

Figure 10.4 – GitHub account settings

2. **Find the security section**: On the sidebar in the **Settings** menu, look for **Password and authentication**. This is where you can manage your account's defenses.

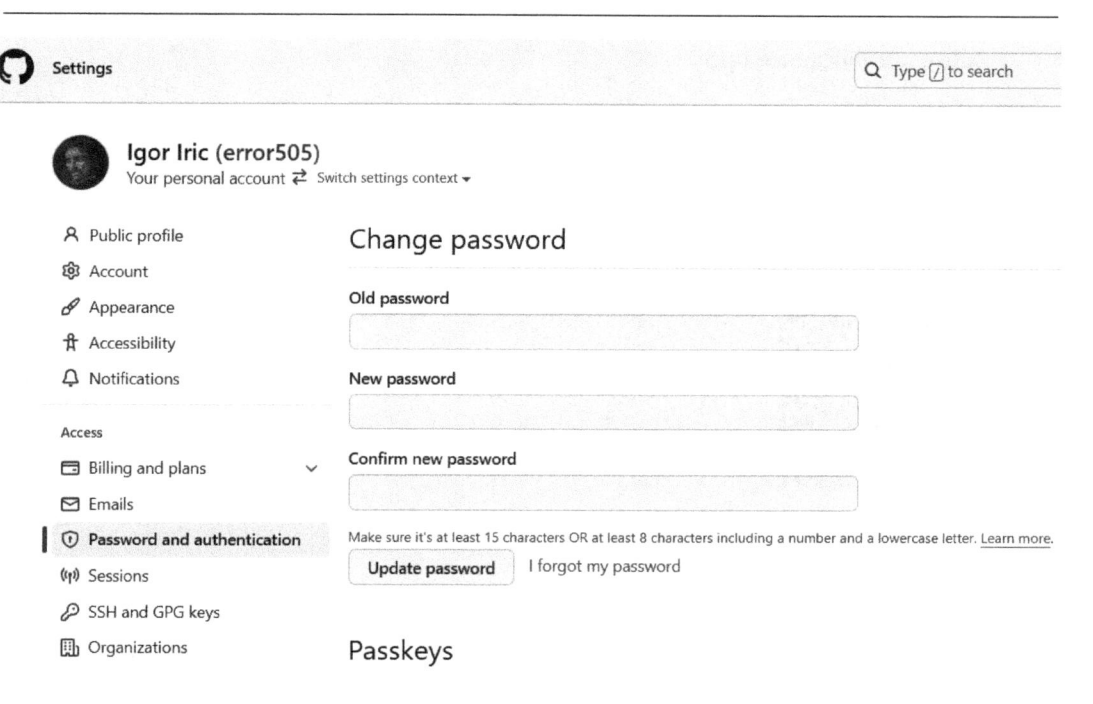

Figure 10.5 – GitHub's Password and authentication section

3. **Start the 2FA process**:

    I.   Find the option for **Two-Factor Authentication** and click on **Enable two-factor authentication**. It's like deciding to add that second lock to your door.

    II. Scan the QR code, which will be displayed to you by your mobile device's app, such as Microsoft Authenticator. After you have scanned the QR code, the app will generate a six-digit code that you can enter on GitHub.

Setup authenticator app

Authenticator apps and browser extensions like 1Password, Authy, Microsoft Authenticator, etc. generate one-time passwords that are used as a second factor to verify your identity when prompted during sign-in.

Scan the QR code

Use an authenticator app or browser extension to scan. Learn more about enabling 2FA.

Unable to scan? You can use the setup key to manually configure your authenticator app.

Figure 10.6 – Setting up the authenticator app

4. **Choose your 2FA method**:

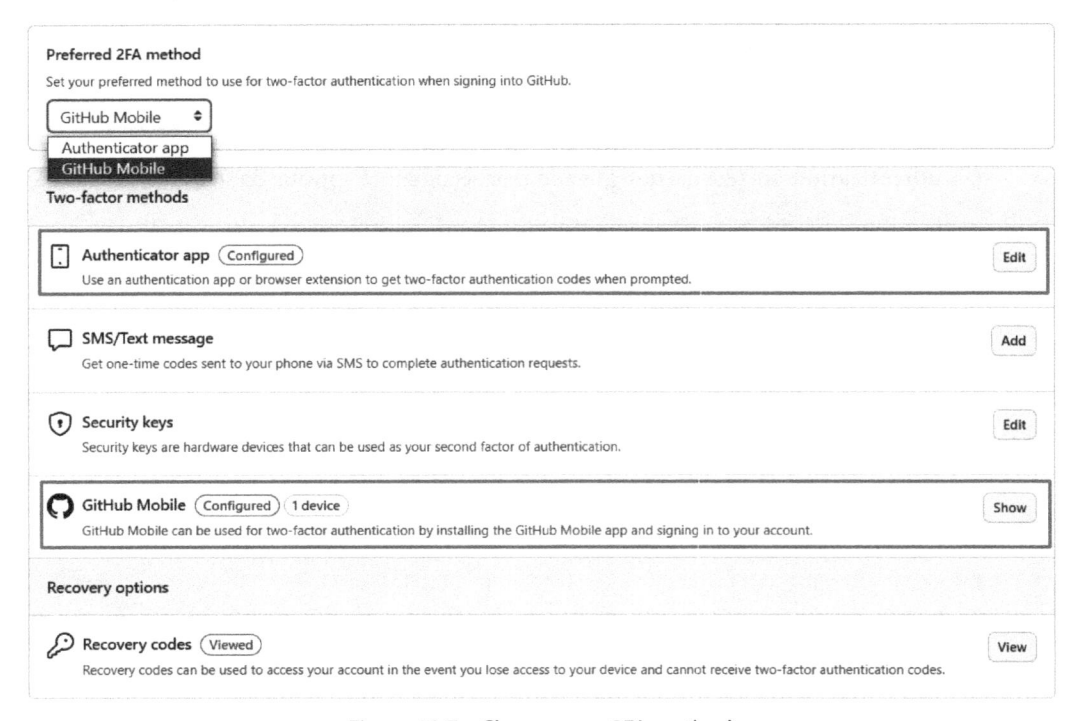

Figure 10.7 – Choose your 2FA method

- **SMS/Text message**: You'll get the second code sent to your phone as a text message. It's straightforward but requires your phone to be handy.
- **Authenticator app**: Use an app such as Google Authenticator, Microsoft Authenticator, or Authy. The app generates the code for you. It works even if your phone isn't connected to the internet.

5. **Follow the setup instructions**: GitHub will guide you through the setup process. It involves verifying your password (your first lock) and then setting up how you'll receive your 2FA codes (your second lock).
6. **Backup codes**: During setup, GitHub gives you backup codes. These are like spare keys. Click **Download** to download your recovery codes to your device. Save them somewhere safe but accessible, just in case you can't use your phone or authentication app.
7. After saving your two-factor recovery codes, click **I have saved my recovery codes** to enable 2FA for your account.
8. **Configure 2FA using GitHub Mobile**: You can have multiple options for 2FA, one of which is using the GitHub Mobile app when signing into your GitHub account.

Figure 10.8 – The 2FA method using GitHub Mobile

You just have to enable 2FA and install the GitHub mobile app from your mobile app store to sign in to your GitHub account.

Setting up 2FA is more like adding a high-tech security system to your house. It's a bit more effort to get in the door, but it means your website project – and all the hard work you've put into it – is much safer. Next, we will learn about another GitHub feature that will help you add security.

Keeping your website safe – vulnerability scanning with Dependabot

To make your website run smoothly and look good, you can use various tools and libraries to develop it, kind of like using different ingredients to bake a cake. But what if one of those ingredients was spoiled? It could ruin your cake, right? In the world of web development, *spoiled ingredients* are vulnerabilities in the tools and libraries you use, which could make your website unsafe. That's where vulnerability scanning and Dependabot come into play on GitHub.

What is vulnerability scanning?

Vulnerability scanning with Dependabot is like having a gadget that scans all your ingredients to make sure they're safe to use. On GitHub, it scans the tools and libraries (called **dependencies**) that your website uses to find any known vulnerabilities.

Why is it important?

Just like you wouldn't want to use spoiled ingredients in a cake, you also don't want to use vulnerable dependencies that could make your website an easy target for bad actors. It ensures your website remains safe for everyone who visits it.

And what's Dependabot?

Dependabot is like a helpful robot in your kitchen that not only tells you when an ingredient is spoiled but also automatically replaces it with a fresh one. On GitHub, Dependabot automatically updates your dependencies if it finds vulnerabilities.

How does it work?

If Dependabot finds a vulnerability in one of your website's dependencies, it will alert you and can automatically create a *pull request* to update it to a newer, safer version. Creating a pull request in this scenario would be like swapping out the vulnerable dependency for a safe one.

Setting up Dependabot

You enable Dependabot in your website's GitHub repository settings. It's like turning on the automatic ingredient checker in your kitchen. To navigate to the Dependabot settings, go to **Code**, and then **Code security and analysis**. You can refer to *Figure 10.9* to check whether you are on the right screen.

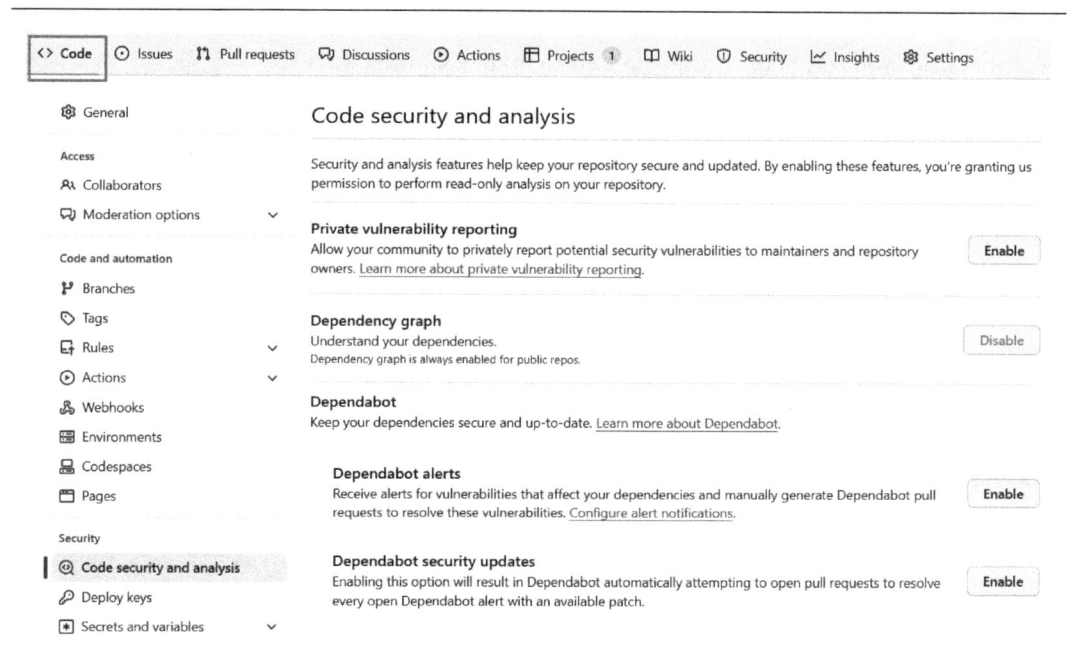

Figure 10.9 – Setting up Dependabot

To enable Dependabot in your repository, you should enable **Dependabot alerts** and **Dependabot security updates**, as shown in *Figure 10.10*.

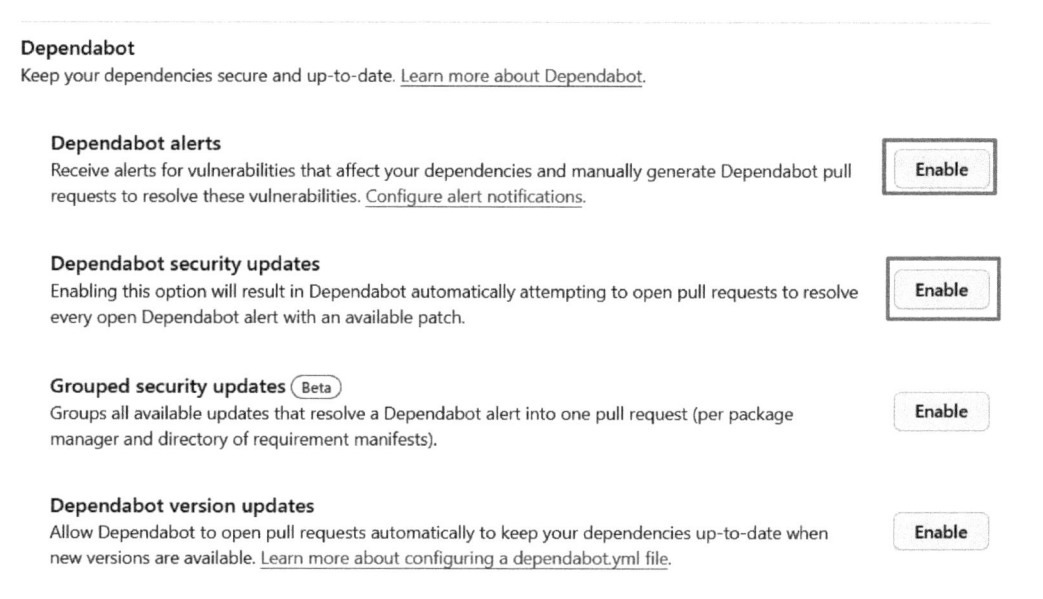

Figure 10.10 – Enabling Dependabot alerts

Dependabot helps you take your security up a notch in the following ways:

- **Scanning for vulnerabilities**: Dependabot regularly checks the dependencies your website uses. When it finds a problem, it alerts you by sending an email to you and a notification on GitHub, kind of like a smoke detector going off when it detects something wrong.

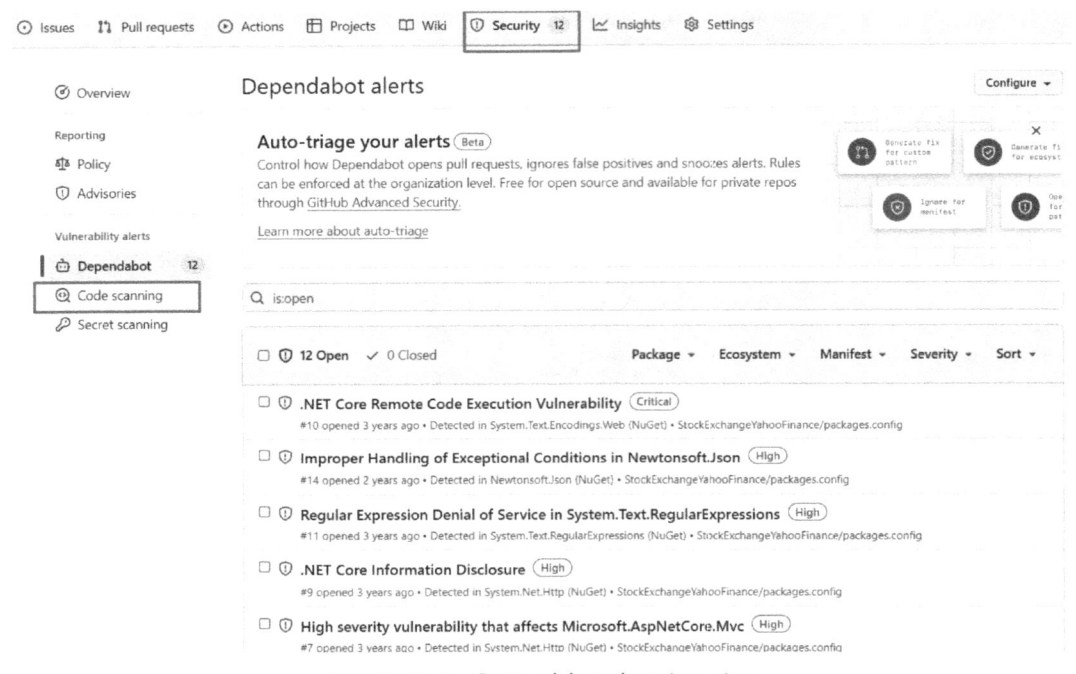

Figure 10.11 – Dependabot alerts in action

- **Automatically fixing issues**: If Dependabot finds a safer version of a problematic dependency, it will automatically prepare an update. It's as if it's not only telling you the milk is spoiled but also going to the store to get fresh milk for you.

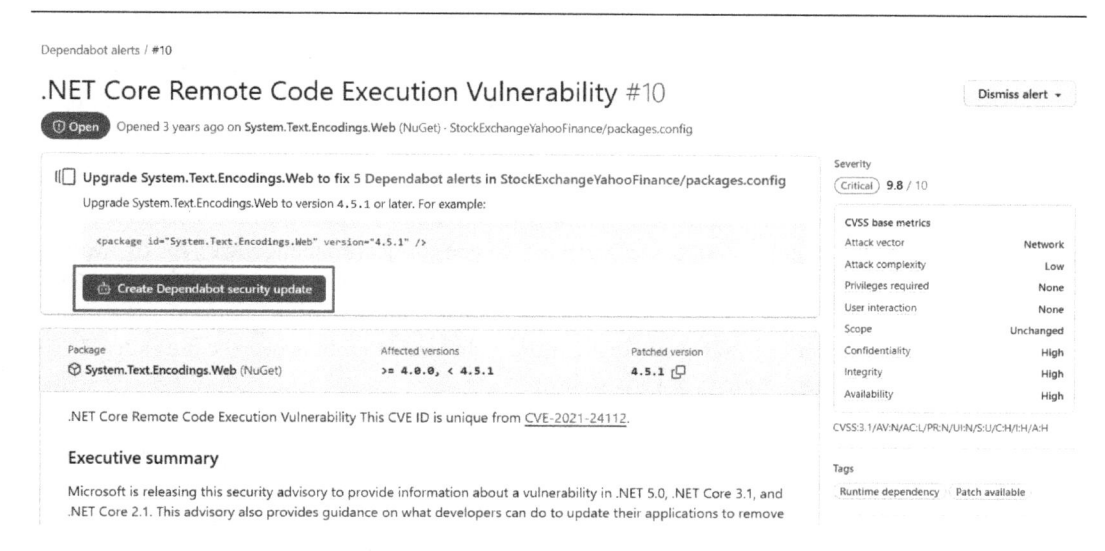

Figure 10.12 – Fixing issues found by Dependabot

- **Reviewing and merging**: You can review the update (the pull request) that Dependabot made. If everything looks good, you merge it into your project, updating the dependency. It's like checking the quality of new milk and then putting it into the fridge for use.

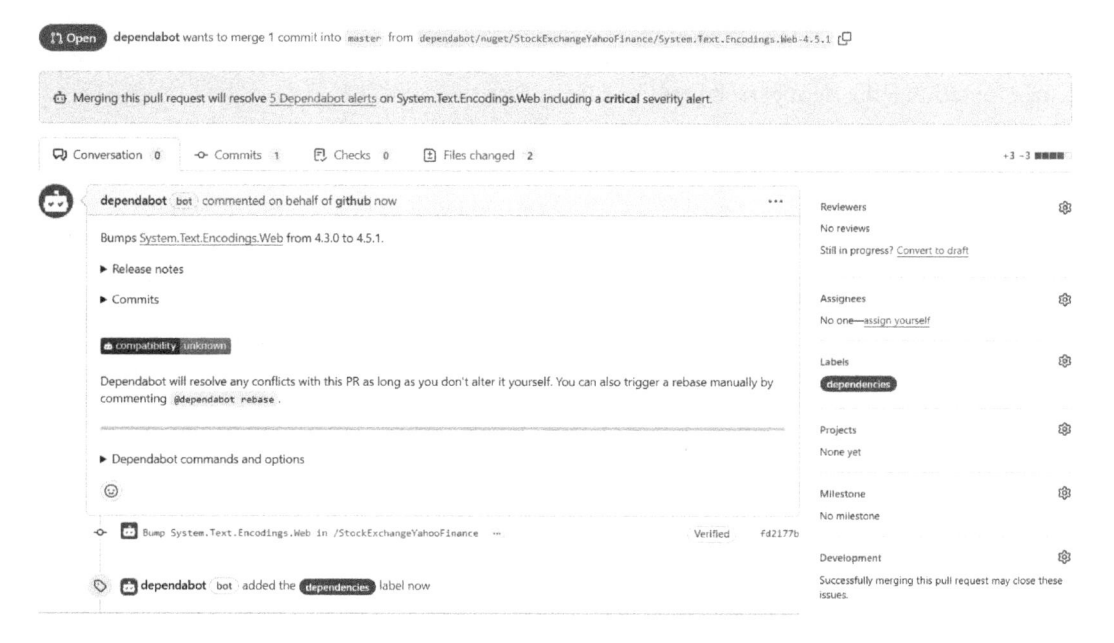

Figure 10.13 – Merging fixed issues

Regular scans and updates mean your website is less likely to have vulnerabilities that could be exploited. Automating the update process saves you the hassle of manually checking and updating each dependency. Knowing that Dependabot is keeping an eye on things lets you focus on other aspects of your website or new projects.

You've learned how to keep your website safe by using vulnerability scanning with Dependabot on GitHub. This tool helps you identify and fix security issues in your project's dependencies, making your website more secure. Next, we'll talk about protecting your one-page website with CODEOWNERS on GitHub. This feature allows you to designate specific people of your team to review and approve changes to certain parts of your project, ensuring that the right team members oversee the code. This helps maintain the quality and security of your website.

Protecting your one-page website with CODEOWNERS

As you're working on your one-page portfolio website with your friends, you've got different parts of the website you're each responsible for. Now, you want to make sure that any changes to these parts are checked by the right person before they become part of the main website. This is where something called **CODEOWNERS** comes into play on GitHub.

What is CODEOWNERS?

Think of CODEOWNERS as a list you put up that says who's in charge of what. For your website, it's a file you add to your project that lists who oversees different parts of your site. So, if someone wants to change something, the right person gets to review it first.

Creating the CODEOWNERS file is easier than you think; you can do it in a few easy steps:

1. First, you create a new file named `CODEOWNERS` and put it in the root of your repository or `.github` folder in your project. It's like putting that list in a place where everyone can see it.
2. The next thing you have to do is to add names to specify who is responsible for what. Inside the `CODEOWNERS` file, you write down which parts of the website each person is responsible for. For example, if your friend Alex is good at styling, you might say that Alex is the owner of the CSS file.

It might look something like this – `*.css @Alex`. This means that Alex is the boss of the `styles.css` file.

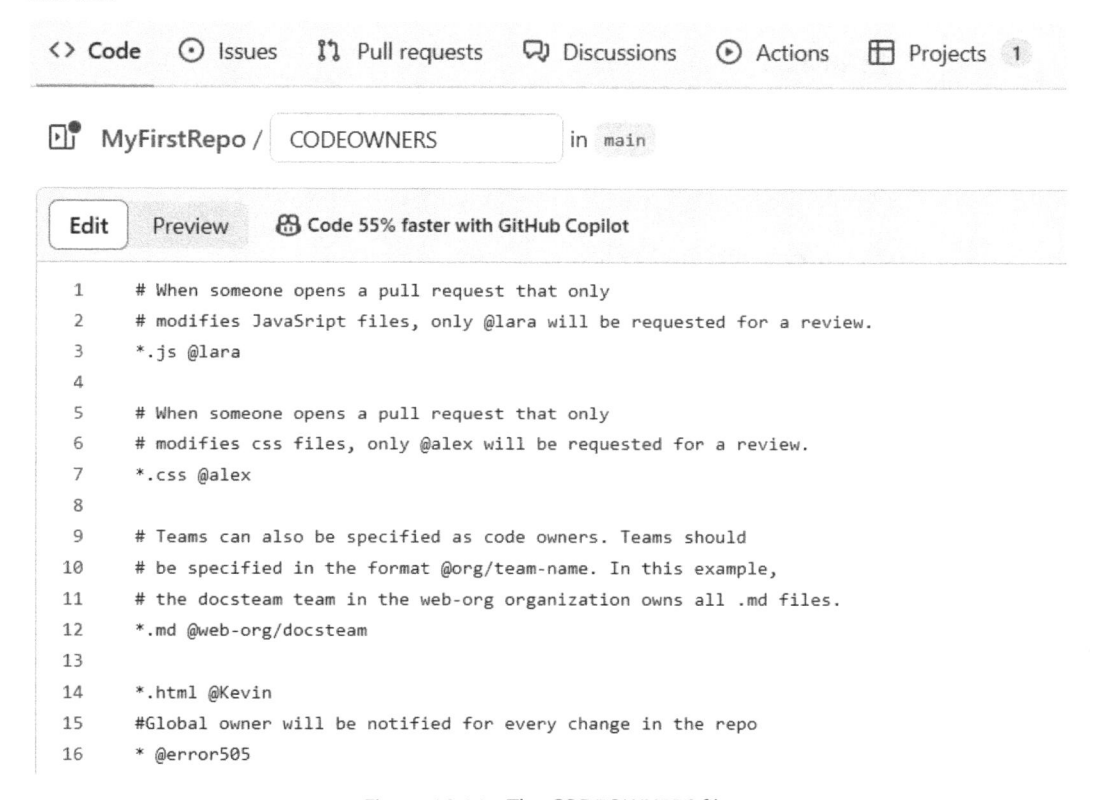

```
# When someone opens a pull request that only
# modifies JavaSript files, only @lara will be requested for a review.
*.js @lara

# When someone opens a pull request that only
# modifies css files, only @alex will be requested for a review.
*.css @alex

# Teams can also be specified as code owners. Teams should
# be specified in the format @org/team-name. In this example,
# the docsteam team in the web-org organization owns all .md files.
*.md @web-org/docsteam

*.html @Kevin
#Global owner will be notified for every change in the repo
* @error505
```

Figure 10.14 – The CODEOWNERS file

How does CODEOWNERS help?

CODEOWNERS adds automatic reviews, meaning that when someone makes a change to a part of the website listed in CODEOWNERS, GitHub automatically knows to ask the right person (such as Alex for the CSS file) to review the changes before they're added to the main project.

CODEOWNERS makes things safe, helping to prevent mistakes or unwanted changes from slipping through because the person who knows a certain part best gets to check it first.

This way, you can organize who looks after what, making sure everyone's working together smoothly without stepping on each other's toes.

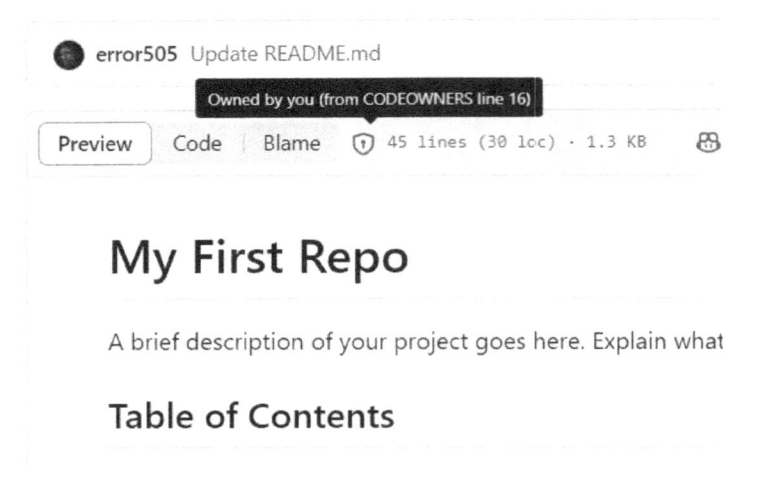

Figure 10.15 – File protected by CODEOWNERS

Let's say you're great at writing documentation, so you're the CODEOWNER for the documentation on the website. Anytime someone wants to update the website's documentation, you'll be asked to review it. It ensures that the website's message stays clear and true to what you envisioned.

You'd add a line to the CODEOWNERS file such as `*.md` or `*.txt @YourUsername`, indicating that you're in charge of the content directory.

By using CODEOWNERS, you're putting up a friendly *guard* on different parts of your website project. It's like saying, "*Hey, I've got this part covered, but I'm happy to look over your ideas before we make it official.*" It keeps your project safe and ensures that changes are always checked by someone who knows the ropes.

So, setting up CODEOWNERS is like creating a safety net for your one-page portfolio website, ensuring that every change is given the thumbs up by the right person, keeping your project on track and looking its best.

Now that you've set up CODEOWNERS to make sure the right people review changes to our website, let's focus on making your site even safer. Next, we're going to learn about applying secure coding practices. This is like learning the best ways to lock up our website at night, making sure it's safe from any unwelcome visitors. You'll make sure your website isn't just well-organized but also well-written and secure.

Applying secure coding practices

After you've built your one-page portfolio website and are using GitHub to manage and share your code, you'll want to keep your personal information safe online. It's essential to protect your website's code from potential threats.

Building a one-page portfolio website on GitHub is exciting, but ensuring it's secure is crucial. It's like inviting guests into your digital home; you want to make sure it's safe and secure for everyone.

Let's dive into how you can apply secure coding practices using GitHub features.

Why secure development matters

Think about your website like a diary with a lock. You wouldn't want strangers reading your secrets, right? Similarly, when developing software, you need to protect it from unwanted access or changes. This means making sure only the right people can see or modify your website's code and content.

Planning a secure development strategy

We're going to look at several key practices that act as layers of security, from keeping sensitive data out of public view to regulating who can make changes to our project:

- **Protecting information using .gitignore**: Just like you wouldn't leave your diary open for everyone to read, you need to keep your website's data safe from those who shouldn't see it. Imagine you have a key to a secret page where you keep sensitive information away from prying eyes. In your GitHub repository, the `.gitignore` file acts like this secret page, telling Git which files or directories to ignore and not upload to GitHub. This could include configuration files with passwords or API keys. This file can be created in the root of your repository to tell Git which files and directories to ignore when you make a commit. *Figure 10.16* shows `.gitignore` in action.

268   Enhancing GitHub Security Measures

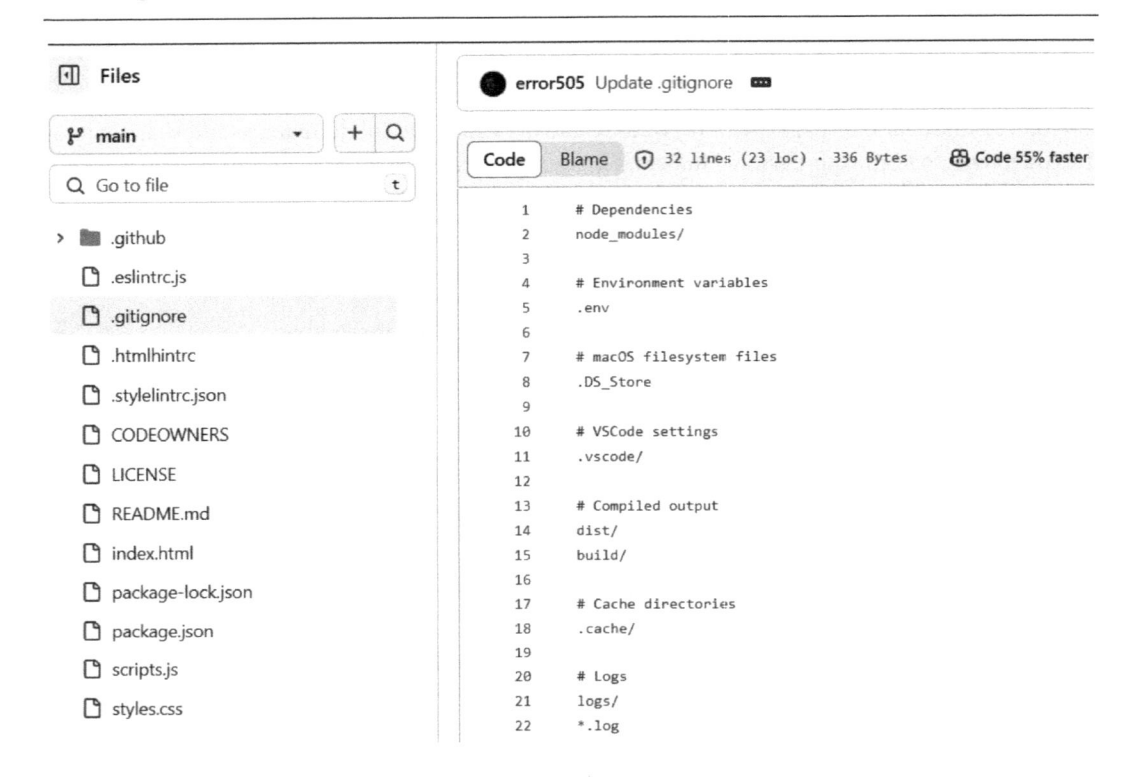

Figure 10.16 – The .gitignore file

GitHub provides a way to store sensitive information such as API keys securely using **secrets**. It's like having a digital safe within your GitHub repository where only you and authorized collaborators can access the combination. To access this secure spot, you just navigate to your repository on GitHub, click on **Settings**, and then find the **Secrets and variables** section on the sidebar. Here, you can add, update, or remove secrets.

Applying secure coding practices   269

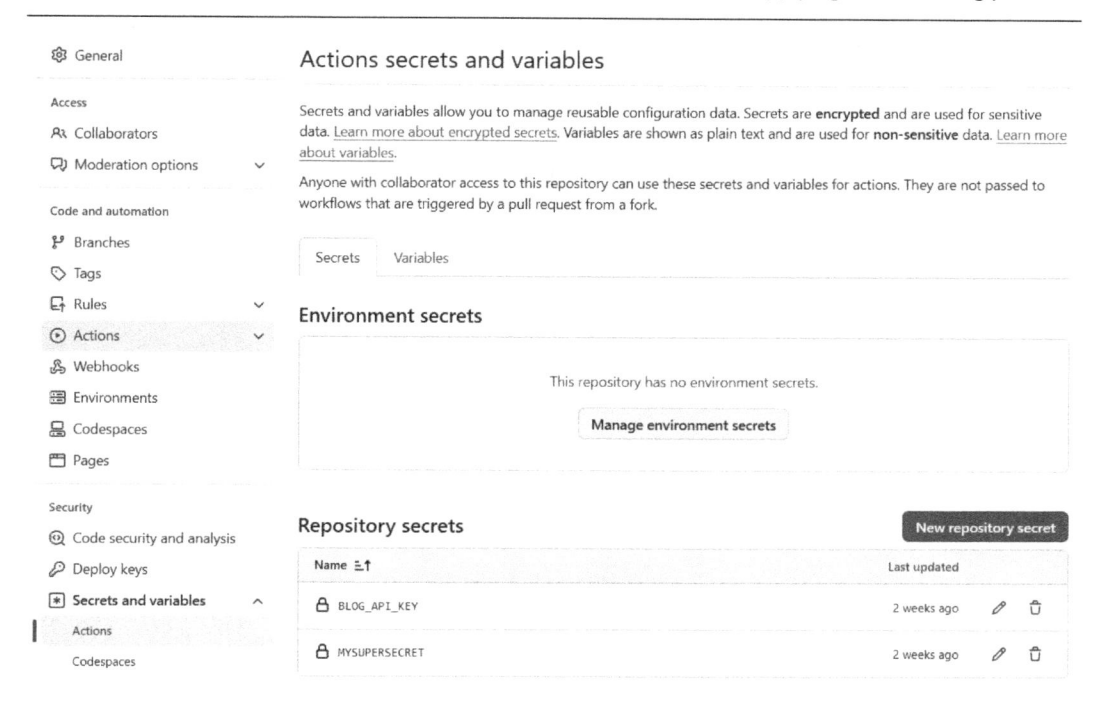

Figure 10.17 – Repository secrets and environmental variables

- **Who's knocking?**: It's important to check who's accessing your website, just like you'd look through the peephole before opening your front door. This involves making sure that anyone who accesses your site has the right to do so and that rules are followed before making any changes:

    - OK, but how can we be sure has the proper rights? It can be easily done by setting up the **branch protection rules**. *Figure 10.18* shows where you can find this setting.

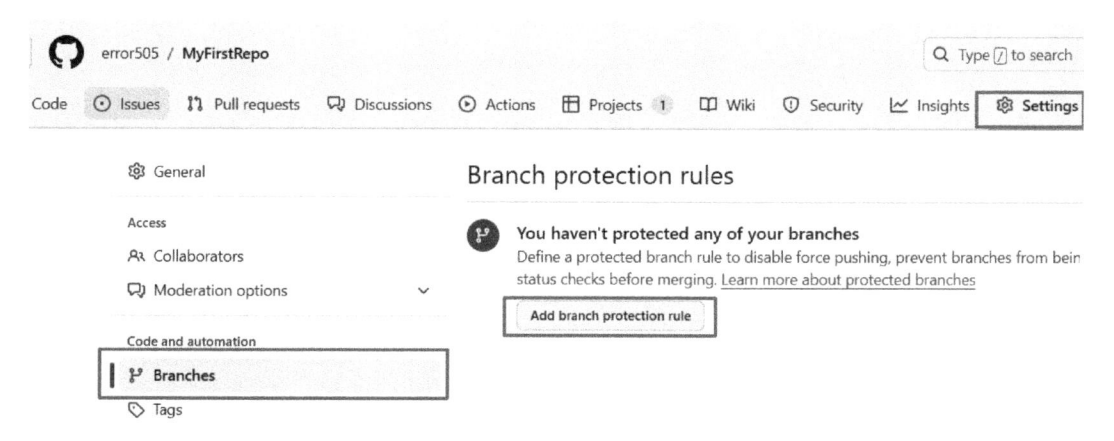

Figure 10.18 – Branch protection rules

- Setting up branch protection rules is like installing a security system on your door. It ensures that changes to your site can only be made by those who have the right permissions. You can enforce rules such as requiring a review before changes are merged and making sure only trusted updates make it to your website.

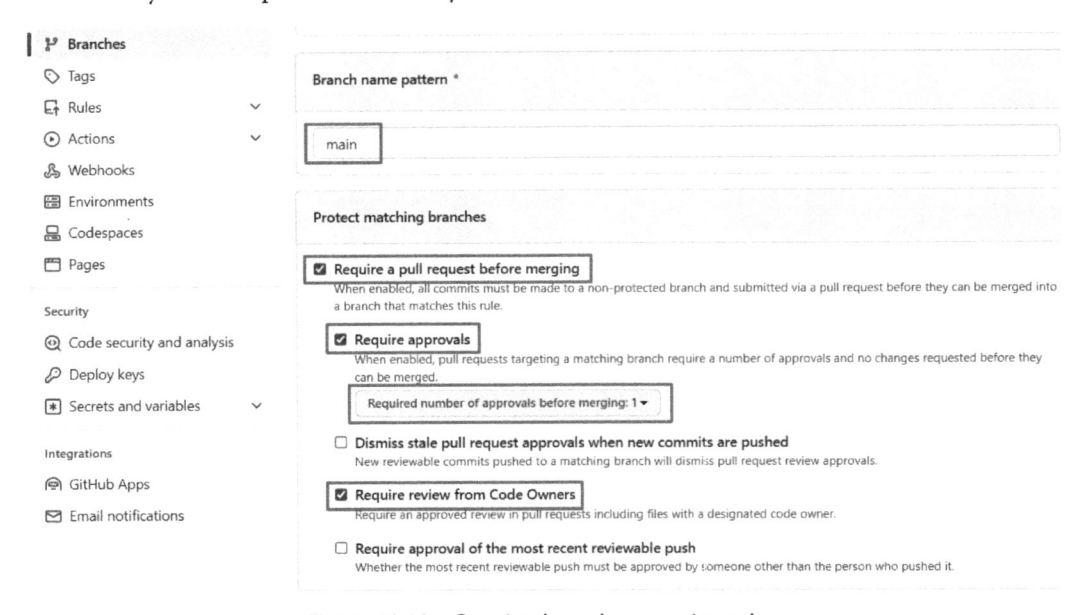

Figure 10.19 – Creating branch protection rules

- Setting up authentication and permissions is crucial for the security of your project. When someone wants to contribute to your site, make sure they are who they say they are. This involves setting up proper authentication mechanisms on your website and within GitHub, ensuring that contributors have the correct level of access.

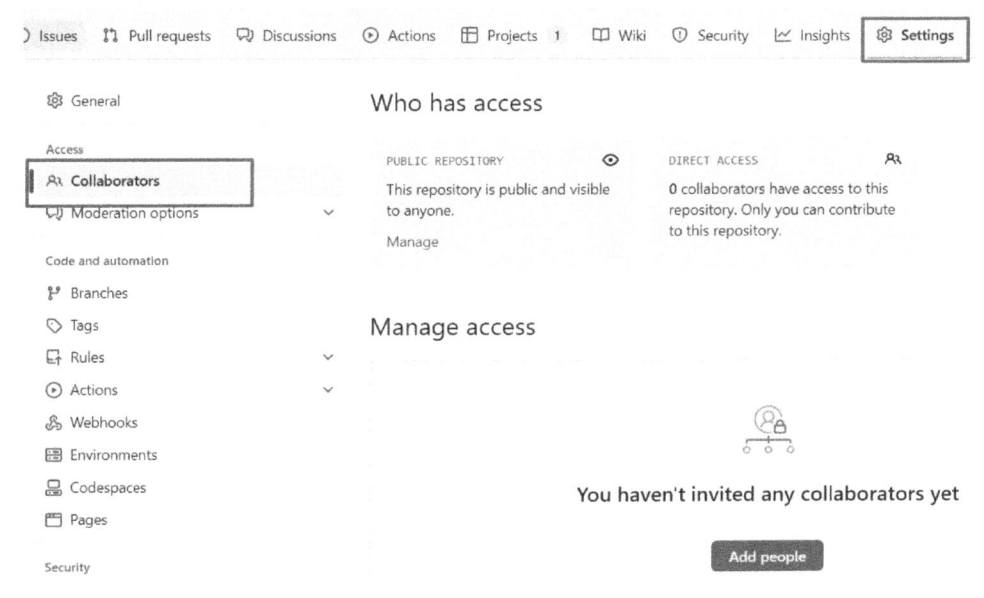

Figure 10.20 – Adding collaborators to the repository

- **Keeping a history**: Imagine you find a page torn out of your diary. You'd want to know when it happened and maybe who did it. Keeping logs or records of what happens on your website can help you spot and understand when something goes wrong.

  Every change made to your website is recorded in the commit history, just like journal entries. It's a detailed logbook of who did what and when. If something goes wrong, you can look through this history to find clues and revert changes if necessary.

Commits

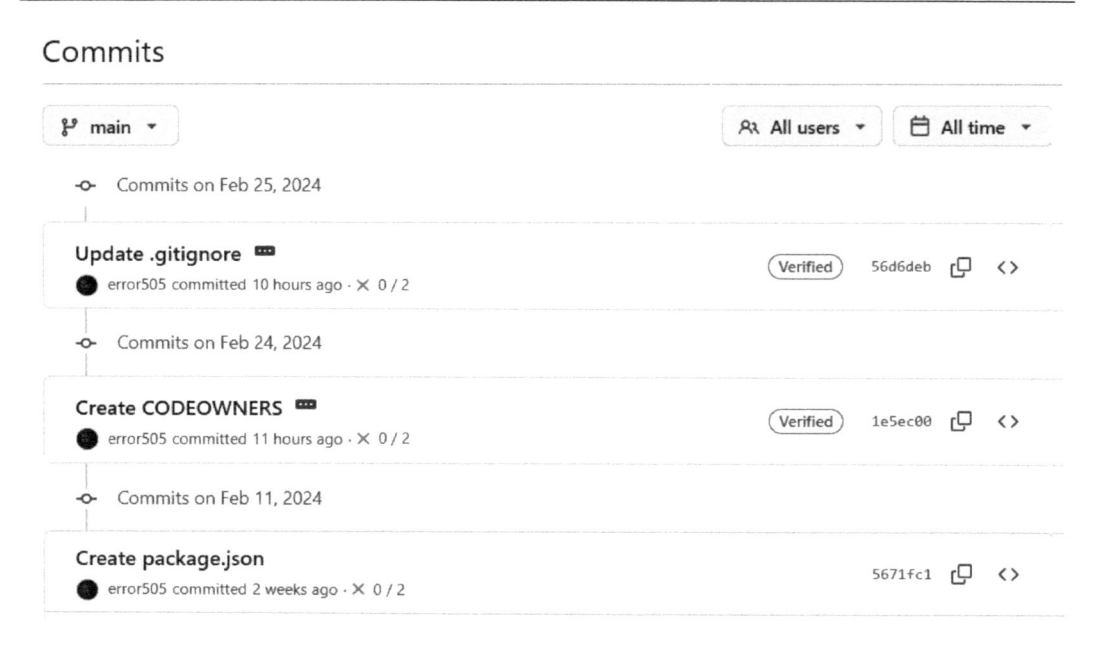

[Screenshot of GitHub commits page showing branch "main", "All users", "All time" filters]

Commits on Feb 25, 2024

Update .gitignore
error505 committed 10 hours ago · ✕ 0 / 2
Verified · 56d6deb

Commits on Feb 24, 2024

Create CODEOWNERS
error505 committed 11 hours ago · ✕ 0 / 2
Verified · 1e5ec00

Commits on Feb 11, 2024

Create package.json
error505 committed 2 weeks ago · ✕ 0 / 2
5671fc1

Figure 10.21 – Git history

- **GitHub's security features**: GitHub has built-in tools to help keep your project secure, like a security system for your digital home:

    - The **Security** tab is where you can manage security features. It's like the control panel for your home security system.

    - With GitHub Actions, you can automate your workflow, including security checks and tests. It's like having a guard that automatically checks visitors (code changes) at the gate, ensuring they're safe before letting them in. You can set up actions to scan your code for vulnerabilities or ensure coding standards are met before changes are merged.

Applying secure coding practices   273

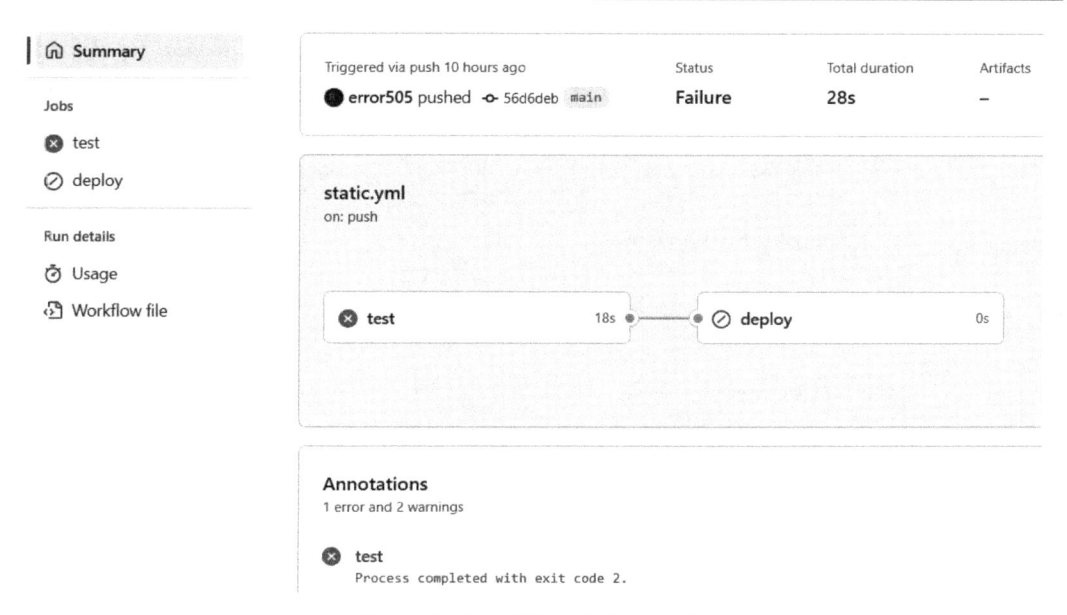

Figure 10.22 – GitHub Actions tests

- Security alerts with GitHub's Dependabot are like a watchtower, scanning the horizon for incoming threats. It alerts you if it finds vulnerabilities in the dependencies your project uses. If Dependabot finds something, it can even suggest or create a pull request to update the vulnerable dependency to a safer version.

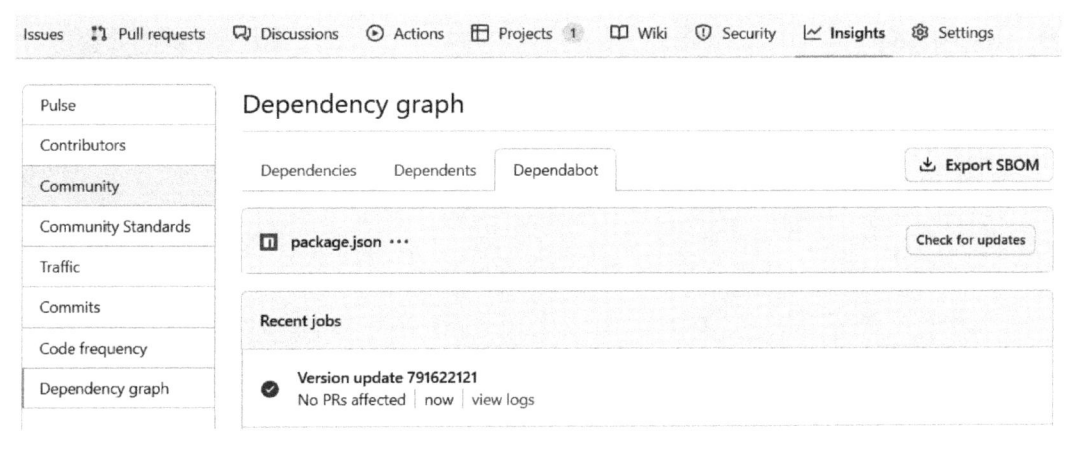

Figure 10.23 – Dependabot alerts

- Improve security by adding *security policies* with a `SECURITY.md` file to your repository. With this file, you tell others how to responsibly report security issues, like putting up a sign that says, "*Lost? Here's how to get help.*"

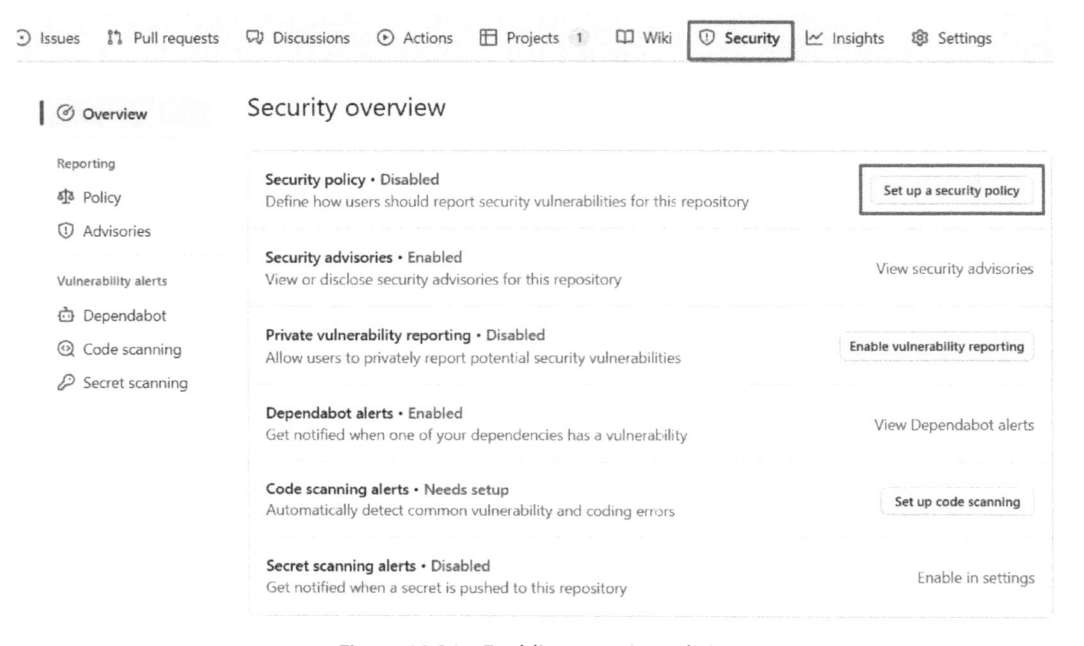

Figure 10.24 – Enabling security policies

Figure 10.25 – Creating a security.md file

- Code scanning with the **CodeQL** feature scans your code for security issues, similar to checking your doors and windows before leaving the house.

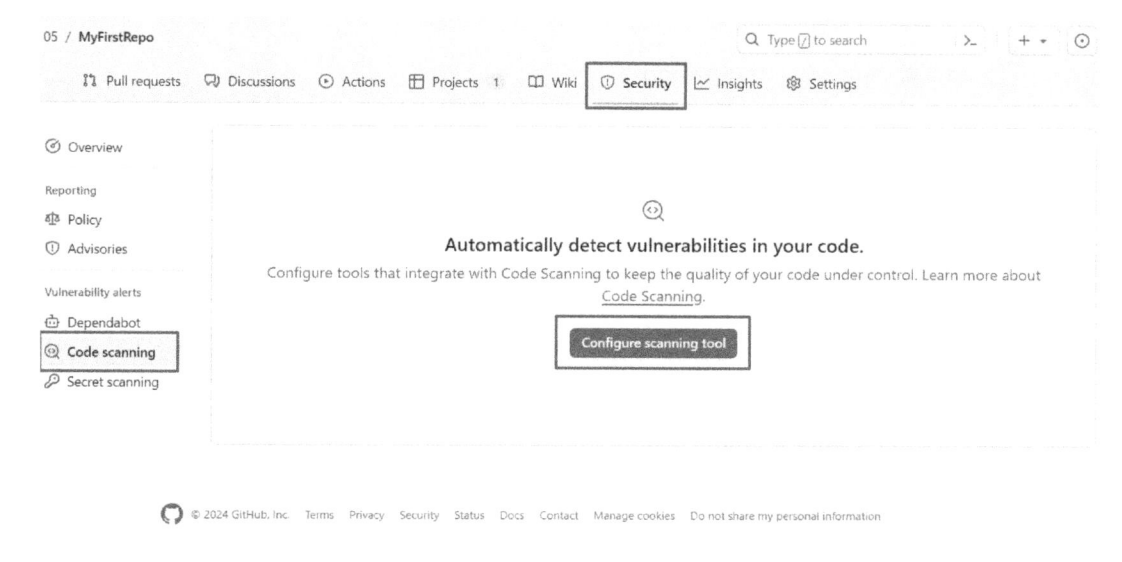

Figure 10.26 – Configuring code scanning

- CodeQL supports compiled and interpreted languages, and it finds vulnerabilities and errors in your code if it is written in the following supported languages:

    - C/C++
    - C#
    - Go
    - Java/Kotlin
    - JavaScript/TypeScript
    - Python
    - Ruby

- Swift

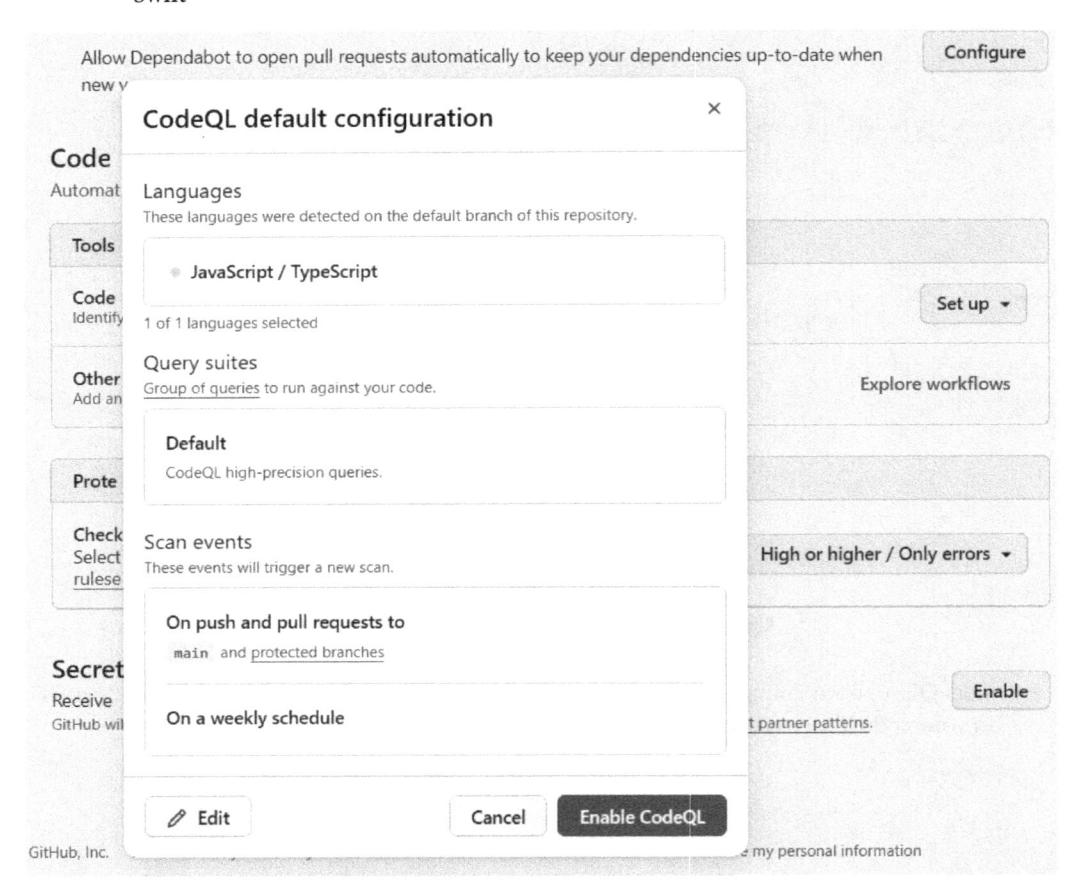

Figure 10.27 – Configuring CodeQL

- The next thing to consider when we talk about security is the *Secret Scanning* feature. What if you've lost your diary, which is locked, but you wrote your secret password on top of it? Not the best idea, as now anyone who finds it can open it without any problems and start reading your diary right there. In the world of web development, especially when building your one-page portfolio website, such keys (e.g., API tokens or passwords) can accidentally be left exposed in your code, posing a risk if someone with bad intentions finds them. Secret scanning on GitHub scans your code for any secrets. If it finds any, it alerts you so that you can retrieve and secure them before they're misused. Just go to the **Secrets** tab of your repository again and enable the **Secret scanning** feature:

Secret scanning

Receive alerts on GitHub for detected secrets, keys, or other tokens. GitHub will always send alerts to partners for detected secrets in public repositories. Learn more about partner patterns.

Figure 10.28 – Enabling Secret scanning

The **Secret scanning** feature will automatically alert you if there are any secrets in your code that you should keep in a safe place. If there aren't any, you may encounter a screen like the one shown in *Figure 10.29*.

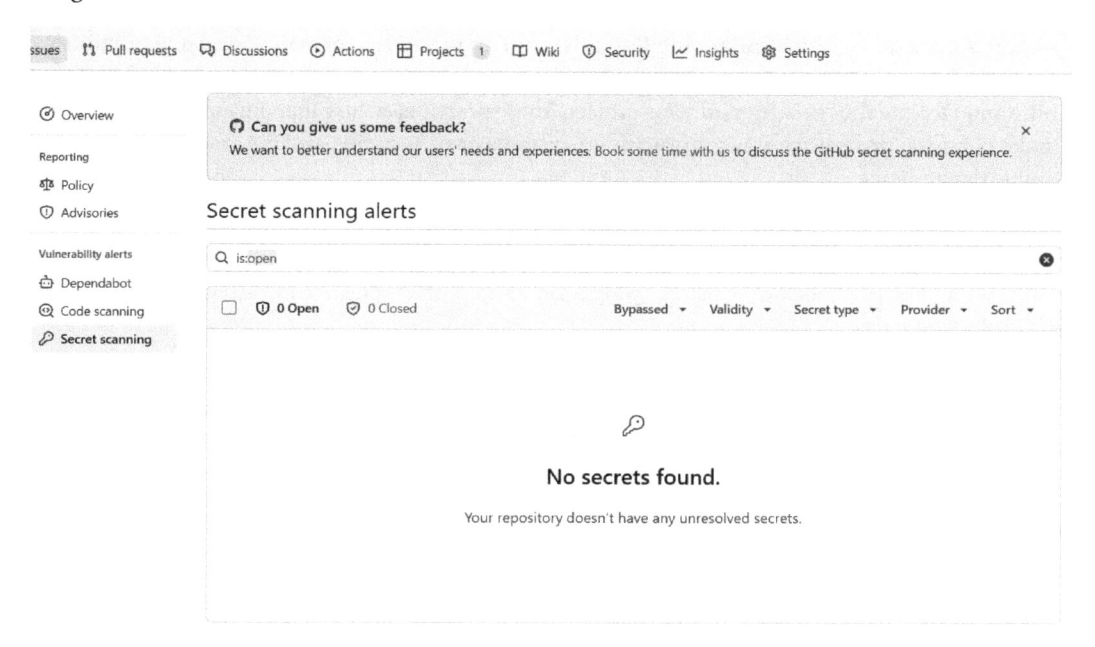

Figure 10.29 – Secret scanning in action

Building and deploying securely

As we go through this chapter, we focus on how to build and deploy your one-page portfolio website securely. This is crucial because it's not just about making a website that looks good; it's also about making sure it's safe for everyone who visits it. Here's what you should keep doing:

- **Continuous learning**: Cybersecurity is like a game where the rules constantly change. You and your team need to keep learning to stay ahead.
- **Writing safe code**: Just like following a recipe ensures your cake doesn't flop, writing code correctly makes sure your website works as intended without opening doors for hackers.

- **Following the rules**: Your website needs to comply with laws and guidelines, much like how certain ingredients need to be declared on food packaging.

By incorporating these security practices into your GitHub workflow for your one-page portfolio website, you're not just building a project; you're also ensuring it's a safe and welcoming place for all your team members.

Summary

In *Chapter 10*, we delved into enhancing GitHub security measures for your one-page portfolio website, focusing on setting permissions, enabling 2FA, and utilizing features such as vulnerability scanning and Dependabot to safeguard your project. These practices ensure that only authorized users can make changes, protect your account with an extra layer of security, and automatically update vulnerable dependencies.

We learned that the introduction of the CODEOWNERS file allows you to specify who reviews changes to certain parts of your project, further securing your work. We also covered the importance of secure coding practices, including using `.gitignore` to protect sensitive information and setting branch protection rules.

In the next chapter, we'll explore setting up **continuous integration** and **continuous deployment** (**CI/CD**) workflows in your project, which is a natural progression from securing your code base to automating and optimizing your development process for efficiency and reliability.

Quiz

Based on the chapter's coverage of securing a one-page portfolio website on GitHub, here are some quiz questions to encapsulate the key points about applying secure development practices:

1. What is the purpose of setting up permissions and access controls in a GitHub repository?

    A. To decide who can contribute code

    B. To lock the repository from everyone

    C. To delete unwanted users

    D. To automate code deployment

    Answer: A. To decide who can contribute code

2. Trur or false – 2FA on GitHub adds an extra layer of security to your account by requiring a second form of verification.

    A. True

    B. False

    Answer: A. True

3. Fill in the blank: _____ automatically updates your dependencies to fix vulnerabilities.

    Answer: Dependabot

4. Which file helps to specify who can review and approve changes to certain parts of your project on GitHub?

    A. `README.md`

    B. `.gitignore`

    C. CODEOWNERS

    D. `.gitconfig`

    Answer: C. CODEOWNERS

5. True or false – branch protection rules can help ensure that changes to your site are only made by individuals with the correct permissions.

    A. True

    B. False

    Answer: A. True

6. What is a key benefit of applying secure coding practices to your GitHub project?

    A. Faster deployment

    B. Increased project size

    C. Enhanced security

    D. Easier collaboration

    Answer: C. Enhanced security

7. Fill in the blank – GitHub's _____ feature scans your project for exposed secrets and alerts you to secure them.

    Answer: secret scanning

8. What GitHub feature allows you to communicate how to report a security vulnerability in your project?

    A. Pull requests

    B. Issues

    C. Security policies

    D. Wiki

    Answer: C. Security policies

9. True or false – it's important to keep a history of changes and logs to spot and understand when something goes wrong with your website.

    A. True

    B. False

    Answer: A. True

10. How does Dependabot help maintain your one-page portfolio website's security?

    A. By writing new content for your website

    B. By automatically updating vulnerable dependencies

    C. By changing your website's design

    D. By promoting your website on social media

    Answer: B. By automatically updating vulnerable dependencies

11
Engaging with the Open Source Community

Welcome to the thrilling journey into the heart of the open source community! In this chapter, we're going to dive into the vibrant ecosystem of open source projects, where innovation, collaboration, and learning intersect. Open source software powers much of the digital world around us, and understanding how to engage with this community can open doors to countless opportunities for growth, learning, and contribution.

By the end of this chapter, you'll know how to explore and select open source projects that align with your interests, make meaningful contributions that are recognized and valued, and navigate the complex landscape of licensing and legal considerations. You'll also learn how to apply open source etiquette and best practices, ensuring that your contributions are not only accepted but also appreciated. Whether you're looking to contribute to your first project or aiming to launch your own, this chapter will equip you with the knowledge and skills needed to thrive in the open source community.

In this chapter, we're going to cover the following main topics:

- Exploring open source projects
- Making meaningful contributions with real examples
- Navigating licensing and legal considerations
- Creating a license for your GitHub repository
- Open source etiquettes and best practices

Technical requirements

Before diving into the chapter, make sure you have a basic understanding of common coding languages (JavaScript and Python) and tools related to the projects you're interested in. You can visit www.w3schools.com to gain quick knowledge about many programming languages used on GitHub. To access the code examples and resources mentioned in this chapter, visit our GitHub repo at https://github.com/PacktPublishing/GitHub-for-Next-Generation-Coders/tree/main/Space%20Explorer%20Game/Game-final.

Exploring open source projects

Welcome to the world of open source! In the world of coding, working together on projects through the internet is not just common; it's also the way most cool stuff gets made! Open source projects are like big, digital playgrounds where coders from all over the world come to play, learn, and create together.

Imagine building a giant LEGO set where each piece is a line of code, and anyone can add a piece, suggest a change, or find a better way to connect them.

Sounds very interesting, but where to start?

First, you would start by finding projects you love. Start by thinking about what you enjoy. Do you like video games, drawing, or maybe helping the planet? There's almost certainly an open source project that fits your interests. GitHub lets you search for projects using keywords, just like searching for your favorite videos online. All you need to do is use the search bar, as shown in *Figure 11.1*.

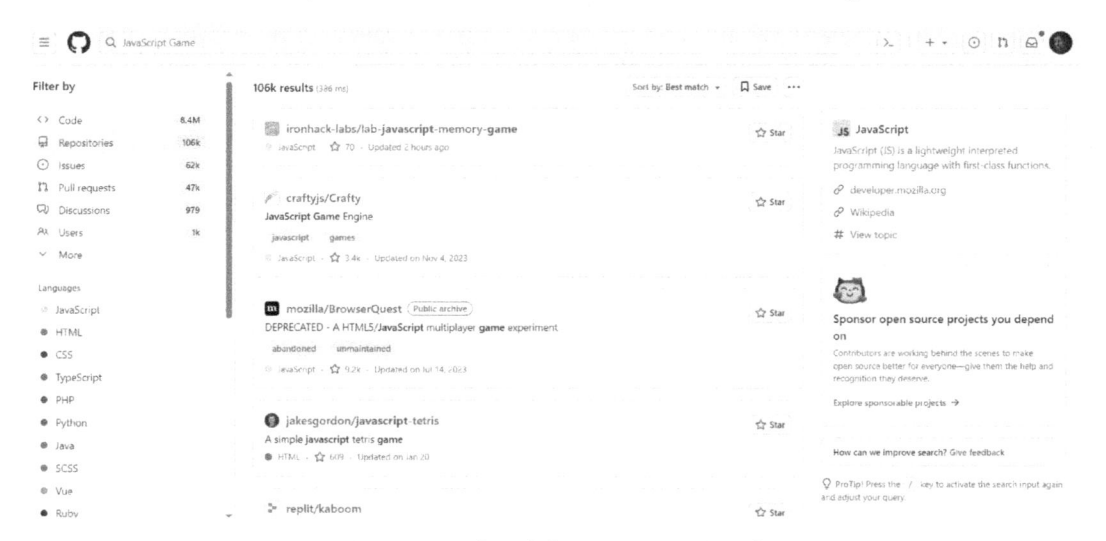

Figure 11.1 – Search for projects on GitHub

Understanding project pages

When you find a project you like, click on it to navigate to its page, which will serve as your map. It shows you everything about the project.

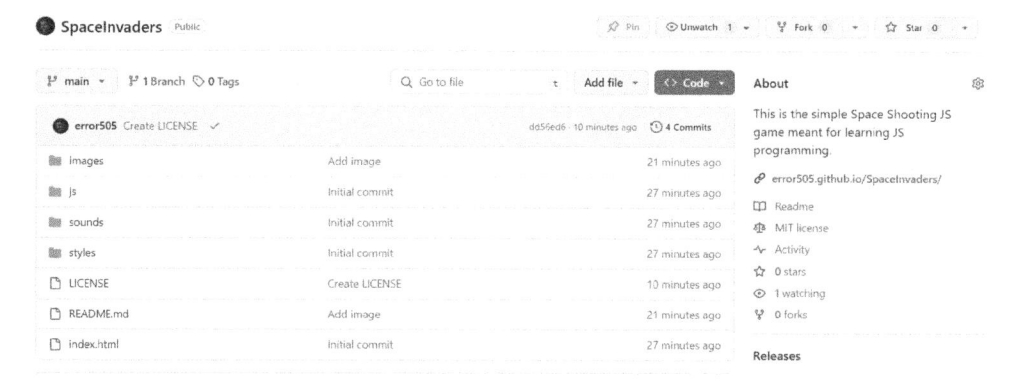

Figure 11.2 – An open source project on GitHub

On the project's page, here's what you'll find:

- The README file, which is like the introductory page of a book. It tells you what the project is about, how you can use it, and how you can help.

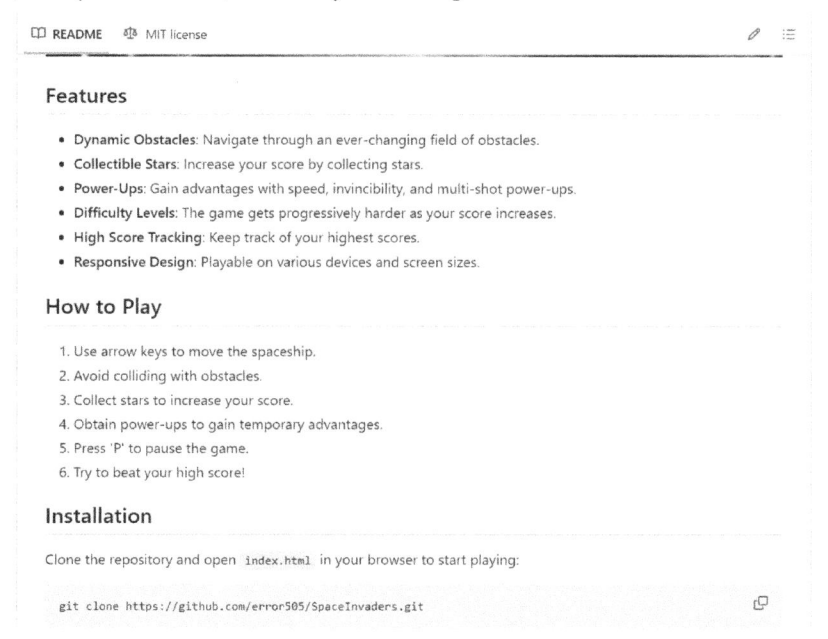

Figure 11.3 – The README file

- The **Issues** tab, which is like a to-do list. It shows things that need to be fixed or added to the project. It's a great place to start if you want to help and contribute to building the open source project further. Some tasks are perfect for beginners and even say **good first issue** to let you know.

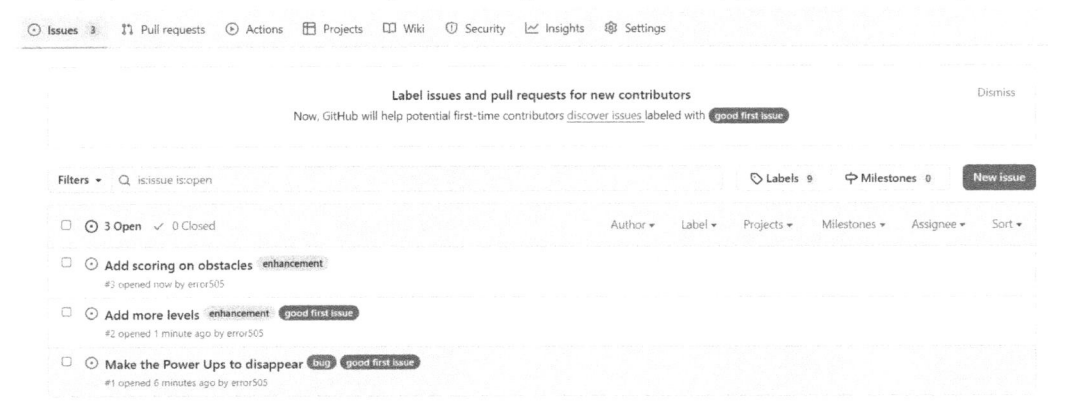

Figure 11.4 – Project Issues

- The *Contribution guidelines*, which are instructions on how to help out. They're like the rules of a game, showing you how to play along with others in the project.

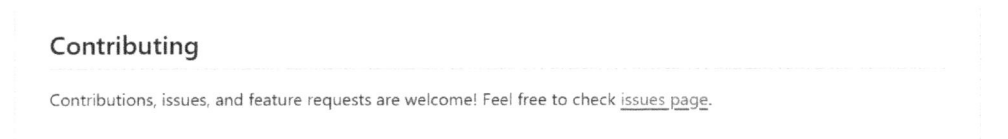

Figure 11.5 – The Contributing guide

Getting started

Here are some things that should be kept in mind when getting started with open source projects:

- **Saying hi!**: Before jumping in, it's nice to say hello to the project owner. Many projects have a place for discussions or a chat. You can introduce yourself, tell everyone what you like about the project, and ask how you can help. It's like joining a new class at school and making friends.

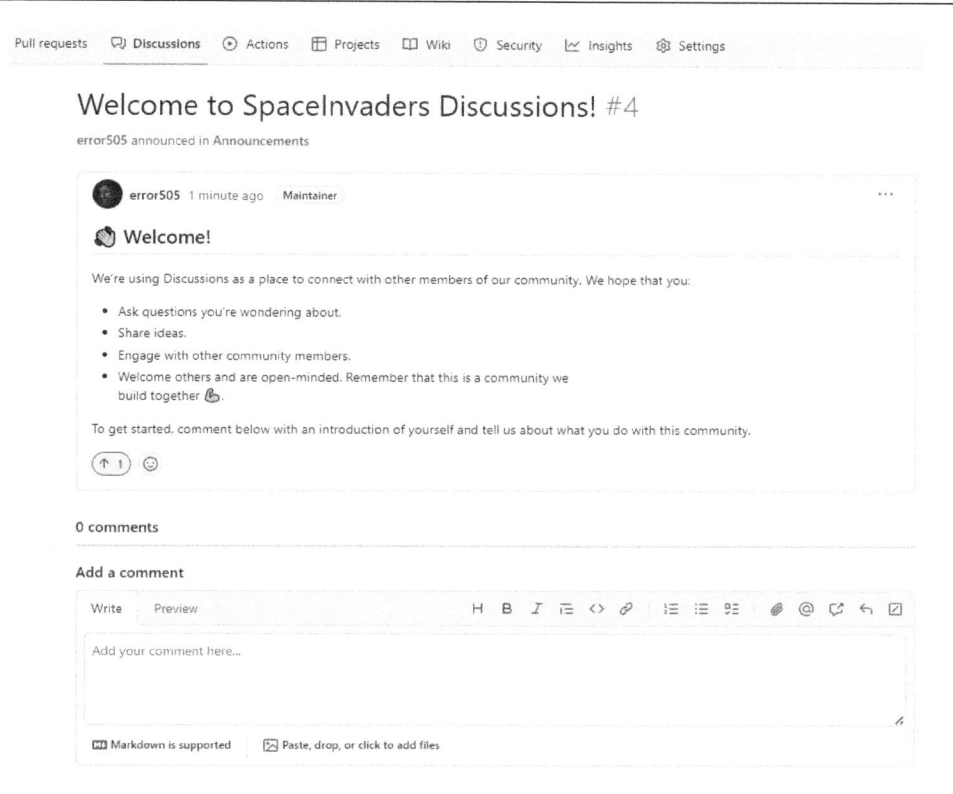

Figure 11.6 – Project Discussions

- **Taking small steps**: Don't worry about starting big. Small contributions are just as important. Maybe you find a typo in some instructions, or you have an idea to make something look nicer. These little suggestions are a great way to get started.

In the example project demonstrated through the preceding figures, you can see that pausing the game is not implemented, so your contribution could be to implement that feature or to update the README file by removing the line **5. Press 'P' to pause the game**.

How to Play

1. Use arrow keys to move the spaceship.
2. Avoid colliding with obstacles.
3. Collect stars to increase your score.
4. Obtain power-ups to gain temporary advantages.
5. Press 'P' to pause the game.
6. Try to beat your high score!

Figure 11.7 – How to play README with missing features

- **Learning and growing**: As you explore and contribute to open source projects, you'll learn new things, make mistakes, and improve. Remember, every coder, even the experts, started just like you. By joining the open source community, you're on your way to becoming a great coder, making cool stuff, and helping others.

You've got a taste of how fun and rewarding it is to explore and contribute to open source projects, so let's take the next step. In the upcoming section, we'll see how you can start making your own contributions that make a difference. You'll see examples of small changes that have a big impact and learn how every little bit helps.

Making meaningful contributions with real examples

Here, you're going to learn how to contribute to a real project on GitHub. We'll use the *Space Invaders* game as our example. Right now, there's a small mistake in the game – the title says *Space Explorer* when it should say *Space Invaders*. Let's fix it together!

Figure 11.8 – Search for the projects on GitHub

Follow these steps to fix the issue:

1. **Find the issue**: First, you need to make sure that the issue is listed on the project's GitHub page. If it's not there, you can create a new issue to let everyone know what you're planning to fix.

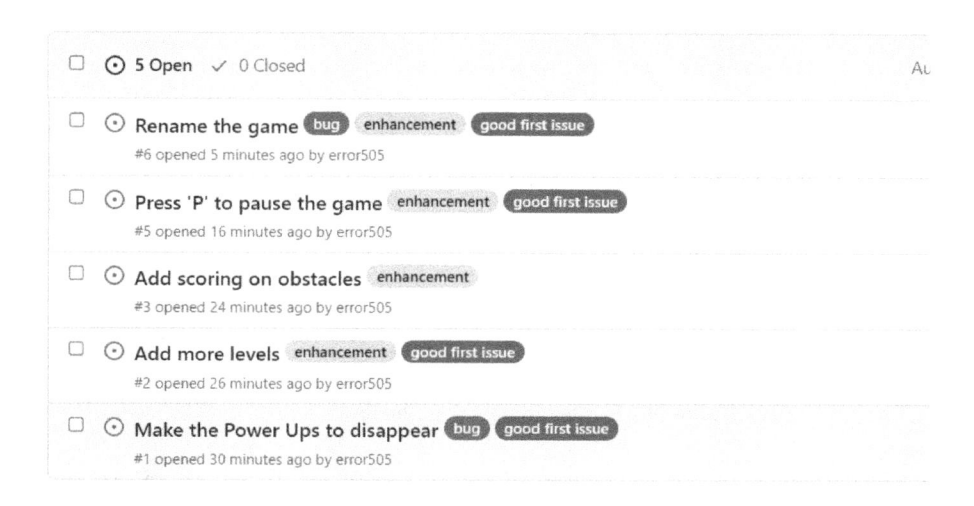

Figure 11.9 – Finding the issue to work on

As you can see, there is already an issue open about renaming the game. Now, you can proceed to the next step.

2. **Fork the repository**: Next, make your own copy of the game's code by *forking* the repository.

Figure 11.10 – Forking the repository

This means you'll have your own version on your GitHub account that you can change however you like.

Create a new fork

A *fork* is a copy of a repository. Forking a repository allows you to freely experiment with changes without affecting the original project.

Required fields are marked with an asterisk (*).

Owner * / Repository name *
Choose an owner / SpaceInvaders

By default, forks are named the same as their upstream repository. You can customize the name to distinguish it further.

Description (optional)
This is the simple Space Shooting JS game meant for learning JS programming.

☑ Copy the `main` branch only
Contribute back to error505/SpaceInvaders by adding your own branch. Learn more.

[Create fork]

Figure 11.11 – Creating a fork of the repository

3. **Clone the repository**: Now, you'll download your forked version of the game to your computer. This is called **cloning**. You do this so that you can make changes to the code on your own computer.

Local | Codespaces

Clone

HTTPS SSH GitHub CLI

https://github.com/error505/SpaceInvaders.git

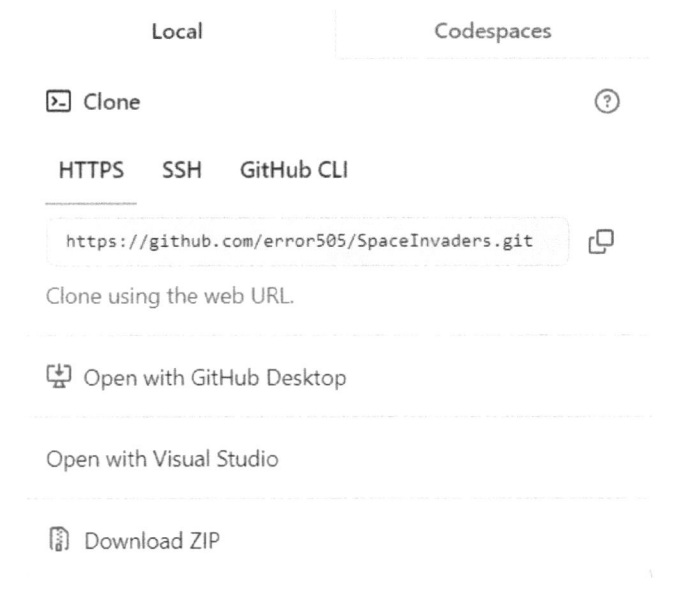

Clone using the web URL.

Open with GitHub Desktop

Open with Visual Studio

Download ZIP

Figure 11.12 – Cloning the repository

4. **Comment on the issue**: It's always nice to comment on the issue so that others know you will be working on it.

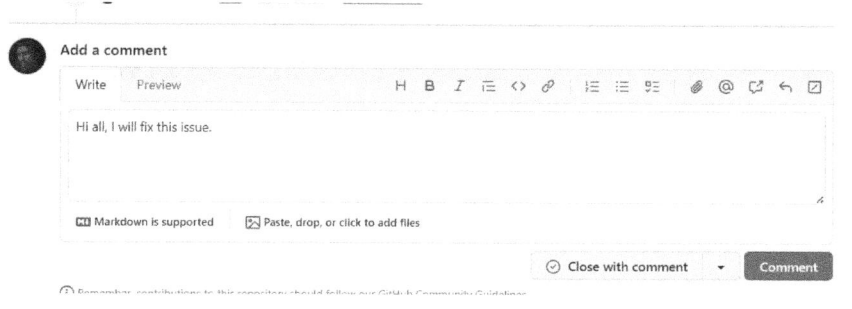

Figure 11.13 – Commenting on the issue

5. **Make the change**: Let's find the part of the code that has the wrong game title.

Figure 11.14 – Making the change

You'll change *Space Explorer* to *Space Invaders*. This might be in a file named something like `index.html`, depending on how the game is made.

6. **Test the change**: Before you share your change with everyone, you need to make sure it works. Let's run the game on our computer by opening the `index.html` file in our browser and see whether the title has changed to *Space Invaders*.

Figure 11.15 – Test whether the change has been applied properly

7. **Commit and push your changes**: After testing, you'll *commit* your changes, which means you save them to your version of the game on GitHub. Then, you *push* these changes, sending them from your computer to GitHub.

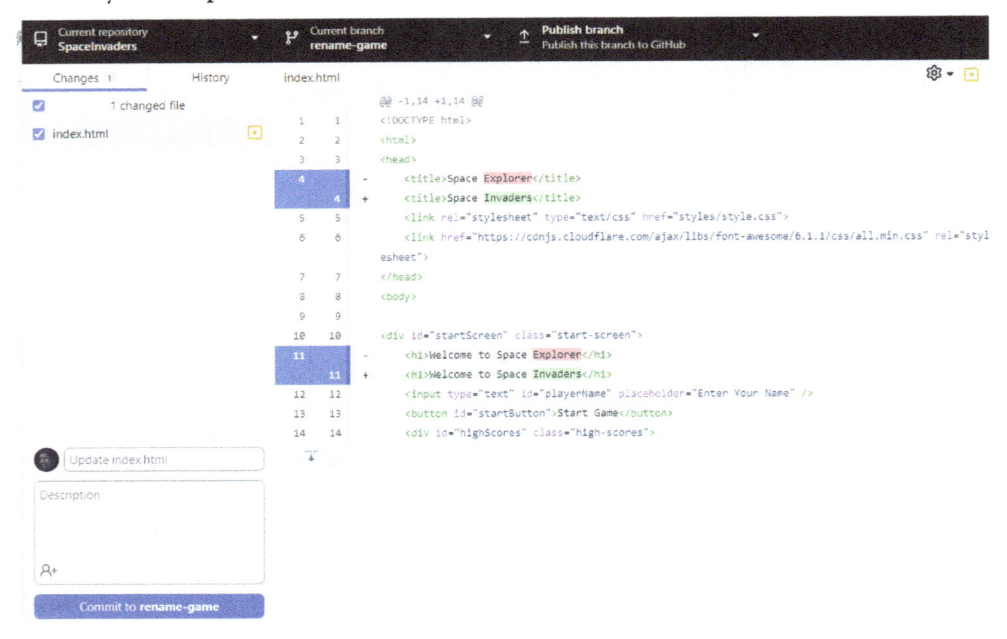

Figure 11.16 – Pushing your changes

8. **Create a pull request**: Finally, you ask the original creators of the game to accept your changes. You do this by creating a *pull request*.

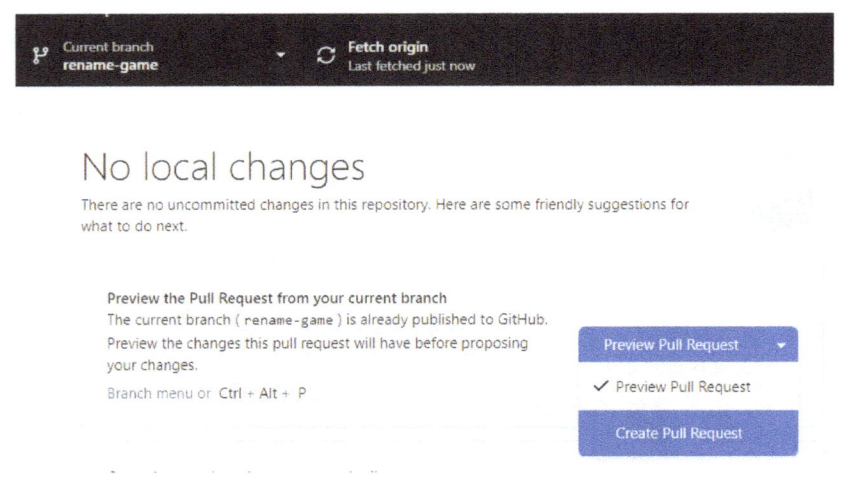

Figure 11.17 – Create a pull request

9. They'll review what you've done, and if they like it, they'll merge your changes into the official game.

Congratulations! You've just made a meaningful contribution to a real project. This is how coders work together to make projects better for everyone.

Now that you've tackled your first real coding collaboration by fixing the title in the *Space Invaders* game, let's move forward. Next, we're going to look into an important aspect of working on projects like these – understanding licensing and legal stuff. It might not sound as exciting as coding, but it's super important to make sure that you can share your work without any legal issues.

Navigating licensing and legal considerations

When you create something cool, such as a game or an app, you might want to share it with the rest of the community to seek contributions and speed up the development. GitHub is a great place to do that, but before you upload your project, it's important to think about licensing and legal stuff. This might sound boring, but it's really about protecting your project and deciding how others can use it.

js	Initial commit	yesterday
sounds	Initial commit	yesterday
styles	Initial commit	yesterday
LICENSE	Create LICENSE	yesterday
README.md	Add image	yesterday
index.html	Initial commit	yesterday

Figure 11.18 – LICENSE on the repository

What is a license?

A **license** is like a set of rules for your project. It tells people what they can and cannot do with your code. For example, can they use it for free? Do they need to give you credit? Can they change it and share their own version? Picking the right license helps you share your work the way you want.

Why are licenses important?

Licenses protect you and your work from any legal issues that could occur. They can also protect people who use your project so that they know what they are and aren't allowed to do. Without a license, your project is like a game without a guide; no one knows what the rules are! This can lead to misunderstandings or people being afraid to use or contribute to your project.

Choosing a license

There are many licenses to choose from, and it can be very confusing to pick the right one that suits your needs. There is a huge list of them on the GitHub documentation page, as shown in *Figure 11.19*.

License	License keyword
Academic Free License v3.0	AFL-3.0
Apache license 2.0	Apache-2.0
Artistic license 2.0	Artistic-2.0
Boost Software License 1.0	BSL-1.0
BSD 2-clause "Simplified" license	BSD-2-Clause
BSD 3-clause "New" or "Revised" license	BSD-3-Clause
BSD 3-clause Clear license	BSD-3-Clause-Clear
BSD 4-clause "Original" or "Old" license	BSD-4-Clause
BSD Zero-Clause license	0BSD
Creative Commons license family	cc
Creative Commons Zero v1.0 Universal	CC0-1.0
Creative Commons Attribution 4.0	CC-BY-4.0
Creative Commons Attribution ShareAlike 4.0	CC-BY-SA-4.0

Figure 11.19 – A list of the licenses

But you are lucky, as there are a few that are friendly for young coders like you. Some of them are as follows:

- **MIT license**: This one is very simple. It lets others do almost anything they want with your project, such as using it, changing it, and sharing it, as long as they give you credit.

```
         README      ⚖  MIT license

        MIT License

        Copyright (c) 2024 Igor Iric

        Permission is hereby granted, free of charge, to any person obtaining a copy
        of this software and associated documentation files (the "Software"), to deal
        in the Software without restriction, including without limitation the rights
        to use, copy, modify, merge, publish, distribute, sublicense, and/or sell
        copies of the Software, and to permit persons to whom the Software is
        furnished to do so, subject to the following conditions:

        The above copyright notice and this permission notice shall be included in all
        copies or substantial portions of the Software.

        THE SOFTWARE IS PROVIDED "AS IS", WITHOUT WARRANTY OF ANY KIND, EXPRESS OR
        IMPLIED, INCLUDING BUT NOT LIMITED TO THE WARRANTIES OF MERCHANTABILITY,
        FITNESS FOR A PARTICULAR PURPOSE AND NONINFRINGEMENT. IN NO EVENT SHALL THE
        AUTHORS OR COPYRIGHT HOLDERS BE LIABLE FOR ANY CLAIM, DAMAGES OR OTHER
        LIABILITY, WHETHER IN AN ACTION OF CONTRACT, TORT OR OTHERWISE, ARISING FROM,
        OUT OF OR IN CONNECTION WITH THE SOFTWARE OR THE USE OR OTHER DEALINGS IN THE
        SOFTWARE.
```

Figure 11.20 – An MIT License

- **GNU General Public License (GPL):** This says that if someone changes your project and shares it, they have to share their explanations of all the changes they made. It's about sharing improvements.

```
                          GNU GENERAL PUBLIC LICENSE
                            Version 3, 29 June 2007

         Copyright (C) 2007 Free Software Foundation, Inc. <https://fsf.org/>
         Everyone is permitted to copy and distribute verbatim copies
         of this license document, but changing it is not allowed.

                                   Preamble

           The GNU General Public License is a free, copyleft license for
         software and other kinds of works.

           The licenses for most software and other practical works are designed
         to take away your freedom to share and change the works.  By contrast,
         the GNU General Public License is intended to guarantee your freedom to
         share and change all versions of a program--to make sure it remains free
         software for all its users.  We, the Free Software Foundation, use the
         GNU General Public License for most of our software; it applies also to
         any other work released this way by its authors.  You can apply it to
         your programs, too.

           When we speak of free software, we are referring to freedom, not
         price.  Our General Public Licenses are designed to make sure that you
         have the freedom to distribute copies of free software (and charge for
         them if you wish), that you receive source code or can get it if you
         want it, that you can change the software or use pieces of it in new
         free programs, and that you know you can do these things.
```

Figure 11.21 – A GNU License

- **Creative Commons licenses**: These are not just for code; these licenses are also for creative work such as pictures, games, and stories. They let you choose how people can use your work. You can find more information in the following figure.

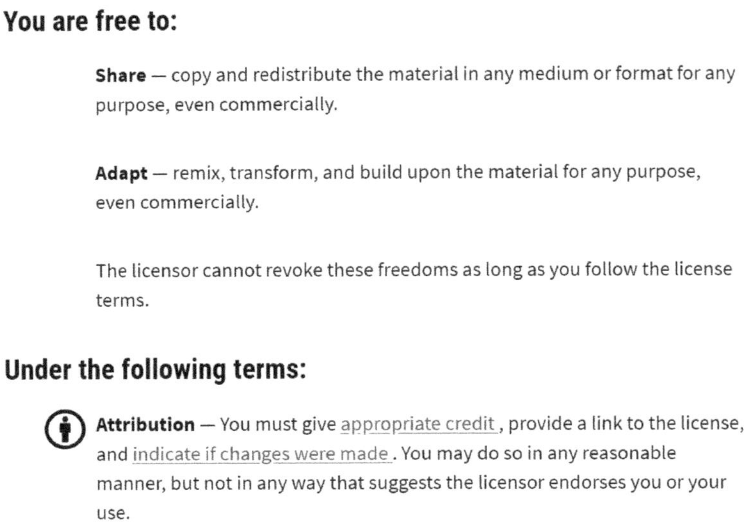

You are free to:

Share — copy and redistribute the material in any medium or format for any purpose, even commercially.

Adapt — remix, transform, and build upon the material for any purpose, even commercially.

The licensor cannot revoke these freedoms as long as you follow the license terms.

Under the following terms:

Attribution — You must give appropriate credit, provide a link to the license, and indicate if changes were made. You may do so in any reasonable manner, but not in any way that suggests the licensor endorses you or your use.

Figure 11.22 – A Creative Commons license

Need more help?

If choosing a license sounds confusing to you, it's okay to ask for help. There is a cool website at `https://choosealicense.com` that can help you choose the appropriate license. There, you can easily see the differences and pick the right one. You can also talk to someone you trust to give you the necessary guidance. They can help you understand the best way to share your project while keeping it safe.

Sharing your work on GitHub is a great way to learn and grow. Just make sure you think about how you want to share it so that everyone knows the rules and your project stays protected.

Creating a license for your GitHub repository

So, now you've created your repository and you want to share it with the community, but how do you create the license for the repository?

In this section, we will explain how to do it in a few simple steps:

1. When creating a new repository on GitHub, after you enter the repository name and description, you'll find an option to add a license.

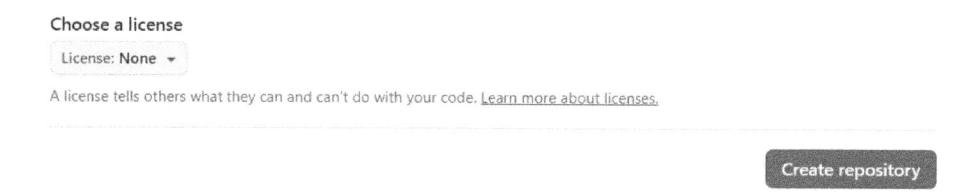

Choose a license

License: None

A license tells others what they can and can't do with your code. Learn more about licenses.

Figure 11.23 – Create a license while creating a repository

2. To choose the license, just click on the **Choose a license** drop-down menu.

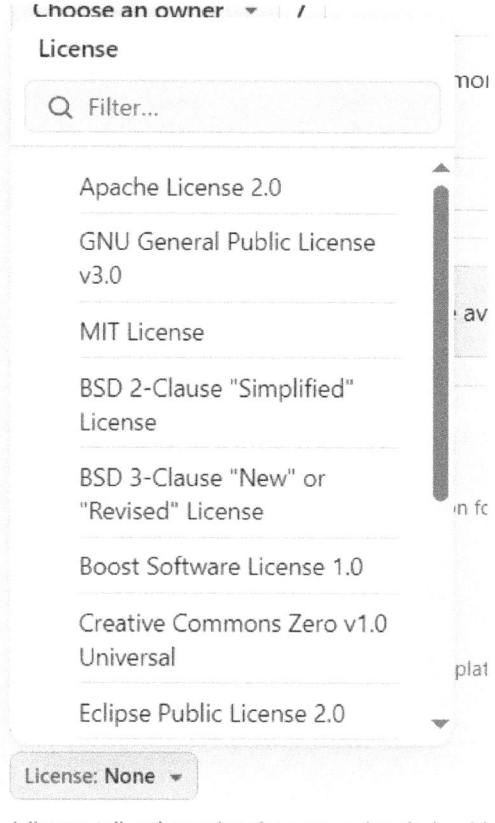

Figure 11.24 – Choosing a license

GitHub provides a list of licenses you can choose from. Select the license that fits your project's needs. If you're unsure, GitHub offers a short description of each to help you decide.

What if you have already created a repository but you have forgotten to add the proper license? No worries – there is a quick trick to do it easily. Let's see how:

1. Go to your GitHub repository, click on the **Add file** button, and select **Create new file**.

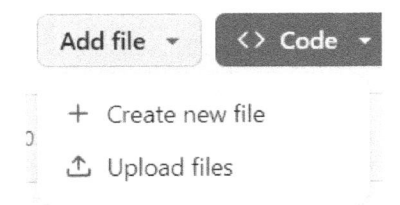

Figure 11.25 – Adding a new license file

2. Name this file `LICENSE` or `LICENSE.md`, and a new option will automatically be displayed where you can choose the license from templates. Choosing this name is key in ensuring that it displays properly in the **License** section.

Figure 11.26 – Choosing a license template

3. Click on **Choose a license template**, and a new page will open for you, where you will have the option to choose the license that best fits your needs.

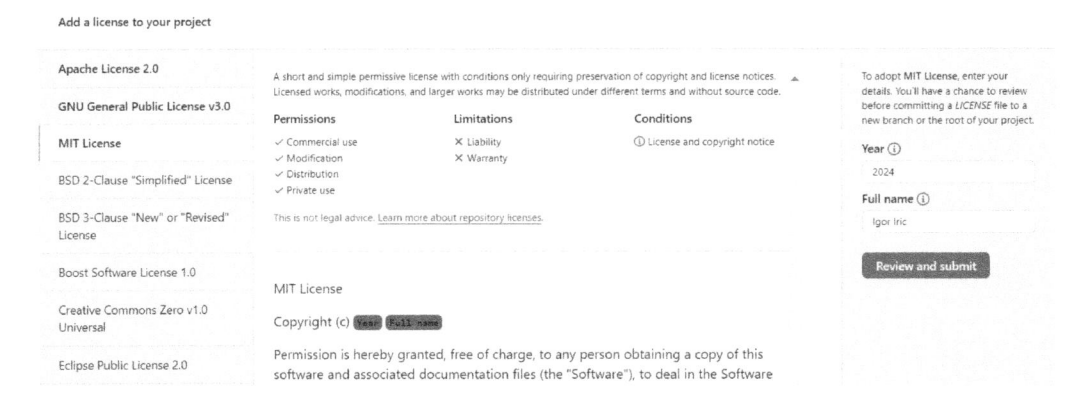

Figure 11.27 – Submitting a license

4. After you have chosen the right one, click on the **Review and submit** button to add it to your repository.

As you can see, adding a new license is very easy and makes it clear what others can and cannot do with your project. Now, let's learn about some best practices associated with open source.

Open source etiquettes and best practices

As you already know, open source is a place where you can share your coding projects with others and help improve the projects they've shared. However, just like in any community, there are ways to be a good and respectful member. In this section, we'll talk about how to be polite and helpful when you're working with others on coding projects.

Here's what you should remember:

- **Be respectful and kind**: When you're talking to other coders, always be nice and respectful. Remember, there's a real person on the other side of the screen. If you're suggesting changes to someone's project, say it in a friendly way. Think about how you'd feel if someone was talking to *you* about *your* project.
- **Ask before you leap**: If you find a project you want to help with, that's great! But first, check whether the project owner wants help and what kind of help they're looking for. Some projects have guidelines for this, so look for a file named CONTRIBUTING in the project's root folder or README file.
- **Make clear changes**: When you make changes to a project, explain why you're making them. This helps the project owner understand your thinking and makes it easier for them to decide whether they want to include your changes. If you're fixing a bug or adding a new feature, say so in your message.
- **Learn from feedback**: Sometimes, the project owner might ask you to change something in your contribution. That's okay! It's all part of the process. Listen to their feedback and try to learn from it. It's a great way to become a better coder.
- **Say thank you**: If someone helps you with your project, don't forget to say thank you. A little kindness goes a long way in making the open source community a friendly and welcoming place.

By following these simple rules, you'll be well on your way to becoming an excellent and respectful member of the open source community. Happy coding!

Summary

In this chapter, we've navigated through essential aspects of engaging with the open source community, including how to explore and contribute to open source projects meaningfully, understand licensing, create a license for your project, and adhere to best practices and etiquette. These lessons equip you with the knowledge to contribute to the open source world responsibly and effectively, enhancing both your skills and the broader coding community.

In the next chapter, we'll shift our focus to crafting your GitHub profile. This will be a natural progression, emphasizing how to present yourself and your contributions in the best light, further engaging with the community and potential collaborators.

Quiz

Answer the following questions:

1. Why is it important to explore open source projects before contributing?

    A. To find the easiest projects

    B. To match your interests and skills with the right project

    C. To see how many stars a project has

    D. None of the previous

    Answer: B. To match your interests and skills with the right project

2. True or false – making meaningful contributions requires understanding a project's needs and community guidelines.

    A. True

    B. False

    Answer: A. True

3. Fill in the blank – understanding _____ is crucial to ensure that your contributions do not violate any legal or ethical standards.

    Answer: licensing and legal considerations

4. What is the first step in creating a license for your GitHub repository?

    A. Choosing a license that restricts all uses (which restricts any use of the code without permission)

    B. Consulting a lawyer

    C. Understanding the types of licenses available

    D. Any of the above

    Answer: C. Understanding the types of licenses available

5. True or false – open source etiquette involves respecting a project's culture and communication norms.

    A. True

    B. False

    Answer: A. True

6. What is a key element of open source contributions?

    A. Keeping your work private

    B. Making large changes to show your skill

    C. Providing clear and helpful documentation with your code

    D. Only contributing code

    Answer: C. Providing clear and helpful documentation with your code

7. Fill in the blank – when navigating licensing, it's important to understand the difference between _____ and permissive licenses.

    Answer: restrictive

8. Which of the following is NOT a good practice when engaging with the open source community?

    A. Submitting code without testing

    B. Reading a project's contributing guidelines

    C. Asking questions when in doubt

    D. Offering constructive feedback

    Answer: A. Submitting code without testing

9. True or false – creating a license for your GitHub repository is optional but recommended.

    A. True

    B. False

    Answer: A. True

10. How can you make a meaningful contribution to an open source project?

    A. By only fixing typos in the README file

    B. By adding new features without discussing them with maintainers

    C. By engaging in constructive discussions and submitting quality pull requests

    D. By observing the project without contributing

    Answer: C. By engaging in constructive discussions and submitting quality pull requests

Part 5: Personalizing Your GitHub Experience

Finally, this part focuses on personalizing your GitHub presence and exploring new tools such as GitHub Copilot. You'll learn how to craft an impactful GitHub profile that showcases your best work, connects with the community, and enhances your professional image. Additionally, discover how GitHub Copilot can aid your coding process.

This part contains the following chapters:

- *Chapter 12, Crafting Your GitHub Profile*
- *Chapter 13, GitHub Copilot Aiding Code Creation*

12
Crafting Your GitHub Profile

In this chapter, we will learn how to craft an impactful GitHub profile that showcases your projects and contributions in a way that highlights your skills, dedication, and versatility as a coder. By optimizing your profile, you can make a strong first impression on visitors, including potential collaborators, employers, and fellow developers. This chapter is designed to provide practical guidance on making your GitHub profile not only a reflection of your coding journey but also a platform for professional opportunities.

In this chapter, we're going to cover the following main topics:

- Optimizing your GitHub profile overview
- Showcasing skills and expertise
- Adding stats and achievements to your profile
- Displaying projects and contributions strategically

Technical requirements

For some activities, we'll be using external tools and services, such as `https://shields.io/` and `https://github-readme-streak-stats.herokuapp.com`. Ensure you're familiar with Markdown, as it's essential for customizing your README profile.

Additionally, download a GitHub folder named `Chapter 12` from `https://github.com/PacktPublishing/GitHub-for-Next-Generation-Coders/tree/main/Chapter%2012`, where you can find a README template that you can use to follow.

Optimizing your GitHub profile overview

Your GitHub profile is like your online coding biography – a place to display your skills, projects, and contributions to the coding world. Just as you would like to make a good first impression in real life, having an optimized and personalized GitHub profile can make a big difference. In this section, we'll learn some cool tips to make your profile look great!

Let us start:

- **Adding a profile picture and bio**: Start with a clear picture of yourself with a nice one-color background. It makes your profile more personal. Write a bio that's short but tells a lot about you. Are you a web developer, a Python developer, or a game design enthusiast? Mention it here!

    Example: **Hi! I'm Igor Iric and I'm fascinated by how coding can bring ideas to life. I love learning new programming languages and collaborating with other coders. Right now, I'm building a fun new game!**

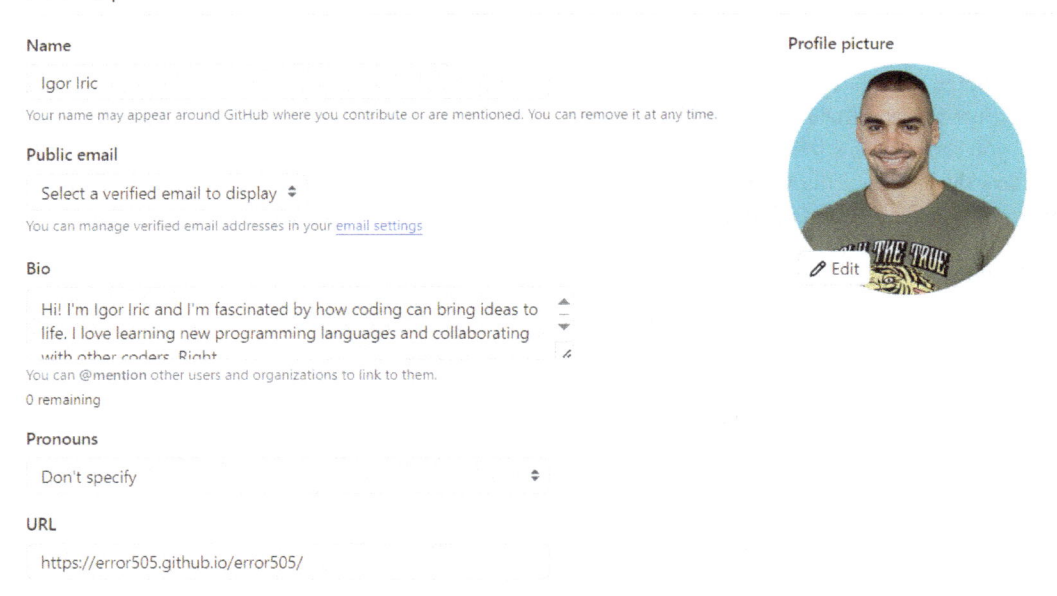

Figure 12.1 – GitHub profile

- **Pinning repositories**: GitHub lets you pin repositories to your profile. Choose projects that showcase your skills and what you enjoy working on. It could be a website you built, a game, or a tool that makes something easier. As you can see in *Figure 12.2*, you can click on **Customize your pins** and choose which repositories you want to showcase:

Optimizing your GitHub profile overview   305

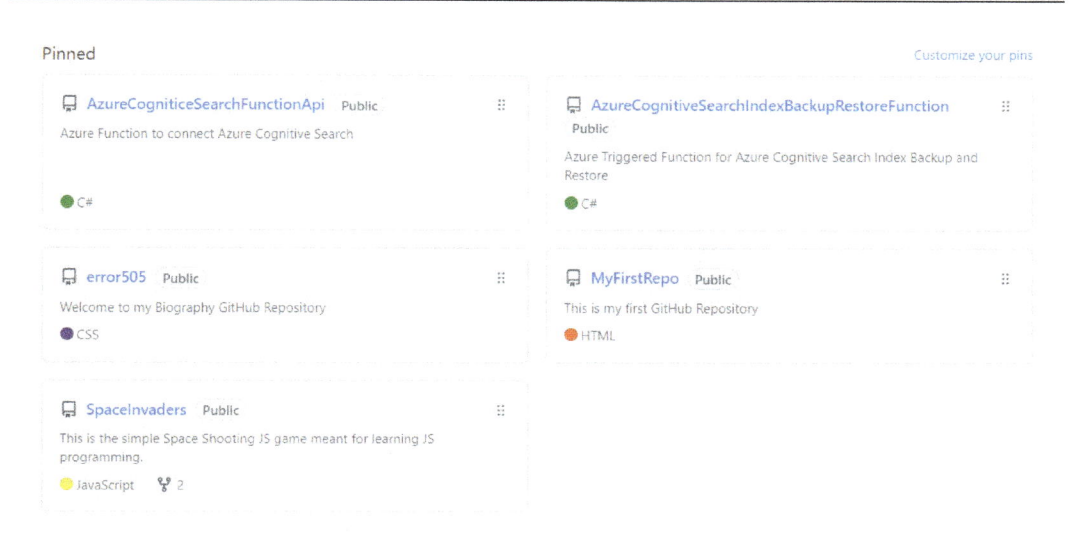

Figure 12.2 – Pinned projects

- **Contribution graph**: This is the place that tells about your activity on GitHub, so stay active! Your contribution graph shows how much you code. Try to contribute to your projects or others' regularly. It shows you're always learning and improving:

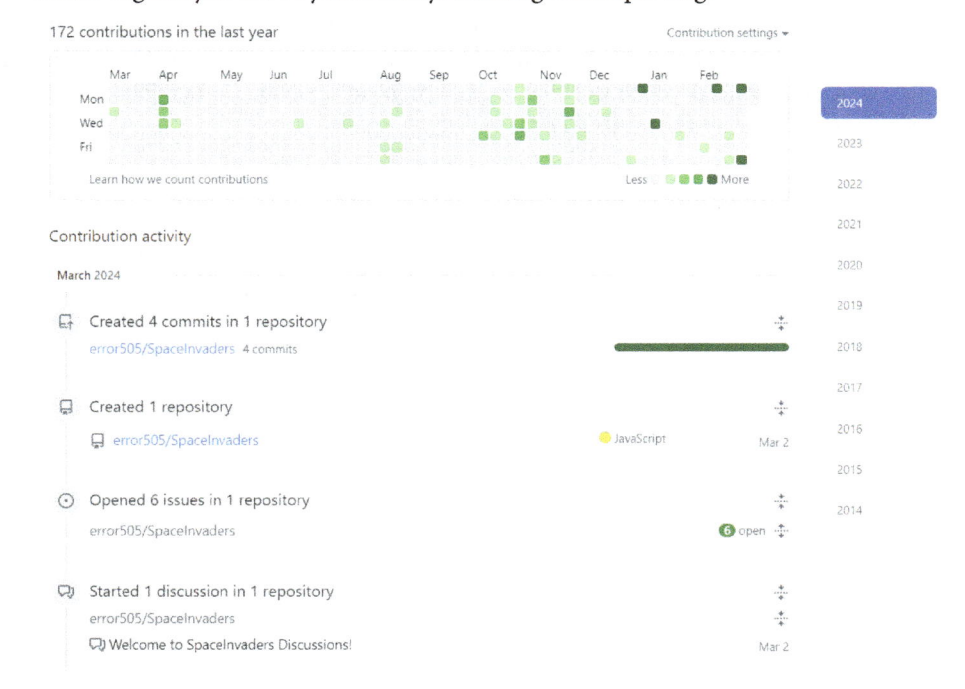

Figure 12.3 – Contribution graph

- **Badges**: These are mini achievements that show off your coding skills and contributions. Try to earn badges by participating in coding challenges or contributing to open source projects. They add a bit of fun and show others you're an active coder:

Figure 12.4 – Badges

Now that you've got the basics down for making your GitHub profile look good and stand out, let's move forward. Up next, we're going to focus on how to really shine a light on your skills and expertise. We'll explore ways to highlight what you're good at and how to make it easy for others to see your coding strengths. This is your chance to showcase the projects and achievements you're most proud of!

Showcasing skills and expertise

In this section, we'll explore a special feature on GitHub that is called a **GitHub README profile**! This is like your complete biography or CV on GitHub – a space where you can tell the community about yourself, what cool projects you're working on, and many other details about yourself. Sounds great, but how can you create your own README profile?

Creating your README profile

It's very simple to make a special repository on GitHub that's all about you. Just create a new repository with the same name as your GitHub username. Yes – it has to be exactly the same. In my case, the repository is named `error505`:

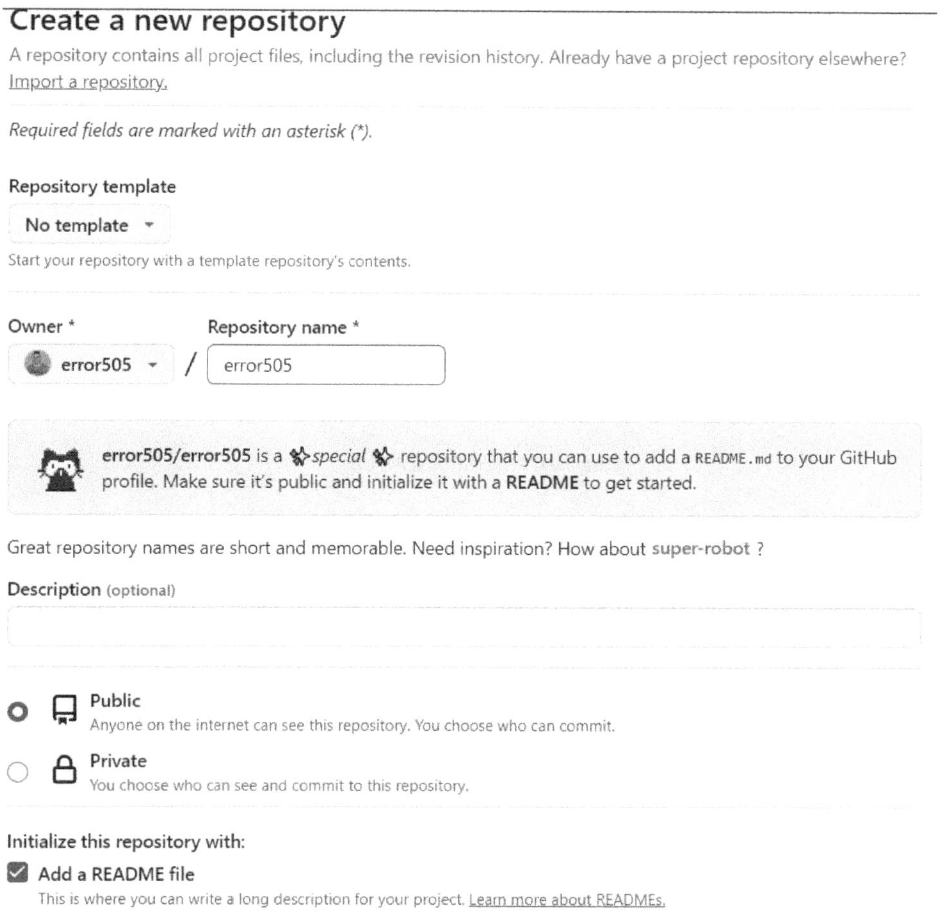

Figure 12.5 – Creating a profile repository

As you can see in *Figure 12.5*, there is a message that automatically displays after typing your username as the name of the repository. The message shows that this repository will be special and asks if you'd like to create a README file and set it as public.

Once you do this, GitHub does something pretty cool. It shows the README file from this repository right on your profile for everyone to see. Now, you can start editing your README file and add some cool information about yourself and your achievements:

Figure 12.6 – Adding README

What to include in the README file?

As usual, you can start by sharing your name, a picture of yourself, and where you live. Additionally, you can add cool badges to showcase your key skills. You can take a look at my profile in *Figure 12.7*:

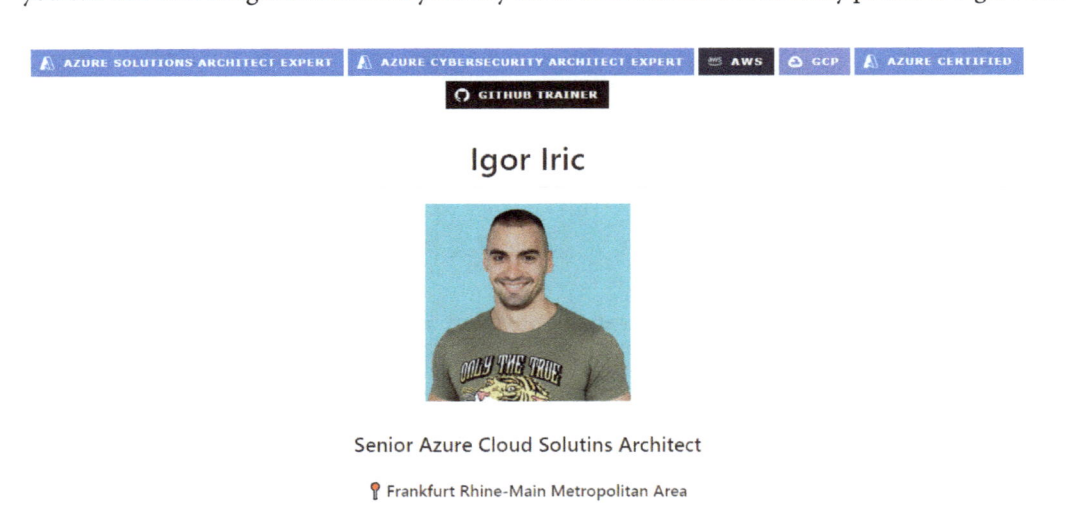

Figure 12.7 – Adding a headline for the profile

The following code block is designed to enhance your GitHub profile, transforming it into a visually appealing and informative page about you, your skills, and your location. It's like setting up a digital business card or resume that anyone visiting your GitHub profile can see right away.

Here is a breakdown of the README template found in the repository (`https://github.com/PacktPublishing/GitHub-for-Next-Generation-Coders/tree/main/Chapter%2012`) and what each part does:

- **Adding a badge**: This part adds a badge to your profile, showcasing your certification as an Azure Solutions architect expert. It's a quick visual cue to visitors about your expertise:

  ```
  <!-- Header -->
  <p align="center">
    <img src="https://img.shields.io/badge/Azure%20
  Solutions%20Architect%20Expert-%230078D4.svg?&style=for-the-
  badge&logo=microsoft-azure&logoColor=white" alt="Azure Solutions
  Architect Expert">
  </p>
  ```

- **Setting up name**: The following code sets up your name in a large font and centers it on the page, making it immediately visible to visitors. Below your name, it adds your job title, **Senior Azure Cloud Solutions Architect**, and your location, 📍 **Frankfurt Rhine-Main Metropolitan Area**, also centered for easy reading. This provides quick, essential information about you:

  ```
  <!-- Name and Location -->
  <h1 align="center">Your Name</h1>
  <h3 align="center">Senior Azure Cloud Solutions Architect</h3>
  <h4 align="center">📍 Frankfurt Rhine-Main Metropolitan Area</h4>
  ```

- **Profile picture**: This includes a space for your profile picture, making your profile more personal and relatable. The image is pulled from a specific GitHub repository and is displayed at a moderate size:

  ```
  <!-- Profile Picture -->
  <p align="center">
    <img src="https://github.com/[your-username]/biographyii/blob/
  main/1674712595713-plava2.jpg" alt="Your Name" width="200">
  </p>
  ```

- **Biography**: Next, you can continue by adding a short biography of yourself, where you can explain what you love about coding and what you're learning right now. You can also write about what are you aiming to learn next, as it shows others your passion for growing. You can check out my **About Me** section in the following figure:

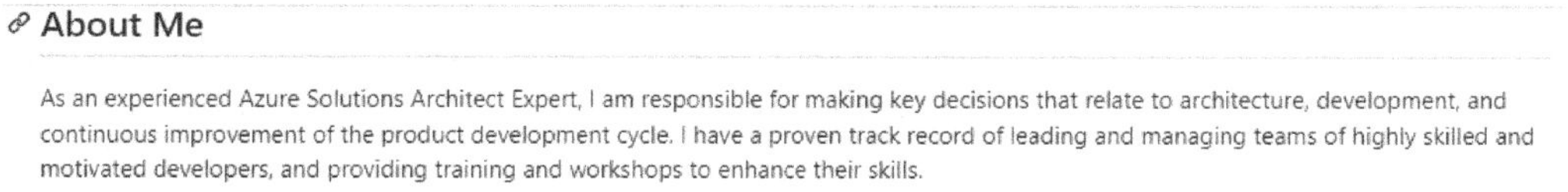

🔗 About Me

As an experienced Azure Solutions Architect Expert, I am responsible for making key decisions that relate to architecture, development, and continuous improvement of the product development cycle. I have a proven track record of leading and managing teams of highly skilled and motivated developers, and providing training and workshops to enhance their skills.

Figure 12.8 – About Me section

- **About Me**: The following code snippet introduces the **About Me** section, where you can share details about your professional background, expertise, and any other personal insights you'd like to highlight. In this section, you might discuss your journey as a young coder, your achievements, what drives you in your career, or how you approach challenges. It's an opportunity to tell your story, connect with visitors, and give context to the work you showcase on GitHub:

    ```
    <!-- About Me -->
    ## About Me
    As an experienced Azure Solutions...
    ```

- **Skills**: You can also add your skills, certificates, and expertise in the form of badges and pictures:

    Core Skills

    Figure 12.9 – Core Skills section

The following code block is designed to showcase your core skills directly on your GitHub profile, highlighting your technical competencies with visually appealing badges. It's like pinning medals to your digital profile for skills in microservices, .NET, C#, and React and Angular frameworks. Aligning these badges in the center creates an eye-catching display that quickly communicates your areas of expertise to visitors. This setup is perfect for developers looking to illustrate their skill set visually, making their GitHub profile not just a place for their code but also a testament to their proficiency in key technologies:

```
<!-- Core Skills -->
## Core Skills

<p align="center">
  <img src="https://img.shields.io/badge/Microservices-%23404D59.svg?&style=for-the-badge" alt="Microservices">
  <img src="https://img.shields.io/badge/.Net%20%7C%20C%23-%23239120.svg?&style=for-the-badge" alt=".Net | C#">
  <img src="https://img.shields.io/badge/React%20%7C%20Angular-%2361DAFB.svg?&style=for-the-badge" alt="React | Angular">
</p>
```

As you can see, the Shields.io website has been used to showcase your most important skills. `https://shields.io/docs/logos` is an open source service for badges, which you can easily include in your GitHub README profile or any other README of your repositories:

Showcasing skills and expertise   311

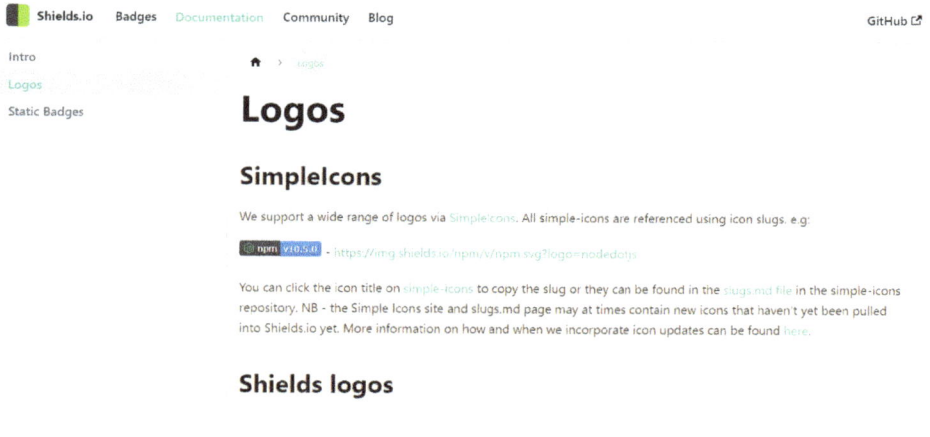

Figure 12.10 – Shields.io

- **Images**: If you want, you can also use any other images from various sources to showcase your certificates, skills, social icons, or anything similar. Just specify the source inside the image tag from where you want to reference your image, and it will be displayed on your profile:

Figure 12.11 – Custom showcase icons

- **Certifications**: The following code block showcases specific certifications on your GitHub profile, serving as a testament to your achievements and expertise. By aligning these badges centrally, the layout ensures they catch the eye of anyone visiting your profile:

```
<!-- Certifications -->
## Certifications

<p align="center">
  <img src="https://img.shields.io/badge/GitHub%20Partner%20Trainer-%23181717.svg?&style=for-the-badge&logo=github" alt="Accredited GitHub Partner Trainer GitHub Admin">
</p>

<p align="center">
  <a href="https://www.credly.com/badges/70f4339c-6104-4d19-b0cf-40471e7cef1f/public_url" target="_blank">
    <img src="https://images.credly.com/size/340x340/images/0ba22331-acf9-4e8a-8ce3-b4cc3d376040/image.png" alt="Microsoft Certified: Cybersecurity Architect Expert"
```

```
            width="100">
          </a>
          <br>
        </p>
```

Certifications

Figure 12.12 – Adding images

Now that you've learned how to customize your GitHub README profile with badges and showcase your skills and certifications, it's time to take it up a notch. Next, we'll look into adding stats and achievements to your profile. This is like putting up your high scores and trophies in your room, showing off what you've accomplished and how active you've been in your coding journey. Let's make your GitHub profile not just a reflection of your skills but also a celebration of your achievements.

Adding stats and achievements to your profile

Your GitHub profile is like your superhero costume. It tells people about your powers (skills), your adventures (projects), and the badges of honor (achievements) you've earned along the way. But how do you make sure people see the cool stuff you've done all at once? That's where widgets come into play!

What are these widgets?

Widgets are like little display windows you can add to your GitHub profile's README file (a special document that introduces you and your work). These widgets can show off your GitHub stats, such as how many projects you've worked on or how many programming languages you've used. It's a way to showcase your coding journey and achievements in a fun and visual way.

Here are some examples of widgets:

- **GitHub Stats**: This widget shows how many contributions you've made, how many people have starred your projects, and so on. It's like a mini-report card of your GitHub activity.
- **Top Languages**: This widget displays the programming languages you use most often, giving viewers insight into your coding preferences.

- **Streak Stats**: This highlights how many days in a row you've made contributions, showcasing your dedication.

Let's add all those **GitHub Stats** widgets to your profile. You can use `https://github.com/anuraghazra/github-readme-stats` API to show the status and top languages used. There, in the GitHub repository of `github-readme-stats`, you can find many demos and configure your own stats widgets very easily:

GitHub Readme Stats comes with several built-in themes (e.g. `dark`, `radical`, `merko`, `gruvbox`, `tokyonight`, `onedark`, `cobalt`, `synthwave`, `highcontrast`, `dracula`).

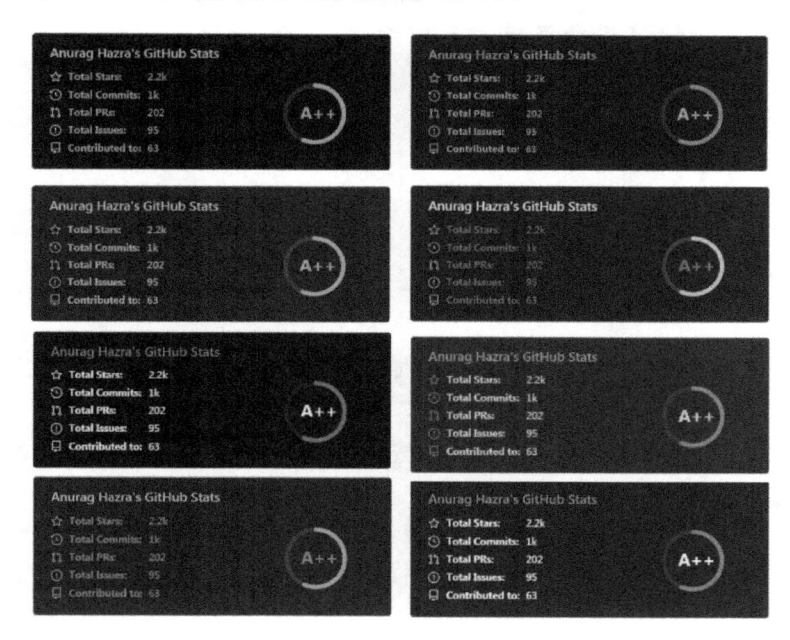

Figure 12.13 – GitHub Stats widgets

Here's how you can let everyone know about your coding preferences and how much you've accomplished on GitHub. By adding these little widgets to your GitHub README profile, you can share which programming languages you use most and highlight your GitHub activity, such as contributions, commits, and more. Just replace `[your-username]` with your actual GitHub username in the URLs, and these charts will automatically update to showcase your stats. It's a fun and dynamic way to show off your skills and hard work to anyone visiting your profile:

```
# 📊 GitHub Stats:

<p><img align="left" src="https://github-readme-stats.
vercel.app/api/top-langs?username=[your-username]&show_
```

Crafting Your GitHub Profile

```
icons=true&locale=en&layout=compact" alt="[your-username]" /></p>
<p> <img align="center" src="https://github-readme-stats.
vercel.app/api?username=[your-username]&show_icons=true&locale=en"
alt="[your-username]" /></p>
```

Similarly, you can use `https://github-readme-streak-stats.herokuapp.com/demo/` to show your streak stats. The **Streak Stats** widget can be easily configured using the configuration UI directly on their website:

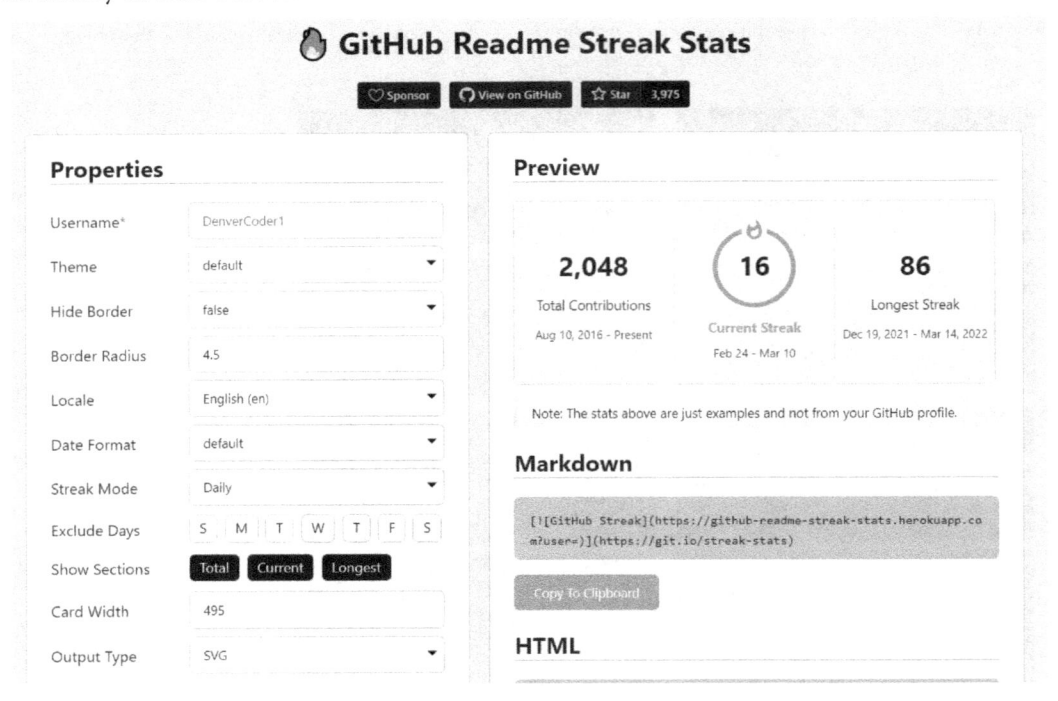

Figure 12.14 – GitHub Streak Stats widget configuration

The following code snippet adds a dynamic chart to your profile, highlighting your GitHub activity streaks. It's like a visual diary of your coding consistency, which gives a glimpse into how regularly you commit to projects and showcases your commitment and progress as a developer. Just replace `[your-username]` with your actual GitHub username to personalize it:

```
<p><img align="center" src="https://github-readme-streak-stats.
herokuapp.com/?user=[your-username]&" alt="[your-username]" /></p>
```

GitHub Stats:

Most Used Languages
- JavaScript 39.53%
- HTML 28.53%
- Visual Basic 13.94%
- C# 9.41%
- CSS 7.73%
- Batchfile 0.86%

Igor Iric's GitHub Stats
- Total Stars Earned: 7
- Total Commits (2024): 107
- Total PRs: 17
- Total Issues: 18
- Contributed to (last year): 4

C+

400
Total Contributions
Jul 10, 2014 - Present

0
Current Streak
Mar 10

6
Longest Streak
May 7, 2017 - May 12, 2017

Figure 12.15 – GitHub Stats section

Why should you add stats and achievements?

Having stats and achievements on your profile has the following advantages:

- It's a quick way to show off your skills and for visitors to see what you've been working on.
- A profile with colorful widgets and stats looks more engaging than just text. It makes your profile stand out from other profiles.
- When people see your achievements and contributions, you might inspire them to start their own coding journey.

Showcasing your GitHub trophies

Think of your GitHub profile as a display shelf with your achievements on it. Each of those trophies on the shelf tells a story about a challenge you've faced and conquered. GitHub trophies are just like that! They're fun, colorful badges that celebrate your achievements and showcase your achievements to anyone who visits your profile.

What are GitHub trophies?

GitHub trophies are like virtual medals you earn for different accomplishments on GitHub. For example, you might get one for working with multiple programming languages or for contributing a lot to projects.

How do they add value to your profile?

They can highlight your versatility with languages (the **MultiLanguage** trophy) or your enthusiasm for contributing to projects (such as the **Commits** trophy). Trophies can inspire you to engage more with the community. When you see a **PullRequest** or **Issues** trophy, it's a nudge to keep collaborating and problem-solving. Getting your first follower or having someone star your project is exciting. Trophies for **Followers** and **Stars** celebrate these social milestones on your coding adventure:

Figure 12.16 – GitHub Trophies section

The following code block introduces a fun way to display your GitHub achievements directly on your profile. By adding the **GitHub Trophies** section, you're essentially showcasing a collection of trophies you've earned for various accomplishments on GitHub. The image link provided generates a set of personalized trophies based on your activity and achievements, styled in a *radical* theme to make your profile stand out. It's a great way to highlight your hard work and contributions in a creative format. Simply replace [your-username] with your actual GitHub username to see your unique trophies:

```
## 🏆 GitHub Trophies
![](https://github-profile-trophy.vercel.app/?username=[your-username]&theme=radical&no-frame=false&no-bg=true&margin-w=4)
```

Adding these trophies to your README profile turns your GitHub page from a simple bio to a storybook of your achievements. It adds a personal touch and a bit of competitive spirit, making your profile not just informative but also fun.

Now that you've boosted your GitHub profile with stats, achievements, and trophies, let's move forward. Next, we're going to learn how to organize your projects and contributions so that anyone visiting your profile can easily see the great work you've done. It's about making your profile not just a collection of impressive stats but a well-organized portfolio that tells the story of your development adventure.

Displaying projects and contributions strategically

To make your profile more professional, you can mention the projects you're most proud of. This way, you can direct visitors to the most important projects you are working on:

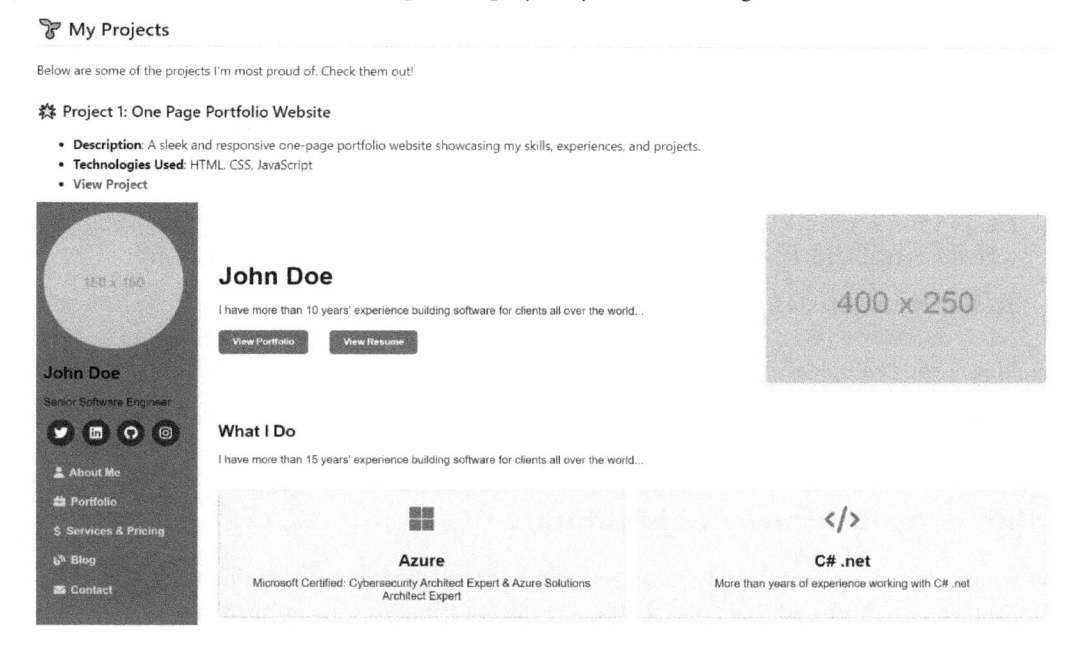

Figure 12.17 – Displaying projects

The following is a snippet to introduce the **Projects** section on your GitHub README profile. This section is where you get to highlight the projects you've worked on and are especially proud of. It's your chance to showcase your skills and creativity, letting others see what you've accomplished. Each project entry gives a brief overview, lists the technologies used, and provides links to view the project and its screenshot. This way, visitors can get a quick understanding of your work and even dive deeper into projects that catch their interest:

```
<!-- Projects -->
## 🌱 My Projects
Below are some of the projects I'm most proud of. Check them out!
### 🌟 Project 1: One Page Portfolio Website
- **Description**: A sleek and responsive one-page portfolio website showcasing my skills, experiences, and projects.
- **Technologies Used**: HTML, CSS, JavaScript
- **[View Project](link-to-your-project)**
```

```
![One Page Portfolio Website](link-to-screenshot-of-your-project)
### 🎮 Project 2: Space Invaders
- **Description**: A dynamic, exciting and challenging shooting game
set in space.
- **Technologies Used**: HTML, CSS, JavaScript, GitHub Actions
- **[View Project](link-to-your-project)**
![Space Invaders](link-to-screenshot-of-your-project)
```

To make your projects stand out on your GitHub profile, follow these simple steps for each project you list:

- For [View Project](link-to-your-project), replace link-to-your-project with the actual URL to your project. This could be a live site or a GitHub repository.

- To display images (such as screenshots of your projects), replace link-to-screenshot-of-your-project with the actual URL to your image. If you're not sure how to host an image, you can upload screenshots directly to your GitHub repository and link to them there.

- Customize the **Technologies Used** section to reflect the actual technologies you worked with on each project.

Adding contributions and highlights

If you've contributed to other projects, make sure those contributions are easy to find on your GitHub profile. You can do this by adding a widget that lists all contributions in your profile's README. It shows you're not just working on your own projects but helping others with their projects too.

The following is a code snippet you can use to include this feature. Just remember to replace [your-username] with your actual GitHub username. This will generate a dynamic image (*Figure 12.18*) displaying your top contributions based on the criteria you set, such as the limit on the number of repos shown and combining all yearly contributions. It's a great way to visually represent your hard work and dedication to your projects:

```
### 🔝 Top Contributed Repo
![](https://github-contributor-stats.vercel.app/api?username=[your-use
rname]&limit=5&theme=default&combine_all_yearly_contributions=true)
```

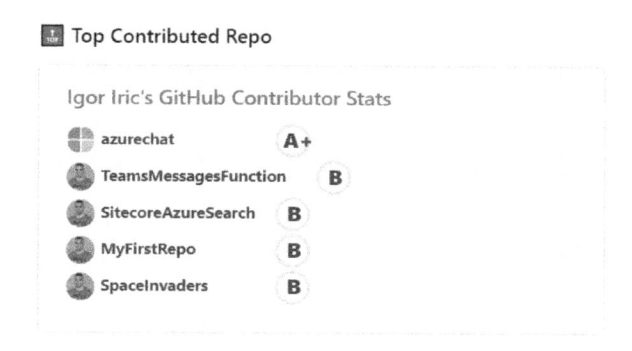

Figure 12.18 – Contribution repos

The next thing you can do is add your personal website and social networks to connect with your community:

Figure 12.19 – Social networks

The following code snippet is designed to enrich your README with a dedicated section for your personal or project website and social media links. This addition not only enhances your profile's visual appeal but also makes it easier for visitors to connect with you across various platforms:

```
<!-- Website -->
## Website
Link : [[your name]](https://[your-username].github.io/[your-username]/)
<!-- Social Media Links -->
## 𝌆 Let's connect!
<p align="center">
  <a href="https://dev.to/[your-username]" target="blank"><img align="center" src="https://dev-to-uploads.s3.amazonaws.com/uploads/logos/resized_logo_UQww2soKuUsjaOGNB38o.png" alt="[your-username]" height="30" width="40" /></a>
  <a href="https://twitter.com/[your-username]" target="blank"><img
```

```
align="center" src="https://www.freepnglogos.com/uploads/twitter-x-
logo-png/twitter-x-logo-png-9.png" alt="[your-username]" height="30"
width="40" /></a>
<a href="https://linkedin.com/in/[your-username]" target="blank"><img
align="center" src="https://www.freepnglogos.com/uploads/linkedin-
logo-design-30.png" alt="[your-username]" height="30" width="40" /></
a>
<a href="https://www.youtube.com/c/[your-username]"
target="blank"><img align="center" src="https://www.freepnglogos.com/
uploads/youtube-circle-icon-png-logo-14.png" alt="[your-username]"
height="30" width="40" /></a>
</p>
```

In the preceding snippet, we have the following:

- The **Website** section allows you to insert a direct link to your personal or project website, inviting visitors to explore more of your work.
- The **Let's connect!** section is designed as a centralized hub for your social media links. Each icon is a clickable link that directs visitors to your profiles on platforms such as dev.to, X (formerly known as Twitter), LinkedIn, and YouTube.

Remember to replace `[your-username]` with your actual username on these platforms and `[your name]` with the name you want displayed for the website link. This personalized approach not only showcases your projects and skills but also opens up channels for professional networking and personal connection.

Using GitHub Pages to show off your work

As you already learned in *Chapter 9*, GitHub Pages lets you turn your repository into live websites for free. This is a fantastic way to show off your profile page as a website. It's like having your own digital portfolio that you can share with friends, family, or even potential employers.

By following these tips, you can make your GitHub profile a great showcase of your coding journey. A well-arranged GitHub README profile not only makes a great first impression but also guides visitors (such as potential collaborators or employers) to see your best work and understand your journey as a developer.

With these steps, you've learned how to turn your GitHub profile into a dynamic portfolio that not only showcases your projects but also connects you with the wider community and potential opportunities. By highlighting your proud achievements, sharing your contributions, and linking out to your social networks and personal website, you create a comprehensive snapshot of your professional self.

Summary

In this chapter, we explored how to strategically display projects and contributions on your GitHub profile to best showcase your skills, experience, and achievements. We covered optimizing your profile overview, showcasing your expertise, and making your profile stand out with stats and achievements. These efforts not only enhance your profile's appeal but also underscore your capabilities to potential collaborators and employers.

In the next chapter, we'll focus on leveraging Copilot for code completion, suggestions, ensuring clean code and best practices, collaborative coding, and optimizing workflows. This subsequent chapter will naturally extend our discussion by introducing tools and strategies to further enhance your coding efficiency and productivity.

Quiz

Answer the following questions:

1. What is the main goal of optimizing your GitHub profile overview?

    A. To display your contact information

    B. To highlight your coding skills

    C. To showcase your hobbies

    D. To list your educational background

    Answer: B. To highlight your coding skills

2. True or false: Adding a custom README to your GitHub profile allows you to showcase your projects and contributions.

    A. True

    B. False

    Answer: A. True

3. Fill in the blank: Showcasing _____ on your GitHub profile helps highlight your expertise in specific areas.

    Answer: skills

4. Which feature can make your GitHub profile stand out by displaying your coding activity?

   A. Profile views counter
   B. **GitHub Stats**
   C. Follower list
   D. Starred repositories

   Answer: B. **GitHub Stats**

5. True or false: It's important to strategically display projects on your GitHub profile to attract potential collaborators and employers.

   A. True
   B. False

   Answer: A. True

6. What does adding a project README include that is beneficial for visitors?

   A. Your resume
   B. Code snippets only
   C. Detailed project description and how to use it
   D. Only screenshots

   Answer: C. Detailed project description and how to use it

7. Fill in the blank: Including _____ achievements on your GitHub profile can demonstrate your contributions to open source projects.

   Answer: coding

8. How can GitHub Pages be used in relation to your GitHub profile?

   A. To create a personal blog
   B. To host a static project portfolio website
   C. To track your daily activities
   D. To sell merchandise

   Answer: B. To host a static project portfolio website

9. The inclusion of what element in your profile can enhance its appeal and provide insight into your personal coding journey?

    A. Animated GIFs of cats
    B. Detailed travel logs
    C. Visual graphs of GitHub activity
    D. Random quotes

    Answer: C. Visual graphs of GitHub activity

10. True or false: Regularly updating projects on your GitHub profile is unnecessary.

    A. True
    B. False

    Answer: B. False

13
GitHub Copilot Aiding Code Creation

In this chapter, we'll talk about how GitHub Copilot can become your ultimate coding assistant, helping you code easily. GitHub Copilot offers real-time code suggestions across various programming languages, significantly improving your coding efficiency and creativity. Through this chapter, you'll learn to use Copilot for generating code snippets, best practices, and even conducting unit testing, all aimed at improving your coding projects. By the end of this chapter, you will be proficient in utilizing GitHub Copilot to streamline your coding process, create cleaner code, and, consequently, speed up your learning curve in new programming languages.

In this chapter, we're going to cover the following main topics:

- Understanding GitHub Copilot, your coding assistant
- Code completion and suggestions with Copilot
- Using Copilot for clean code and best practices
- Prompt engineering with GitHub Copilot
- Is GitHub Copilot free for coders in school?

Technical requirements

Before going deeper into this chapter, be sure you have *Visual Studio Code* installed, as it's the primary IDE we'll use with GitHub Copilot. A GitHub account is also required to access Copilot. Begin by installing the GitHub Copilot extension for Visual Studio Code, which can be found at https://marketplace.visualstudio.com/items?itemName=GitHub.copilot. Download the GitHub folder named Chapter 13 from https://github.com/PacktPublishing/GitHub-for-Next-Generation-Coders/tree/main/Chapter%2013 to follow the hands-on parts.

Understanding GitHub Copilot, your coding assistant

Imagine you're trying to build a big and beautiful LEGO castle, but you are not so good with LEGO. Luckily, you've got this friend who's very experienced at LEGO. They can suggest to you what piece to use to start building or come up with cool design ideas you hadn't thought of. That's pretty much what GitHub Copilot is, but instead of building LEGO, it is for coding in almost any programming language.

Figure 13.1 – GitHub Copilot

GitHub Copilot is like a super smart friend you always wanted to have who helps you write code. You tell it what you want to do in simple English, and it suggests chunks of code that might help you do just that. It's a tool developed by GitHub that uses a lot of data from code available on the internet to give you suggestions. It uses **Artificial Intelligence** (**AI**) to suggest lines of code, complete functions, and even write entire blocks of code based on the comments and code you've already written.

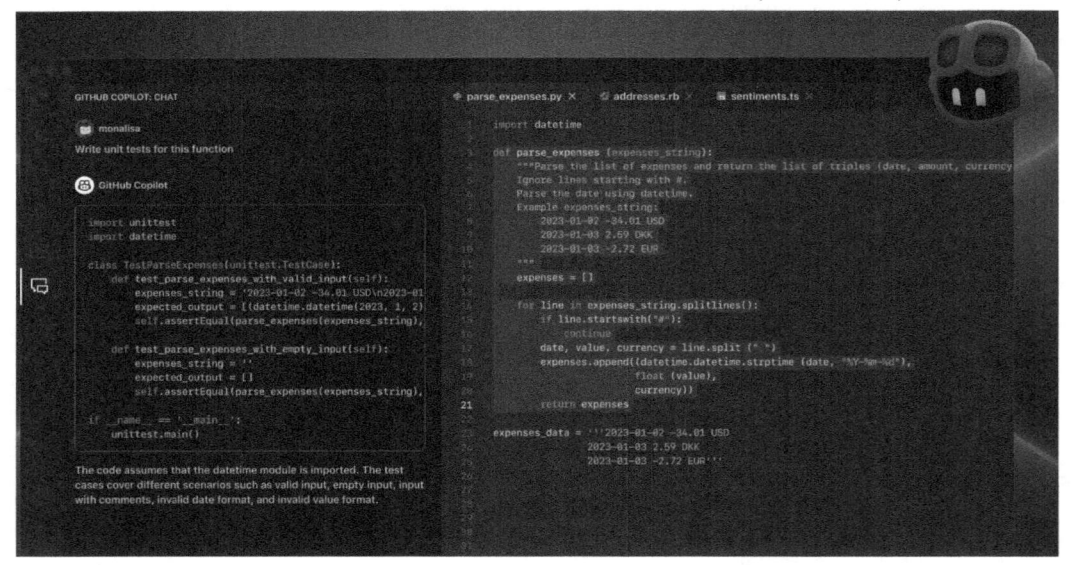

Figure 13.2 – A GitHub Copilot presentation

How does it work?

GitHub Copilot is powered by a technology called **Large Language Models** (**LLMs**). Copilot has been trained on a vast amount of publicly available code from the internet. This means it has seen many different coding styles, languages, and solutions to common problems.

When you start typing in your code editor, Copilot looks at the context of what you're writing. It understands the language you're using, the libraries you've imported, and the functions you're working on.

Based on the context, Copilot suggests lines of code or entire functions that might fit what you're trying to do. You can choose to accept, modify, or ignore these suggestions.

Let's say you're working on your one-page portfolio website and you just want to add a new feature, such as a slideshow of your projects or contact page. You start typing a comment in your code editor, such as `Create a slideshow of projects`. Then, GitHub Copilot jumps in and suggests a block of code that could create that slideshow of projects or contact page, as shown in *Figure 13.3*. It's like it just talked to you and handed you the exact LEGO piece you wanted.

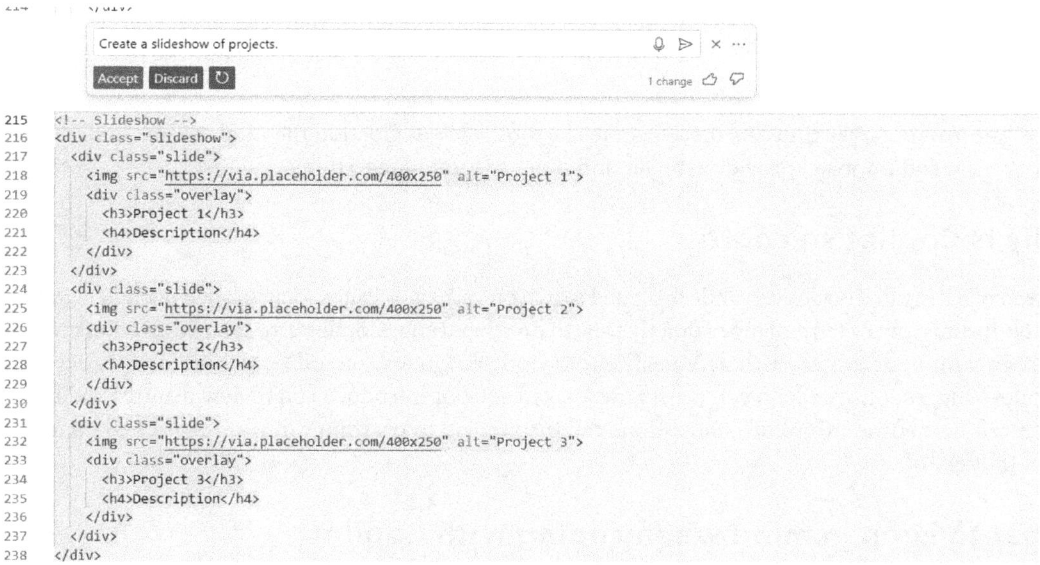

```html
<!-- Slideshow -->
<div class="slideshow">
    <div class="slide">
        <img src="https://via.placeholder.com/400x250" alt="Project 1">
        <div class="overlay">
            <h3>Project 1</h3>
            <h4>Description</h4>
        </div>
    </div>
    <div class="slide">
        <img src="https://via.placeholder.com/400x250" alt="Project 2">
        <div class="overlay">
            <h3>Project 2</h3>
            <h4>Description</h4>
        </div>
    </div>
    <div class="slide">
        <img src="https://via.placeholder.com/400x250" alt="Project 3">
        <div class="overlay">
            <h3>Project 3</h3>
            <h4>Description</h4>
        </div>
    </div>
</div>
```

Figure 13.3 – GitHub Copilot in action

You look at the suggested code snippet, and if it looks good, you can accept it and continue coding another part. If not, you can ask Copilot for another suggestion or to enhance it the way you think is best. It's like collaborating with your LEGO friend, deciding together how to build the coolest castle.

A quick introduction to AI and LLMs

AI is a way to make computers perform tasks that usually require human intelligence. LLMs are a type of AI that understands and generates human-like text. They are trained on large datasets and can predict what comes next in a sentence or a piece of code.

Here's how LLMs work:

- **Training**: LLMs are trained on a massive amount of text data, including books, articles, and code. During training, a model learns to predict the next word in a sentence or the next line in a piece of code.
- **Generating text**: Once trained, the model can generate text based on the input it receives. For example, if you start writing a function, the model can predict and suggest the next part of the function.
- **Improving over time**: As more people use tools such as Copilot, the models can continue to learn and improve, providing better and more accurate suggestions.

Why is Copilot so cool?

Instead of having to Google every little thing and search on various websites, look through documentation, ask on forums, or try to remember exactly how to do something, Copilot gives you suggestions right there in your code editor, such as Visual Studio Code or Codespaces, directly on GitHub. Seeing Copilot's suggestions can help you learn new ways to code or introduce you to new features and best practices. Sometimes, Copilot's suggestions can inspire you to try things in your project you hadn't even thought of.

What to keep in mind when coding with Copilot

Just like your LEGO friend might sometimes hand you a piece that doesn't quite fit, Copilot's suggestions might not always be exactly what you need. You still need to review and maybe tweak them. Copilot is there to help, not to do your complete job for you. The best projects come from your creativity and problem-solving, with Copilot as your assistant.

How to get started

To start using GitHub Copilot for personal projects, you need to sign up for a trial or subscription.

First, head over to the GitHub Copilot page. You can find it by searching for `GitHub Copilot` online or going directly to GitHub Copilot's site: `https://github.com/features/copilot`.

Understanding GitHub Copilot, your coding assistant   329

You can also access GitHub copilot from your GitHub account. Here's a quick guide on how to do so:

1. Click your profile photo on GitHub and select **Your Copilot**.

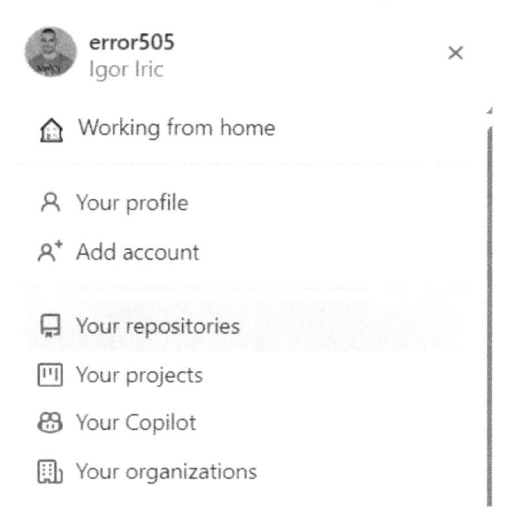

Figure 13.4 – Accessing GitHub Copilot from the profile menu

2. On the settings page, choose **Start free trial**.

Figure 13.5 – Starting a GitHub Copilot free trial

3. Decide between a monthly or yearly payment and click **Get access to GitHub Copilot**.

Figure 13.6 – Getting access to GitHub Copilot

4. Enter your payment details and submit them.

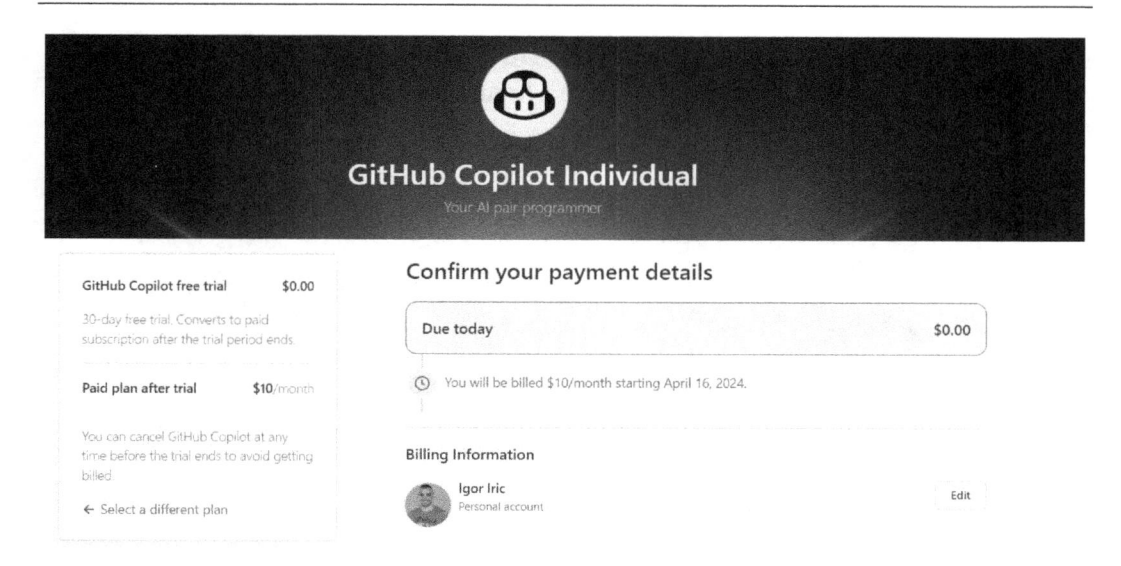

Figure 13.7 – Adding payment details

5. Set your preferences, and then click **Save and complete setup**.

Figure 13.8 – Selecting your preferences

Next, make sure you have **Visual Studio Code** (**VS Code**) installed on your computer so that you can add the GitHub Copilot plugin, which will enable you to utilize the capabilities of GitHub Copilot effectively.

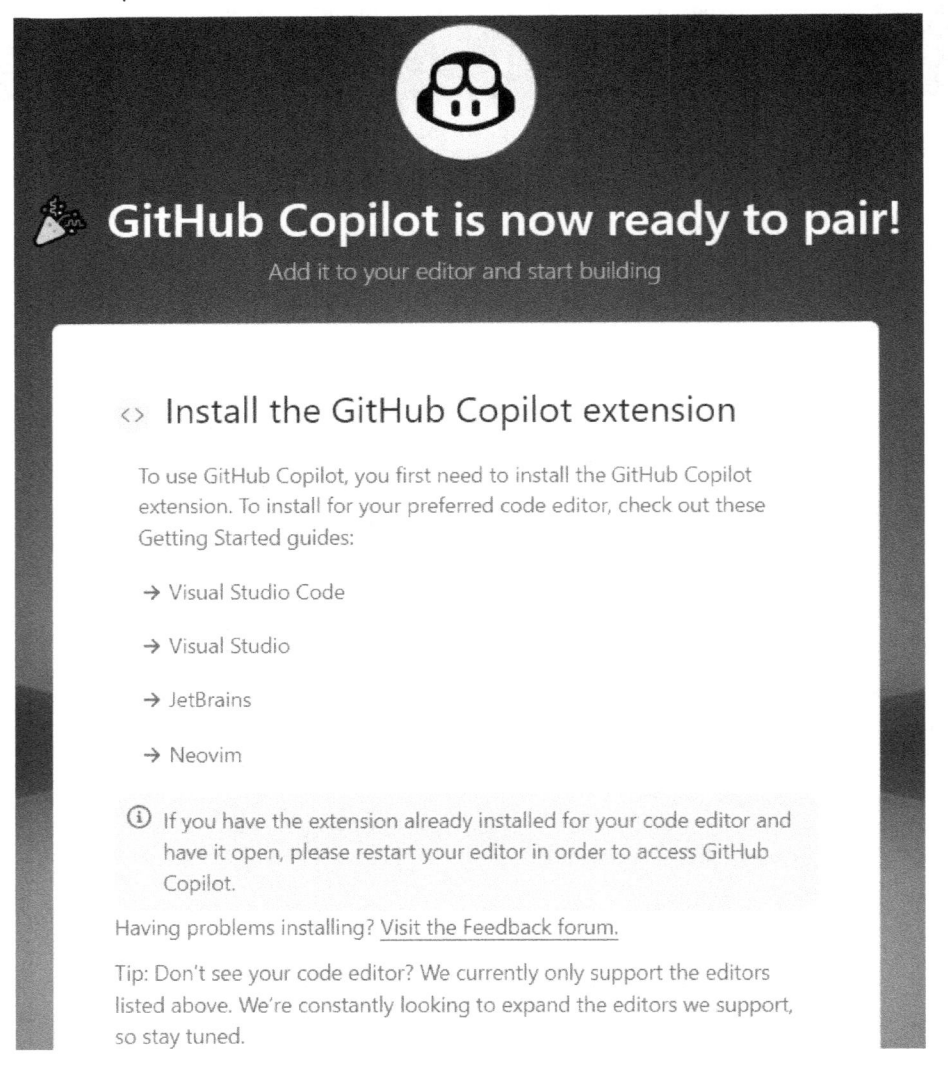

Figure 13.9 – Installing a supported IDE

Installing the GitHub Copilot extension

Open VS Code and find the **EXTENSIONS** view by clicking on the *square* icon on the sidebar, or you can press *Ctrl + Shift + X*.

In the search bar at the top, type `GitHub Copilot` and look for the GitHub Copilot extension. It should be the first one to pop up, as shown in *Figure 13.10*.

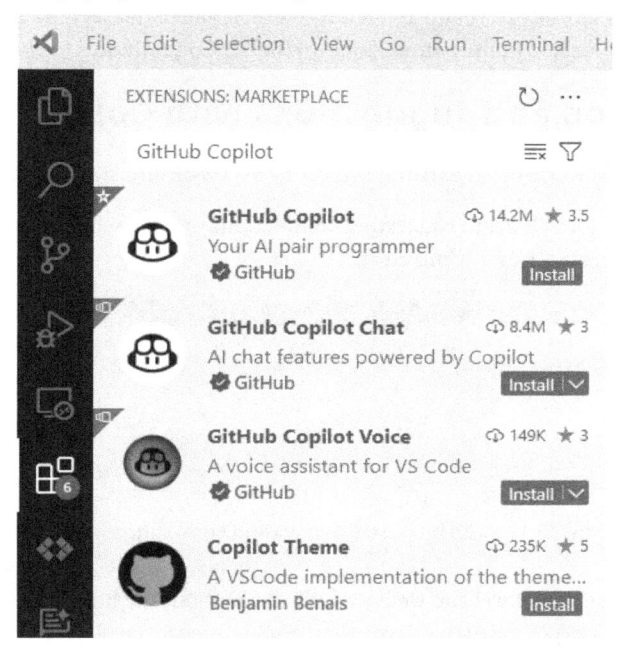

Figure 13.10 – Installing GitHub Copilot VS Code Extension

Click on the blue **Install** button. This is like telling VS Code, "*Hey, I want to add this smart assistant!*" Visual Studio will install the GitHub Copilot extension, and then you will see the screen shown in *Figure 13.11*.

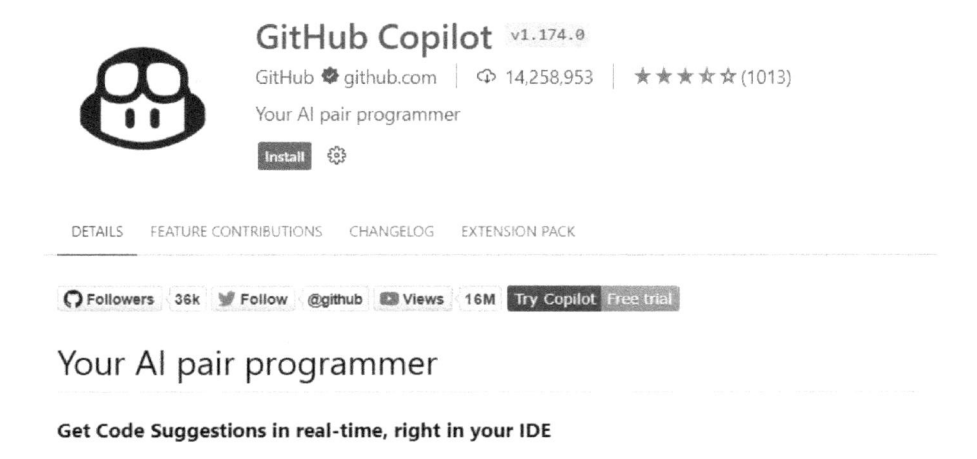

Figure 13.11 – GitHub Copilot with the code extension after installation

Once Copilot is installed in VS Code, you'll be asked to sign into GitHub within VS Code if it hasn't been linked to your GitHub account yet. If you've already linked VS Code with your GitHub account, GitHub Copilot will automatically be set up and ready to use. If you're not seeing a prompt to authorize, you can click the bell icon located in the bottom panel of the VS Code window to check for notifications.

Code completion and suggestions with Copilot

After signing in, open a new file or an existing project in VS Code and start typing your code.

As you type, GitHub Copilot will begin suggesting completions or even whole chunks of code. It's like having a friend who suggests, "*Hey, try this next!*"

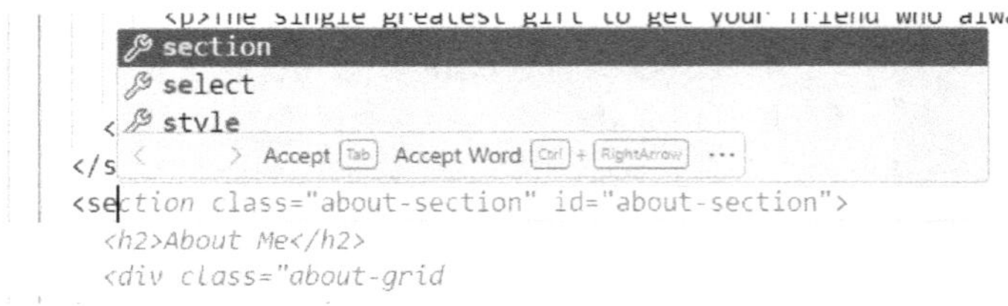

Figure 13.12 – GitHub Copilot providing code suggestions

Let's try it on the one-page profile website we have built throughout the book (`https://github.com/PacktPublishing/GitHub-for-Next-Generation-Coders/blob/main/Chapter%2013/index.html`) and ask Copilot to create the **Get In Touch** section for you. Open the Copilot chat using the *Ctrl* + *Enter* keyboard shortcut (which opens a separate panel showing 10 suggestions). You can use a prompt like `I would like you to create new contact sections with google maps, address, phone, email, and contact form. Use similar coding like previous section and add CSS styling.`

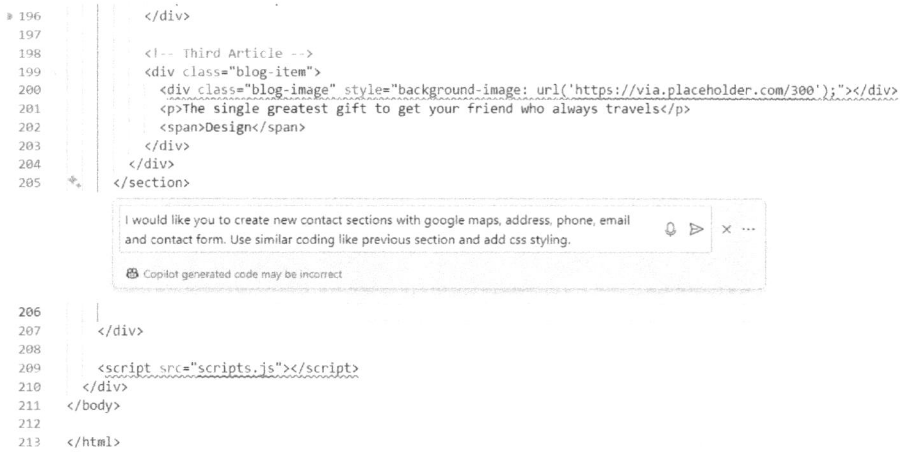

Figure 13.13 – Chatting with GitHub Copilot

After hitting *Enter* or clicking on the arrow in the chat box, you'll see how Copilot starts writing code for your new **Get In Touch** Section. You will have the option to accept, discard, or try again.

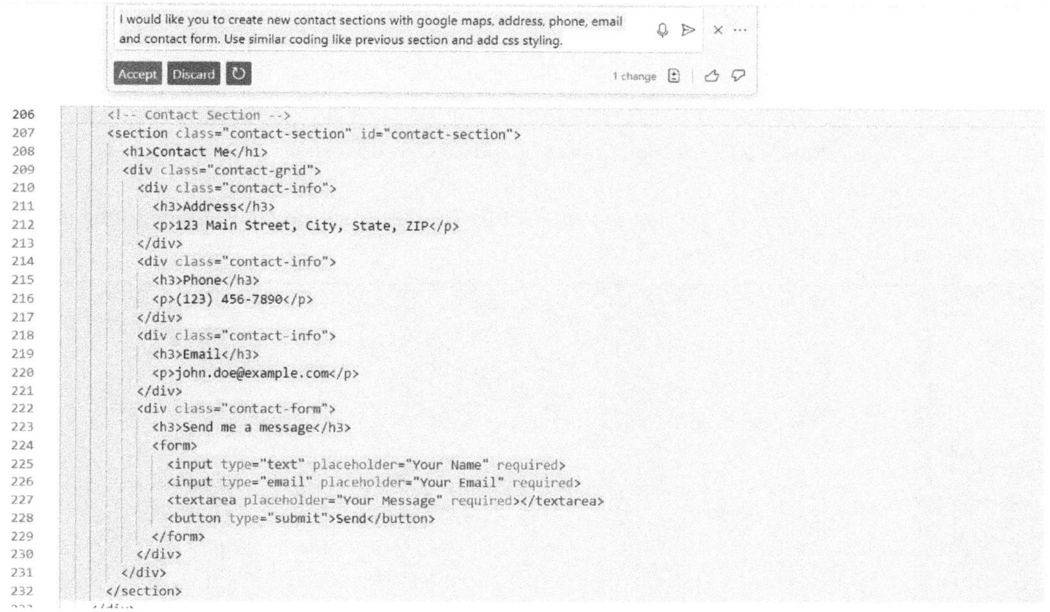

Figure 13.14 – GitHub Copilot providing code snippets

You can also hover your cursor over a piece of code that has been suggested to you, and Copilot will display alternative suggestions for that specific code block. You can then choose the one that best fits your needs.

Now, you can accept the code that Copilot provides you, and you can enhance it further by starting to type to get suggestions. Additionally, you will have to add CSS stylings for each of the classes we created.

Inside the `Chapter 13` directory, go to your `styles.css` file and start typing to get suggestions regarding the styles you would like to create. When you see a suggestion that you like, you can accept it by pressing the *Tab* key. This is like saying *Good idea!* to your friend's suggestion. If you don't like the suggestion, just keep typing, and Copilot will try to offer something else. It's pretty flexible and wants to help you in the best way possible.

```css
.map-container {
  position: relative;
  overflow: hidden;
  padding-bottom: 56.25%; /* 16:9 aspect ratio */
  width: 100%;
}
```

Figure 13.15 – GitHub Copilot providing CSS suggestions

After a few more prompts, you will have a beautiful **Get In Touch** section on your one-page portfolio website. You can see the same in *Figure 13.16*.

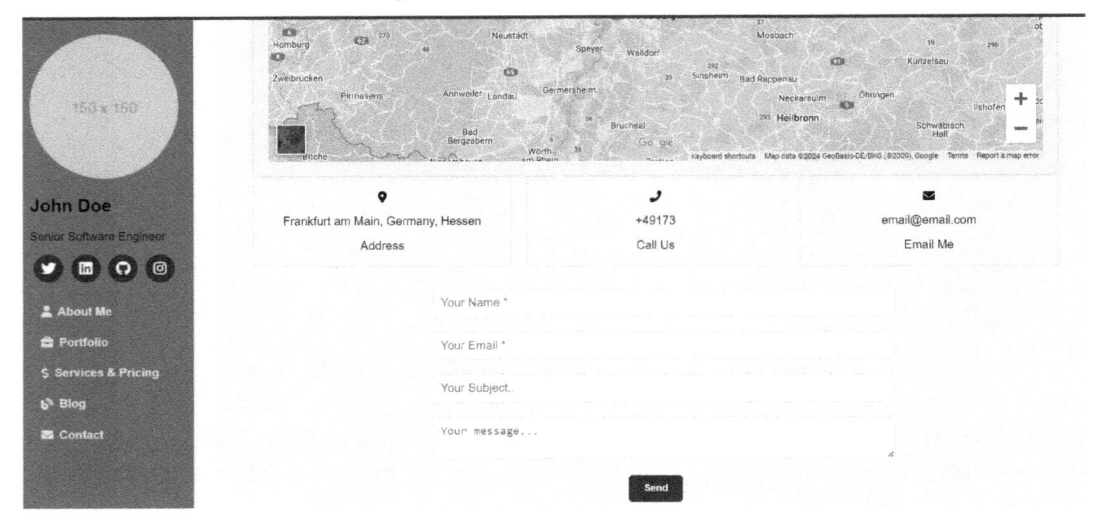

Figure 13.16 – The Get In Touch section on your one-page portfolio website

GitHub Copilot is like having a helper who's seen a lot of code before and can offer suggestions to make your coding faster, and even teach you some new tricks. It's great for the following:

- Speeding up your coding by suggesting code as you type
- Helping you learn by showing different ways to code something
- Making coding a bit less of a lonesome process, especially if you're just starting out

Now that we've looked at how Copilot can help with code completion and suggestions, let's move forward. In the next section, we'll see how Copilot can be used for more than just filling in code. We'll explore how it supports clean coding practices, helps generate unit tests, encourages test-driven development, and assists in refining and improving your code through refactoring.

Using Copilot for clean code and best practices

As well as helping you write better code, GitHub Copilot can give you suggestions on following best practices and guidelines, help you create tests to check whether your code works properly, and even offer tips on making your code cleaner and more efficient. Let's learn about some of these best practices now.

Unit testing generation

What's unit testing?

Unit testing is like giving each piece of your website its own little quiz to make sure every part knows its job. For example, if your website is a puzzle, each unit test checks whether an individual puzzle piece fits right where it should. This is important because it catches mistakes early, ensuring that every section of your website, such as the **MY ARTICLES** section of your `index.html` file, is working properly before you put it all together.

GitHub Copilot can help you to write these quizzes (unit tests). It suggests questions (tests) to ask each part of your code to make sure that they're ready for the final exam (your finished website). Copilot helps you write these tests faster by predicting what you might need to check, based on your code.

Figure 13.17 shows that you're working with a JavaScript function related to creating power-ups in a game. You've asked GitHub Copilot to write some unit tests for the *powerUpTypes* function. Clicking on the GitHub Copilot icon in the **Activity** bar on the left side of VS Code will open the **Copilot** sidebar, where you can interact with Copilot. GitHub Copilot could suggest tests that, for example, confirm that properties are set and that behaviors occur within the expected time, ensuring that each power-up enhances the game as planned.

Figure 13.17 – Asking GitHub Copilot to add unit tests for selected code

After you ask it to help you write unit tests for the selected code, GitHub Copilot will start doing so.

Now, you can copy those unit tests and add them to the new file, and later on, you can run them using GitHub Actions. The following figure demonstrates how Copilot suggests tests for *powerUpTypes* for the *Spaceship* object, which control certain aspects of the spaceship's behavior, such as speed and invincibility in a game.

GitHub Copilot offers to create unit tests using Jest, a popular JavaScript testing framework. It suggests tests that check if the *effect* and *endEffect* methods of the *powerUpTypes* object function as intended. For example, when the `speed` power-up effect is applied, the spaceship's speed should double, and when the effect ends, the speed should return to normal.

You can examine the suggested tests to see whether they align with your expectations and the specifications of the *spaceship* object's functionality. If they match your needs, you can incorporate these tests into your test suite.

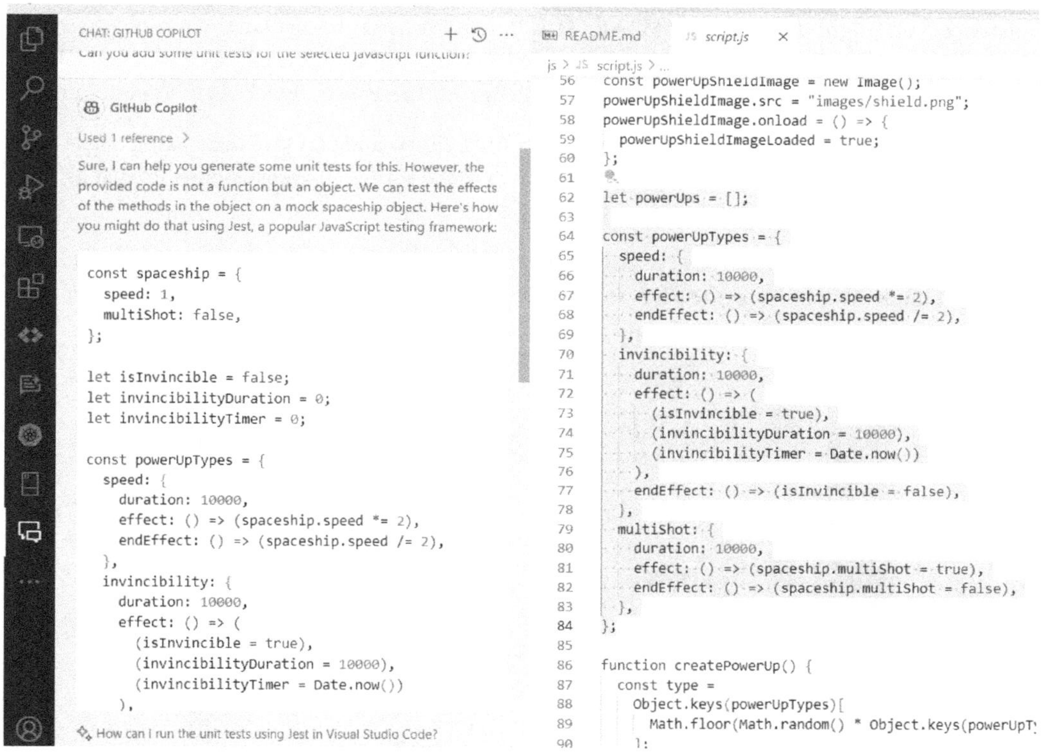

Figure 13.18 – GitHub Copilot writing unit tests

In essence, GitHub Copilot acts as a coding assistant, automatically writing tests that help ensure your game's features work correctly. By writing unit tests, you're making sure every feature, such as the power-ups in your game, works independently as it should. With Copilot's assistance, you can write these tests quickly and more efficiently, maintaining a solid, error-free gaming experience for your players.

Test-driven development practice

Test-Driven Development (**TDD**) is like creating a detailed recipe for your meal before even stepping into the kitchen. You decide what your finished dish should look like and the steps needed to make it happen. This way, you're setting up a plan to ensure your final dish is exactly what you want it to be.

In practical website-building terms, TDD means you first write out the checks (tests) for a new feature on your site, such as the **MY ARTICLES** section, before writing the actual code for that section. By doing this, you know exactly what you need to code to pass the tests.

For instance, you might decide your **MY ARTICLES** section needs to display the five most recent articles and load within two seconds. TDD would have you write tests for these requirements first. Then, as you build the feature, you can continuously check it against these tests, which serve as your guides.

This method is super useful because it focuses your coding efforts and can save time fixing bugs later, since you're checking as you go. It's a systematic way to make sure your website ends up just as you envisioned, working smoothly for anyone who visits it.

Now that you know what TDD is, you can tell Copilot about the feature you would like to code and ask it to help you implement it, using the TDD technique. After prompting Copilot, it will start generating the code snippets for you with the tests and possible code implementation.

Figure 13.19 – GitHub Copilot coding with the TDD approach

If you would like to learn more about TDD, I would recommend some resources such as the following:

- *Test-Driven Development by Example* by *Kent Beck*: A classic book that introduces the fundamentals of TDD with practical examples and a step-by-step approach

- *Clean Code: A Handbook of Agile Software Craftsmanship* by *Robert C. Martin*: While not exclusively about TDD, this book covers many principles that complement TDD practices, such as writing clean, maintainable code
- *Growing Object-Oriented Software, Guided by Tests* by *Steve Freeman and Nat Pryce*: This book provides an in-depth look at TDD in the context of object-oriented design, with a focus on creating robust and maintainable software

Code refactoring

So, you've cooked your dish, but now you want to make it look prettier on the plate. **Refactoring** is like rearranging your cooked dish to make it more appealing without changing its taste.

Sometimes, your code works, but it could be neater or easier to understand. Copilot can suggest ways to tidy up your code, making it cleaner and more organized, just like a chef might help you plate your dish more beautifully.

You can select the part of your code and ask Copilot to refactor your code by following the clean code best practices that we have discussed in this section so far. Then, it will start generating much cleaner and organized code suggestions that you can then use.

Figure 13.20 – GitHub Copilot refactoring selected code

Using GitHub Copilot while building your website is like having a helpful guide by your side, offering advice, helping you test your creations, and making sure your code is the best it can be. It's a tool that not only helps you solve problems but also teaches you to become a better coder, just like cooking with a chef teaches you to become a better cook. Next, we're going to explore prompt engineering with GitHub Copilot, which is key to making the most of this tool, by learning how to effectively communicate with it to improve your coding workflow.

Prompt engineering with GitHub Copilot

Think about a magic lamp and a genie that lives inside who knows how to code. That genie is a GitHub Copilot for you. However, instead of rubbing the lamp and telling the genie your wish, you use *prompts* or special instructions to tell them what kind of coding magic you want them to perform. You wouldn't just say, *make a sandwich*. You'd give the genie step-by-step instructions, such as *get two slices of bread*, *spread peanut butter on one slice*, and so on.

This process of giving the genie specific instructions and telling them exactly what you wish for is known as **prompt engineering**.

Figure 13.21 – GitHub Copilot as a lamp genie

How does prompt engineering work?

Let's say you're building your one-page portfolio website and you want to add a new feature, such as a photo gallery. Instead of writing all the code yourself, you ask GitHub Copilot for help. But here's the trick – you need to be clear about what you're asking.

Be more specific with your wishes, as the more details you provide, the better Copilot can fulfill your request. For example, instead of saying `I want a website gallery`, you could say, `I want a responsive gallery section where you will display the first four pictures with hover-over effects. The pictures are stored in the database, for which we need to have the connection string and a query that takes the first four photos.`

In coding terms, these *magic words* are actually bits of code or descriptions that help GitHub Copilot understand what you're trying to build. For instance, `Create a function that calculates the average of three numbers` is a clear and direct prompt.

By giving these specific instructions, Copilot understands what you're trying to achieve and can suggest code that fits your needs.

This is why clear instructions matter:

- **You avoid misunderstandings**: Without clear instructions, Copilot might guess wrong. It's like ending up with peanut butter on both bread slices because you weren't specific about where to spread it.
- **You achieve better results**: The clearer you are, the more likely Copilot will generate the code you're actually looking for. It turns your ideas into reality more efficiently.

Cool examples of prompt engineering with GitHub Copilot

Imagine you want to create an app that finds recipes based on the ingredients you have. You might prompt Copilot with `Write a Python function that takes a list of ingredients and searches for recipes online that use those ingredients`. Copilot might generate a result such as the one shown in *Figure 13.22*.

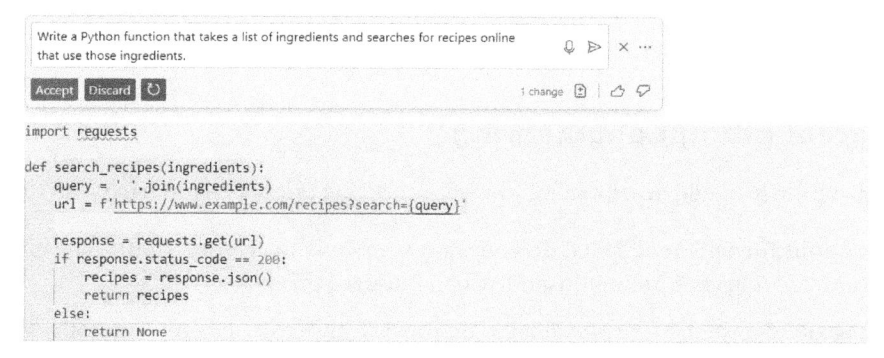

```python
import requests

def search_recipes(ingredients):
    query = ' '.join(ingredients)
    url = f'https://www.example.com/recipes?search={query}'

    response = requests.get(url)
    if response.status_code == 200:
        recipes = response.json()
        return recipes
    else:
        return None
```

Figure 13.22 – GitHub Copilot writing a Python function

Another cool example of prompt engineering would be to ask Copilot, `Design a JavaScript program that tracks my spending and categorizes expenses into 'Food', 'Bills', 'Entertainment'`, etc. Copilot will start drafting the code based on your detailed prompt, as you can see in the following figure.

```javascript
// Define an object to store the categories and their expenses
const expenses = {
    Food: [],
    Bills: [],
    Entertainment: [],
    // Add more categories as needed
};

// Function to add an expense to a specific category
function addExpense(category, amount) {
    // Check if the category exists
    if (category in expenses) {
        // Add the expense to the category
        expenses[category].push(amount);
        console.log(`Expense of $${amount} added to ${category}`);
    } else {
        console.log(`Category '${category}' does not exist`);
    }
}

// Function to calculate the total expenses for a specific category
function calculateTotal(category) {
    // Check if the category exists
    if (category in expenses) {
        const total = expenses[category].reduce((acc, curr) => acc + curr, 0);
        console.log(`Total expenses for ${category}: $${total}`);
    } else {
        console.log(`Category '${category}' does not exist`);
    }
}
```

Figure 13.23 – GitHub Copilot writes a JavaScript app

Tips for great prompt engineering

Here are some tips that should be kept in mind while practicing prompt engineering with Copilot:

- **Think about the end goal**: What do you want your code to achieve? Start there and work backward, crafting your prompt to guide Copilot toward that goal.
- **Be clear and detailed**: The more details you provide, the easier it is for Copilot to generate useful code.
- **Experiment and refine**: Not getting the code you expected? Tweak your prompts, try different phrasings, or add more details. It's like refining your wishes based on the genie's responses.

Prompt engineering with GitHub Copilot turns coding into a collaborative adventure with a digital companion that's eager to help bring your ideas to life. By mastering the art of crafting prompts, you're essentially learning to communicate effectively with a genie who's ready to make your coding wishes come true. Now that you understand the basics of prompt engineering, let's shift focus to practical insights, including whether GitHub Copilot is complimentary for student coders and how to access GitHub's educational benefits.

Is GitHub Copilot free for coders in school?

If you're diving into the world of coding while studying, GitHub Copilot comes as a friendly companion at no cost. Here's what you need to have to buddy up with Copilot:

- **Be a student**: You have to be at least 13 years old and actively learning in a school or college.
- **Have school email**: A school-issued email address proves that you're part of an educational institution. If you don't have one, don't worry! You can show some documents that confirm you're currently studying.

How to apply for free GitHub benefits

This is very easy. Just visit `https://github.com/edu` and click on **Join GitHub Education**, as shown in *Figure 13.24*.

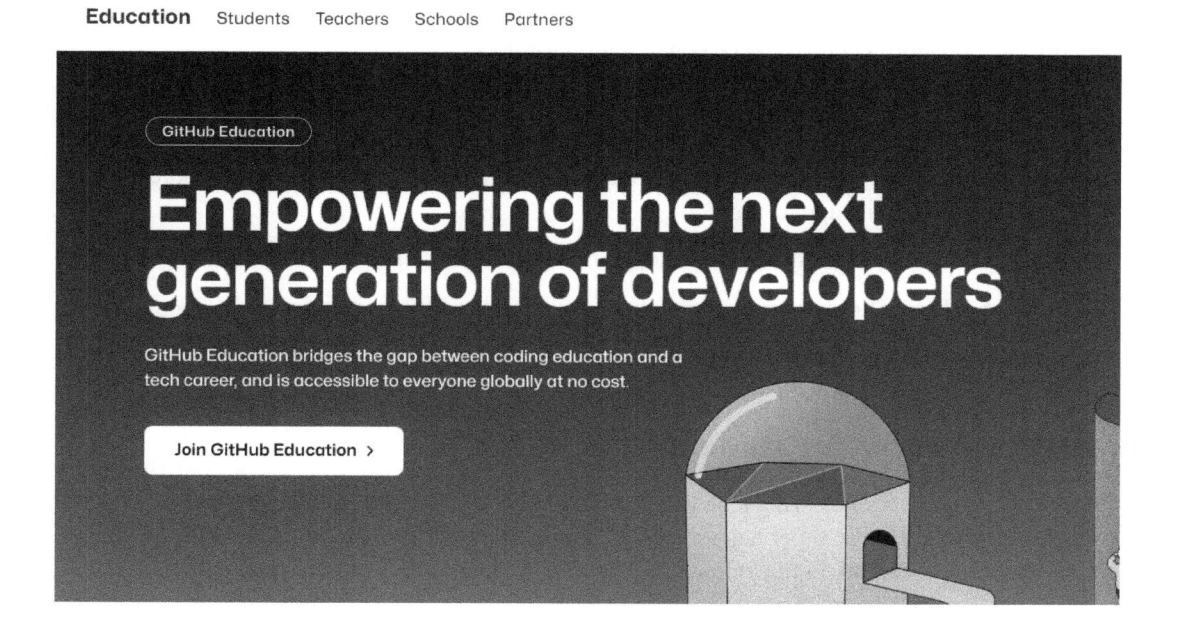

Figure 13.24 – The GitHub Education page

On the next screen, you will have to select that you are a student, provide your school name and school email, and then click **Continue** to finish setting up your student benefits.

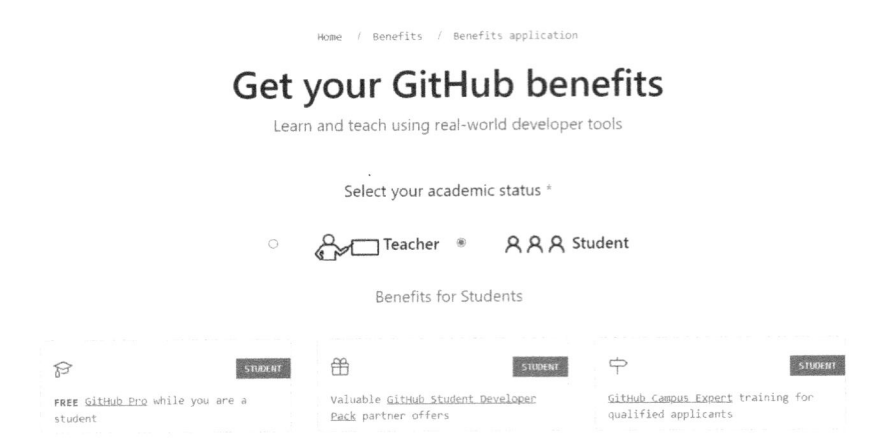

Figure 13.25 – Selecting your role on the GitHub Education page

As a student, not only can you access GitHub Copilot for free but also the Student Developer Pack, which provides a wealth of resources and tools to enhance your learning and development journey.

Here are some of the free resources included:

- **Free GitHub Pro**: While you're a student, you get free access to GitHub Pro, which includes private repositories and advanced collaboration features
- **GitHub Student Developer Pack partner offers**: Unlock discounts and free access to various tools and services, such as cloud hosting, productivity tools, and developer platforms
- **Azure credits**: You'll receive credits for Microsoft Azure services, allowing you to explore cloud computing and build applications
- **Codespaces**: Access cloud-based development environments directly from your browser
- **Student gallery**: Showcase your projects and collaborate with other students
- **GitHub Campus Expert training**: Qualified applicants can become GitHub Campus Experts and contribute to their local developer communities

A word of wisdom for students

While GitHub Copilot is an amazing tool, it's important to use it wisely. It's like learning to ride a bike with training wheels. They're super helpful, but you also need to understand how to balance on your own. Similarly, make sure to do the following:

- **Understand the code**: Don't just copy and paste what Copilot suggests. Try to understand why and how the suggested code works.

- **Stay curious**: Use Copilot as a learning tool. Ask questions, explore the suggestions, and dive deeper into the concepts.

In a nutshell, GitHub offers a free, smart, and supportive learning experience for students starting with their coding journey, and in this section, we learned about how to take advantage of its services.

Summary

In *Chapter 13*, we explored GitHub Copilot, a tool that suggests code in real time, making coding more efficient and less lonely, especially for beginners. We learned how to use Copilot for code completion, best practices, and even unit testing. This knowledge is useful because it can speed up coding projects and help you learn new coding methods. Now, you understand how to make Copilot your coding partner, which can improve your coding skills and project quality.

Quiz

Answer the following questions:

1. What is GitHub Copilot primarily used for?

    A. Designing websites

    B. Suggesting code in real time

    C. Managing team projects

    D. Automating email responses

    Answer: B. Suggesting code in real time

2. True or false – GitHub Copilot can only help with JavaScript coding.

    A. True

    B. False

    Answer: B. False

3. Fill in the blank – GitHub Copilot suggestions can introduce you to new _____ and best practices.

    Answer: features

4. How can GitHub Copilot aid in writing cleaner code?

    A. By suggesting code refactoring

    B. By changing the IDE theme

    C. By deleting unused files

    D. By organizing your bookmarks

    Answer: A. By suggesting code refactoring

5. True or false – to start using GitHub Copilot, you need a paid subscription from the start.

    A. True

    B. False

    Answer: B. False

6. Which IDE is required to use GitHub Copilot effectively?

    A. Microsoft Word

    B. VS Code

    C. Notepad++

    D. Eclipse

    Answer: B. VS Code

7. What is prompt engineering with GitHub Copilot?

    A. Designing computer hardware

    B. Writing clear instructions to improve code suggestions

    C. A method for electrical engineering

    D. Programming video games

    Answer: B Writing clear instructions to improve code suggestions

8. True or false – GitHub Copilot can generate unit tests for your code.

    A. True

    B. False

    Answer: A. True

9. GitHub Copilot can be accessed for free by whom?

    A. Only professional developers

    B. Only GitHub employees

    C. Students

    D. Everyone automatically

    Answer: C. Students

10. How does GitHub Copilot improve coding productivity?

    A. By increasing internet speed
    B. By suggesting code snippets as you type
    C. By cleaning your computer screen
    D. By organizing your desk

    Answer: B. By suggesting code snippets as you type

Index

A

access control 255
actions 233
Actions section 25
action workflows
   creating 233
Activity page 179-181
advanced Git commands 200
   git alias 200
   git blame 201
   git fetch --all 202
   git log follow 200
   git rebase 201
   git remote 202
   git stash 201
   git stash pop 201
   git tag 201
Artificial Intelligence (AI) 326, 328
automated testing 239

B

branch 5, 98, 179
   creating 124-126
branch protection rules 269, 270

C

centralized version control system (CVCS) 6, 7
   flow 7
   for coding 7
cherry-picking 193-195
cloning 288
CODEOWNERS 264, 265
   benefits 265, 266
CODEOWNERS file
   creating 264
CodeQL feature 275
Code section 22, 23
Codespaces 215, 218
   deleting 218, 219
   launching 216, 217
collaboration
   setting up, in website repo 252-255
collaborators, inviting 64
   guest list, checking 65
   invitation spot, finding 64, 65
   RSVP, awaiting 65
commits 11, 100, 194
   reverting, to previous version 188-190

Index

Contact Me form
  adding, to website 99
Creative Commons licenses 294

D

deltas 8
  working, in coding 8, 9
Dependabot 260, 273
  in action 262-264
  setting up 260, 261
  working 260
dependencies 260
deployment 239
DevHub 219
  features 221
  setting up 221
  URL 220
diff command 190
  git diff <REF-1> <REF-2> command 192
  git diff HEAD command 192
  git diff origin/main main command 192, 193
  git diff --staged command 191
direct editing 142
Discussions section 24
distributed version control system (DVCS) 10, 11
draft
  sharing 128

E

environment variables 243
  benefits 243
  using, for MY ARTICLES 243-246
ESLint 239
events 231

F

feature branch
  creating 98
  need for 99
files
  adding, to Git 109-111
  Untracked Files, checking 111
Files changed tab 134, 135
filters
  used, for reducing notification flood 75

G

Git 13-15, 174
  configuration levels 44-46
  downloading 43
  history (or repository) 115, 116
  installing 43, 44
  interaction, with GitHub 32
  network interaction 39
  staging area (index) 114
  working area 113, 114
git alias command 200
git bisect 182, 183
  missing CSS styling case 183-187
  using 187
git blame command 201
Git cherry-picking
  feature branch 194
  main branch 194, 195
Git commands
  git add command 34, 35
  git clone command 32-34
  git commit command 37
  git fetch command 38
  git pull command 39
  git push command 38

git fetch --all command 202
Git history
 parallel universes 179
GitHub 13-15
 benefits 4
 files, creating 35, 36
 neighborhood 40
 repositories 39, 40
 repository, creating 40, 41
 repository, managing 42
 working area 118
GitHub account
 organization account 16, 21, 22
 setting up 16, 17
 user account 16, 18-20
GitHub Actions 230, 272
 common issues 247
 Marketplace 230
 one-page portfolio website, deployment and testing 240-242
 optimizing 248
 starter page 233
 steps 232
 troubleshooting 247
GitHub Actions Marketplace 234
GitHub branch
 deleting 139
 deleting, significance 139
 local branch, deleting after PR approval 140
 local features branch, deleting 141
 local main branch, upgrading 140
 unneeded branches, pruning 141
GitHub Copilot 326
 code completion and suggestions 334-337
 code refactoring 341, 342
 features 328
 for coders, in school 345-347
 for personal projects 328-331

 prompt engineering 342
 test-driven development practice 339, 340
 unit testing generation 337-339
 working 326, 327
GitHub Copilot extension
 installing 332-334
GitHub Desktop 208
 benefits 214
 features 211-215
 setting up 209, 210
 URL 209
github.dev 143
 activating 143, 144
 need for 144
GitHub Discussions 77, 78
 Announcements 78
 General 77
 Ideas 77
 Polls 78
 Q&A 78
 Show and tell 78
 starting 78-80
 stronger team, building 81
GitHub Education 345, 346
 URL 345
GitHub Flow 98, 103
 actual main, with feature branches 106
 branch, creating 103
 branch, pushing 104
 changes, making 99, 100
 changes, pushing 100
 feature branch, creating 98, 100, 104
 files, adding 105
 files, committing 105
 pull request, approving 102
 pull request, creating 101
 pull request, creating on GitHub 104, 105
 pull request, discussing 101

pull request, merging 102, 104, 105
workspace, cleaning up 102, 103
GitHub history 174
GitHub interface
  Actions section 25
  Code section 22, 23
  Discussions section 24
  Insights section 26
  Issues section 23
  navigating 22
  Projects section 25
  Pull requests section 24
  Security section 26
  Settings section 27, 28
  Wikipedia section 26
GitHub Pages
  using, to show off work 320
  workflow syntax, to deploy website to 231-233
GitHub PR interface 128, 129
GitHub profile 303
  badges 306
  contribution graph 305
  contributions and highlights, adding 318-320
  picture and bio 304
  pinned projects 304
  projects, displaying 317, 318
  stats and achievements, adding 315
GitHub project 53
GitHub repository
  license, creating 294-297
GitHub repository insights 89
GitHub Stats widgets 312-314
  reference link 313
GitHub trophies 316
  showcasing 316

GitHub version control 5
  centralized version control system (CVCS) 6, 7
  deltas 8
  deltas, working in coding 8, 9
  distributed version control system (DVCS) 10, 11
  snapshots 11, 12
.gitignore file 267
git log 175
git log follow command 200
git log --oneline command 176
git log --oneline --graph command 176
git log --oneline --graph --decorate --all command 177
git log --oneline --graph --decorate command 177
git push command
  changes, pushing to remote GitHub repository 116, 117
  --set-upstream 117
git rebase command 200, 201
git reflog 178
git remote command 202
Git repository
  Activity page 179-181
  history, viewing locally 174-178
git reset
  hard reset 193
  mixed reset 193
  soft reset 193
git revert 188-190
Git staging area
  file changes, adding 112, 113
git stash command 201
git stash pop command 201
git tag command 201
GNU General Public License (GPL) 293

H

HTMLHint 239

I

Insights 89
  benefits 92
  code frequency 91
  Contributors 89
  pulse 91
  traffic 90
Insights section 26
issues 66
  assigning 68, 69
  creating 66, 67
  discussing 67
  linking, to pull request 69
Issues section 22, 23

J

jobs 232

K

Kanban 82
Kanban board
  need for 86
Kanban project
  setting up, for one-page website on GitHub 82

L

Large Language Models (LLMs) 326
  working 328

license 291
  creating, for GitHub repository 294-296
  Creative Commons licenses 294
  GNU General Public License (GPL) 293
  legal considerations 291
  MIT license 292
  reference link 294
  selecting 292
local copy, of repository
  creating 106
  project, cloning 106-108
  switching, to correct branch 108

M

main branch 98
Marketplace 230, 234
merge conflicts 150-153
  addressing 153
  file, removing 164
  files, restoring 165
  in action 153-155
  multiple merge conflicts 159
  resolving 155-158
  resolving, with command line 161-163
  solving, with removed files 164
MIT license 292
multiple merge conflicts 159
  conflicting file, resolving 160
  PR, completing 160
  resolving, with command line 160, 161
  right changes, selecting 160
MY ARTICLES section 230
  secrets and environment variables, using 243-246

N

notification flood
   reducing, with filters 75
notifications 72
   alerts, customizing 76
   channels, tuning 73, 74
   marking 74
   spotting 73
   updates, emailing 76, 77

O

open source
   best practices 297
   etiquettes 297
open source projects 282-286
   project pages 283, 284
organization account 16, 21, 22
ownership 254

P

personal access token (PAT) 47
   creating 48-51
   keeping safe 51
   using, with Git 51
prebuilt actions 234-237
projects
   creating, on GitHub 82-85
   setting up 81
   setting up, for one-page portfolio website 81
   significance 81
project settings
   modifying 86
Projects section 25
prompt engineering 342
   examples 343, 344
   tips 344
   working 342
prompts 342
Pull Request page 161
pull requests 5, 70, 101, 124, 133
   changes, merging 138
   creating 124, 130-133
   issues, linking to 69
   merging 72
   quality, ensuring 136, 137
   reviewing 71, 133, 137
   significance 72, 130
   using, in website project 70
Pull requests section 24

R

README file 52
   editing, on GitHub 55-57
   example 52
   tips 53, 54
   using 53
README profile 306
   About Me section 310
   badge, adding 309
   biography 309
   certifications 311
   creating 306, 307
   images 311
   name, setting up 309
   profile picture 309
   skills 310
real project, on GitHub
   contributing to 286-291
refactoring 341

Index

repositories, GitHub 39, 40
  creating 40, 41
  managing 42
repository 3, 40
  local copy, creating 106
repository permissions
  collaborators 252
  owner 252
reusable workflows 238
runners 232

S

secrets 243, 268
  benefits 243
  using, for MY ARTICLES 243-246
Secret Scanning feature 276
secure development strategy 267
  planning 267-277
security features 272-275
SECURITY.md file 274
Security tab 272
Settings section 27, 28
SHA code 178
Shields.io
  URL 310
snapshots 11, 12
Sourcetree 203
  branches, creating 206
  branches, merging 206
  branches, switching 207
  downloading 204
  Fetch button 207, 208
  project timeline, viewing 205
  Pull button 207, 208
  Push button 207, 208
  quick commands 206

repository, setting up 204
  stage and commit changes 207
  URL 204
staging area 113
steps 232
Streak Stats 313
Stylelint 239

T

Test-Driven Development (TDD) 339, 340
Top Languages 312
Two-Factor Authentication (2FA) 251, 255
  benefits 255
  setting up 255-259

U

unit testing 337
Untracked Files 111
  checking 112
  checking, reason 112
user account 16-20
uses keyword 237
  working 237

V

version control systems (VCSs) 5
Visual Studio Code (VS Code) 215, 332
  URL 109
vulnerability scanning 260

W

website repo
  collaboration, setting up 252-255

widgets 312
   GitHub Stats 312-314
   Streak Stats 313
   Top Languages 312
wiki 87
   creating, for one-page website on GitHub 87-89
Wikipedia section 26
workflow 231

Y

YAML Ain't Markup Language (YAML) 231

‹packt›

packtpub.com

Subscribe to our online digital library for full access to over 7,000 books and videos, as well as industry leading tools to help you plan your personal development and advance your career. For more information, please visit our website.

Why subscribe?

- Spend less time learning and more time coding with practical eBooks and Videos from over 4,000 industry professionals
- Improve your learning with Skill Plans built especially for you
- Get a free eBook or video every month
- Fully searchable for easy access to vital information
- Copy and paste, print, and bookmark content

Did you know that Packt offers eBook versions of every book published, with PDF and ePub files available? You can upgrade to the eBook version at packtpub.com and as a print book customer, you are entitled to a discount on the eBook copy. Get in touch with us at customercare@packtpub.com for more details.

At www.packtpub.com, you can also read a collection of free technical articles, sign up for a range of free newsletters, and receive exclusive discounts and offers on Packt books and eBooks.

Other Books You May Enjoy

If you enjoyed this book, you may be interested in these other books by Packt:

GitHub Actions Cookbook

Michael Kaufmann

ISBN: 978-1-83546-894-4

- Author and debug GitHub Actions workflows with VS Code and Copilot
- Run your workflows on GitHub-provided VMs (Linux, Windows, and macOS) or host your own runners in your infrastructure
- Understand how to secure your workflows with GitHub Actions
- Boost your productivity by automating workflows using GitHub's powerful tools, such as the CLI, APIs, SDKs, and access tokens
- Deploy to any cloud and platform in a secure and reliable way with staged or ring-based deployments

DevOps Unleashed with Git and GitHub

Yuki Hattori

ISBN: 978-1-83546-371-0

- Master the fundamentals of Git and GitHub
- Unlock DevOps principles that drive automation, continuous integration and continuous deployment (CI/ CD), and monitoring
- Facilitate seamless cross-team collaboration
- Boost productivity using GitHub Actions
- Measure and improve development velocity
- Leverage the GitHub Copilot AI tool to elevate your developer experience

Packt is searching for authors like you

If you're interested in becoming an author for Packt, please visit authors.packtpub.com and apply today. We have worked with thousands of developers and tech professionals, just like you, to help them share their insight with the global tech community. You can make a general application, apply for a specific hot topic that we are recruiting an author for, or submit your own idea.

Share Your Thoughts

Now you've finished *GitHub for Next-Generation Coders*, we'd love to hear your thoughts! Scan the QR code below to go straight to the Amazon review page for this book and share your feedback or leave a review on the site that you purchased it from.

`https://packt.link/r/1835463045`

Your review is important to us and the tech community and will help us make sure we're delivering excellent quality content.

Download a free PDF copy of this book

Thanks for purchasing this book!

Do you like to read on the go but are unable to carry your print books everywhere?

Is your e-book purchase not compatible with the device of your choice?

Don't worry! Now with every Packt book, you get a DRM-free PDF version of that book at no cost.

Read anywhere, any place, on any device. Search, copy, and paste code from your favorite technical books directly into your application.

The perks don't stop there, you can get exclusive access to discounts, newsletters, and great free content in your inbox daily

Follow these simple steps to get the benefits:

1. Scan the QR code or visit the following link:

    https://packt.link/free-ebook/9781835463048

2. Submit your proof of purchase.
3. That's it! We'll send your free PDF and other benefits to your email directly.

Printed in the USA
CPSIA information can be obtained
at www.ICGtesting.com
LVHW061257301124
797961LV00008B/1125